3rd Jordan International Oil Shale Symposium and the 36th CSM Symposium 2016

Madaba, Jordan
21 - 22 November 2016

ISBN: 978-1-5108-5395-9

Printed from e-media with permission by:

Curran Associates, Inc.
57 Morehouse Lane
Red Hook, NY 12571

Some format issues inherent in the e-media version may also appear in this print version.

Printed by Curran Associates, Inc. (2018)

Ministry of Energy & Mineral Resources of the Government of Jordan (MEMR)
PO Box 140027
8th Circle, Abdel-Rahman
Al-Jamal Street / Passage, Building No. 18
Amman, Jordan 11814

Phone: 962-6-5803060
Fax: 962-6-5865714

memr@memr.gov.jo

Additional copies of this publication are available from:

Curran Associates, Inc.
57 Morehouse Lane
Red Hook, NY 12571 USA
Phone: 845-758-0400
Fax: 845-758-2633
Email: curran@proceedings.com
Web: www.proceedings.com

Monday - 21st November 2016 — Day One

08.30	Registration and morning refreshments
09.00	National Anthem and Reading from the Quran
09.10	Opening remarks by the Chairman: **Dr. Ghaleb Maabreh, Chairman, National Petroleum Company**
09.20	Official opening speech by the Minister of Energy and Mineral Resources
09.40	**World Energy Council** (TBC)
Distinguished Panel 1 10.20	**Government Efforts to Encourage Oil Shale Investment** • Jordan is 96% dependent on imports of oil from countries in a politically unstable region. The continuity of supply may be not assured. • Jordan's top priority is Energy Security • The Government's perception of economic and geopolitical risk • Jordan's Energy Road Map to overcome the current challenges • Jordan's policy incentives for oil shale investment projects • The Government's outlook for opportunities of future cooperation between government and international companies in Jordan and in ME Region • The oil shale industry needs a commercial mindset and political will in these critical situations. • Long term energy partnerships through contractual agreements are healthful, building the trust and confidence that are needed during the current crisis. The agreements are clear messages to political & business leaders on the mutual benefits derived from sustainable energy trade flows. **Panellists:** **H.E. Dr. Jawad Anani, Deputy Prime Minister for Economic Affairs** **Dr. Ibrahim Saif, Minister of Energy and Mineral Resources** **Ahamd Qatarneh, Secretary General, Ministry of Environment** **H.E. Thabit Al Wir, President, Jordan Investment Commission**
10.50	**Continued Evolution of our Oil Market – Within the Context of the IEA 2016 World Energy Outlook** 1 **Speaker:** Nathan Frisbee, **Senior Analyst, Global Energy Economics, International Energy Agency 'IEA'**
11.10	Exhibition Opening and Morning Networking Break & Refreshments
11.40	Oil Shale Project Development Journey. 7 **Speaker:** Andres Anijalg, **Chairman of the Management Board, Attarat Power Company**

Monday - 21st November 2016 — Day One

12.05	Jordan Oil Shale Company Production Pilot - Jordan Field Experiment 25 **Speaker:** **Yusuf Siddiqui, Shell Country Chairman & General Manager, Jordon Oil Shale Company**
12.30	KIO Oil Shale Project: Progress & Challenges 34 **Speaker:** **John Fraser, Director, Karak International Oil**
12.55	SACOS Leads a New Era of Oil Shale Exploitation **Speaker:** **Dr. Trad Mohammed Al-Ahmed, Saudi Arabian Corporation for Oil Shale** 42
13.20	Enefit - Operating the World´s Largest Oil Shale Industry 49 **Hando Sutter, Chairman of the Management Board, Enefit**
13.40	Lunch & Networking

	TRACK A Resource Assessment Oil Shale Chemistry, Geology and Mining	TRACK B Oil Shale Processing, Retorting Firing and Upgrading
14:40	The Isfir-Jafr deposit in Southern Jordan and The Resource Assessment of the Isfir-Jafr Oil Shale Deposit in Southern Jordan **Speaker:** **John Brodylo, Questerre Energy Corporation** 56	Extractive De-sulfurization of High Sulphur Shale Oil by Thio-phenes Extraction 91 **Speaker:** **Abdelhafid Foughalia, TTU Galoter**
15:00	Oil Shale Sulfur Content Variations with Different Analytical Methods 63 **Speaker:** **Birgit Maaten, Tallinn University of Technology**	Appraising the Jordan Oil Shale , an example of novel subsurface disciplines integration **Speaker:** **Wasfi Omari, Jordon Oil Shale Company** 98
15:20	Characterisation of Shale Oil Produced by Retorting Sultani Oil Shale 70 **Speaker:** **Dr. Omar Ayed, Al-Balqa Applied University**	Design Considerations of Oil Shale Semi Coke Combustion in Circulating Fluidized Bed **Speaker:** **Dr. Yiqun Huang** 106 **Tsinghua University**
15:40	Assessing the Geochemical Variability of Jordan Oil Shale: Laboratory Geochemistry Vs. Multivariate Statistics 79 **Speaker:** **Prof. Alvar Soesoo, University of Tartu, Estonia**	Sensitivity Analysis of Oil and Gas Production in the In-Situ Pyrolysis of Maoming Oil Shale **Speaker:** **Dr. Peng Guo, SINOPEC** 115

Monday - 21st November 2016 — Day One

Time	Track C	Track D
16:00	**Afternoon Networking Break**	
	TRACK C **Environmental Impact and Mitigation Measures**	**TRACK D** **Oil Shale Strategies, policies and Challenges**
16:30	EIF Onshore Discharges Tool. A Modelling Tool to Assess Environmental Risk to Soil, Groundwater and Surface Water from In-Situ Kerogen Oil Production 126 **Speaker:** **Randi Hagemann**, **Statoil**	Engaging in Public-Private Partnerships with Oil Shale Developers 133 **Speaker:** **Osama Sulieman**, **Director PPP Unit Ministry of Finance**
16:50	Geophysical Alterations And Groundwater Flow Implications Of In Situ And Ex Situ Conversion Of Oil Shale In The Piceance Basin, Colorado, With Applications For Other Oil Shale Basins Via Semi-Advanced Groundwater And 3D Geologic Computer Models **Speaker:** **Paul Quinn**, **International Consultant, Ambet LLC** N/A	**Jordan Oil Shale Industry - Making Progress in Difficult Times** **Speaker:** **Eyas Shuaibi**, **Partner Trowers & Hamlins LLP** 144
17:10	End of Symposium Day One	
19:30	Dinner	

Countries Participating

Tuesday - 22nd November 2016 — Day Two

08.30 — Registration and morning refreshments

09.00 — Opening comments from the Chairman: **Dr. Ghaleb Maabreh, Chairman, National Petroleum Company**

Panel Discussion 2
09.10

Government Involvement and Support to Investment Projects

- Support of a host government can improve the feasibility of an oil shale project and help it secure financing.
- Types of support range from tax concessions through purchase agreements to partial ownership of the assets.
- The support of Jordan's government is presently limited to facilitating project implementation.
- What procedures are in place for obtaining licenses, permits, and concessions?
- What other types of involvement are needed and desired?
- How are these potential initiatives constrained by Jordan's laws and policies?
- What regulatory changes is the Government working on?
- What is the schedule for completing these revisions?

Panellists: Secretary Generals of Relevant Governmental Authorities
H.E. Eng. Amani Azzam, Secretary General, Ministry of Energy and Mineral Resources
Eng. Wijdan Al Rabadi, Commissionaire, Energy and Mineral Resources Commission
His Excellency Eng. Abdelfattah Al Daradkeh, Managing Director, National Electric Power Company
Dr Khair Hadidi, Water Authority of Jordan
Hussain Al Mansi, Director of Exemptions Affairs, Jordan Customs
Dr. Waleed Bawaneh, Income and Sales Tax Department Jordan

09.40 — Challenging Oil Shale Geology of Central Jordan 153
Dr. Vaino Puura, Professor Emeritus, University of Tartu

Panel Discussion 3
10.00

Financing Oil Shale Projects and the Current Challenges

- Keys to successful oil shale investment project financing
- Appetite and capacity of Jordanian banks for lending to support oil shale mega-projects
- How to accurately determine if a project is still feasible and commercially viable
- Success stories of oil shale project financing
- How World Bank Group involvement helps enhance oil shale investment projects
- What is the World Bank Group Strategy for developing countries and how do oil shale projects align with World Bank Group's approach to the energy sector?
- How oil shale companies can assure compliance with environmental and social best practices and World Bank Group international performance standards.

Panellists:
Nathan Frisbee, Senior Analyst, Global Energy Economics, International Energy Agency 'IEA'
Andres Anijalg, Chairman of the Management Board, Attarat Power Company
Dr. Adli Kandah, Director General, Association of Banks in Jordan
Chris Nurse, Director, CEO, Karak International Oil

10.40 — Morning Networking break and Refreshment

11.10 — Oil Shale Economics - Updating a Legacy 163
Dr. Tom Sladek, Director, Ockham Energy Services

Tuesday - 22nd November 2016 — Day Two

11.40	The Ways to Achieve Independence from the Oil Price Volatility for Oil Shale Projects in Jordan **Ruslan Salikhov, Deputy Chief Engineer Designer, TTU Galoter, Russia** 167
12.05	Red Leaf Resources Update: TOTAL Update, Technology Update, Future Plans and Reusable Capsules **Lance Lehnhof**, Director and Vice President of Corporate Affairs, **RedLeaf Resources Inc** 188
12.35	Financing Oil Shale Projects in the Current Macro Environment US$50/Barrel Oil: Challenges and Path Forward **David Argyle, Managing Director, Global Oil Shale Holdings Inc (GOSH)** N/A
13.00	Lunch & Networking

	Track A Financing Oil Shale Projects and enhancement of Projects Economics	Track B Commercialisation and Industry Developments
14:00	APCO Oil Shale Power Project and Chinese Financing **Speaker: Jason Pok, APCO** 196	The Isfir-Jafr Oil Shale Project Status **Speaker: Pete Coldham, Questerre Energy Corporation** 226
14:20	Integrated Development of Renewable Energy and Oil Shale **Speaker: Dr. Jamal Jaber,** Faculty of Engineering Technology, **Al-Balqa' Applied University** 289	Oil Shale Development in Jordan, HSSE/SP and Operational Excellence to Achieve Goal Zero **Speaker: Malek Abu Rumman, JOSCO** 232
14.40	Ways and Means for Oil Shale Economic Enhancements **Speaker: Munther Bseiso, Al-Qamer for Energy and Infrastructure Ltd** 203	Al-Qamer Oil Shale Project at Attarat Umm Ghudran **Khaled Khraisat, Geology and Mining Manager Al-Qamer for Energy and Infrastructure Ltd** 239
15:00	Afternoon Break and Refreshment	

Closing Panel 15.30

Outlook for the Oil Shale Industry

- The world has plentiful supplies of energy resources, and technologies are available to sustain economic growth and satisfy increasing populations.
- However the time of easy oil & gas has come to an end, and hydrocarbon production is becoming increasing complex, whether in geology or petroleum engineering and technology.
- What role will fossil fuels play in the future global energy mix?
- What role will oil shale play?
- How can technologies be improved to provide greater efficiency throughout the oil shale cycle, from mining to consumption, and to reduce the carbon intensity of the industry?
- Well-developed frameworks and policies are needed to build a balanced energy profile, including oil shale

Panellists:
H.E. Eng. Amani Azzam, Secretary General, Ministry of Energy and Mineral Resources
Yusuf Siddiqui, Shell Country Chairman & General Manager, Jordon Oil Shale Company
Munther Akroush, Vice Chairman, Karak International Oil
Andres Anijalg, Chairman of the Management Board, Attarat Power Company
Dr. Trad Mohammed Al-Ahmed, Saudi Arabian Company for Oil Shale

16.00 Close of Symposium

ADDITIONAL PAPERS

FROM THE MINISTER'S DESK

Ministry Of Energy & Mineral Resources

Ref. No: PoD /4415
Date : 24-5-2016

The 3rd International Jordan Oil Shale Symposium and the 36th Colorado School of Mines Oil Shale Symposium / Mövenpick Hotels & Resorts, Jordan-Dead Sea, 21st-22nd November 2016

The **Hashemite Kingdom of Jordan** represented by **The Ministry of Energy and Mineral Resources "MEMR"** along with **Colorado School of Mines** "CSM" and the **Oil Shale Companies investing in Jordan**, are honoured to announce the launching of the joint event of **The 3rd International Jordan Oil Shale Symposium and the 36th Colorado School of Mines Oil Shale Symposium** to be held in the Mövenpick Hotels & Resorts -Dead Sea during the period 21st to 22nd of November 2016.

The 'MEMR' and 'CSM' for and on behalf of the Symposium Co-organizers cordially invite the various international, regional and local sectors in the field of Oil Shale research and investment to participate in this prestigious event. The upcoming symposium is expected to shed the light on the latest updates of the oil shale development projects in Jordan. This important knowledge and business platform will also present global technologies and development strategies in the oil shale industry as also the combined efforts to overcome the current challenges facing the oil shale industry.

We extend our invitation to the global business and finance community from all sectors to participate in reviewing the targeted responses to the unique challenges of the soaring demand for energy.

Your participation in this event as our esteemed speakers, event sponsors, delegates or as media partners will demonstrate you as a valuable asset in strengthening the symposium, achieving its goal and showcasing your organisation in this business.

For all enquiries concerning the symposium please contact:

Mr. Arshed Hussain
Project Director MCI Middle East
Tel: +9714 3116300
Mobile: +97150 5593922
Email: arshed.hussain@mci-group.com

Thank you for your cooperation and together we look forward to a successful symposium.
Best regards

Dr. Ibrahim Saif
Minister of Energy and Mineral Resources

The Hashemite Kingdom Of Jordan
TEL: +962 6 5803060 - FAX: +962 6 5865714 P.O.Box: 140027 AMMAN 11814 JORDAN www.memr.gov.jo

Industry Overview

With global energy demand expected to increase by 44% over the next two decades, there is an urgent need to develop and utilize alternative forms of energy. Oil Shale is growing as a global phenomenon around the world, as more companies and governments increase their investment in oil shale exploration. The global integrated efforts were dedicated to develop the resource exploitation process in the direction of scientific and engineering research, information management, education, and networking related to development and production of energy sources from oil shale. In Jordan, almost 96 per cent of oil and gas supplies were imported and this adds a strain on the economy and finances. With the increasing oil consumption and prices internationally, Jordan began to consider reducing its dependence on oil and gas by exploring and developing its vast oil shale resources. These deposits are located in about 60 per cent of the country, with an estimated resource base of 70 billion tons: The 4th largest in the world.

Estimations of the "near-surface, exploitable reserves are estimated at more than 50 billion metric tons, with extractable crude oil equivalent to about 50 billion barrels of oil." Even though exploration and tests were conducted since the 1980s, the interest in oil shale only developed in 2006 when several companies considered both shale oil extraction and oil shale combustion for power generation. Jordan has signed four concession agreements, one in-Situ oil project and three retorting oil projects through surface mining and one oil shale firing project to produce electricity.

Event Overview

The 3rd Jordan International Oil Shale Symposium (3rd JIOSS) and the 36th CSM Symposium will discuss and evaluate global and domestic developments and challenges in the oil shale investments and an opportunity gathering key players in the energy industry, oil shale working groups, policy makers, academics, researchers, geologists and engineers.

This gathering will transfer you to the deep of the oil shale industry and connects you to the research and development world partners and exchange information about the latest developments, the benefits and opportunities of oil shale industry as well as issues related to technical, environmental and economic risks. Under the patronage of the Minister of Energy & Mineral Resources, the "3rd JIOSS and 36th CSM" Symposium will be a prestigious meeting of oil shale stakeholders working towards a beneficial oil shale development in the region.

Topics of Discussion

- Financing Oil Shale Investment Projects and Enhancement of Projects Economics
- Country Policies, Current Challenges and Oil Shale in the Energy Mix
- Oil shale Industry Developments
- Oil shale chemistry, Geology and Mining
- Oil shale processing, Retorting technologies, Combustion Technologies and Upgrading technologies
- Environmental Impacts and Mitigation Measures

Steering Committee

- Ministry of Energy and Mineral Resources
- Colorado School of Mines
- Jordon Oil Shale Company (JOSCO)
- Attarat Power Company (APCO)
- Karak International Oil (KIO)
- Whitehorn Resources
- INTEK Inc
- Global Oil Shale Holdings (GOSH)
- Al Qamer for Energy & Infrastructure
- Shale Energy (JOSECO)
- Saudi Arabian Company for Oil Shale (SACOS)

Who will attend

- Oil Shale Industry
- Ministry of Energy & Mineral Resources, Jordan
- Representatives from Regional Governments
- Colorado School of Mines Representatives
- International Oil Shale Exploration Companies
- Oil Shale Companies Investing in Jordan
- Service Providers to the Industry
- Mining Companies
- Oil Shale Technology Providers
- Financial institutions and private investors
- Senior Academics and Research specialists
- Environmentalists

Key Reasons to attend

Proven and successful Oil Shale Event encompassing senior decision makers and stake holders in the Oil Shale industry will come together and present their thoughts, their plans and the way forward

- A meeting place for Oil Shale Companies investing in Jordan
- Conduct business
- One to one meetings
- Listen to the best on technology in oil shale processing
- Meet with your partners, clients and prospects
- Opportunity to showcase your company offerings and its services
- Connect with the investors

Keynote Presenters

Hando Sutter
Chairman of the Management Board
Enefit, Estonia

Dr. Thomas Sladek
Director, **Ockham Energy Services, USA**

Nathan Frisbee
Senior Analyst, Global Energy Economics, **International Energy Agency, France**

Lance Lehnhof
Director and Vice President of Corporate Affairs, **RedLeaf Resources Inc**

Prof. Vaino Puura
Professor Emeritus
University of Tartu Estonia

Ruslan Salikhov
Deputy Chief Engineer Designer, **TTU Galoter, Russia**

Speakers & Panellists

H.E. Dr.Jawad Anani
Deputy Prime Minister for Economic Affairs

Dr. Ibrahim Saif
Minister of Energy and Mineral Resources

H.E. Mr. Ahmad Al-Qatarneh
Secretary General
Ministry of Environment

H.E. Thabit Al Wir
President, **Jordan Investment Commission**

H.E. Eng. Amani Azzam, Secretary General, **Ministry of Energy and Mineral Resources**

Dr. Ghaleb Maabreh
Chairman
National Petroleum Company

Andres Anijalg
Chairman of the Management Board
Attarat Power Company

His Excellency Eng. Abdelfattah Al Daradkeh, Managing Director
National Electric Power Company

Hussain Al Mansi
Director of Exemptions Affairs,
Jordan Customs

Osama Sulieman
Director PPP Unit
Ministry of Finance

John Brodylo
Vice President Exploration,
Questerre Energy Corporation, Canada

Eng. Wijdan Al Rabadi
Commissionaire, **Energy and Mineral Resources Commission**

Jason Pok
Chief Financial Officer
APCO, Estonia

Yusuf Siddiqui
Shell Country Chairman & General Manager
Jordon Oil Shale Company, Jordan

Dr. Trad Mohammed Al-Ahmed
Saudi Arabian Company for Oil Shale

Pete Coldham
VP Engineering,
Questerre Energy Corporation, Jordan

Malek Abu Rumman
Health, Safety Environment Manager Jordon Oil Shale Company

John Fraser
Director, **Karak International Oil**

Wasfi Alomari
Head of Geology and Geochemistry
Jordon Oil Shale Company Jordan

Dr. Adli Kandah
Director General
Association of Banks in Jordan

Speakers & Panellists

Dr. Khair Hadidi
Water Authority of Jordan

Dr. Jamal Jaber
Faculty of Engineering Technology, **Al-Balqa' Applied University**

Munther Akroush
Vice Chairman
Karak International Oil

Khaled Khraisat
Geology and Mining Manager, **Al-Qamer for Energy and Infrastructure Ltd**

Dr. Yiqun Huang
Doctor, **Tsinghua University, China**

Chris Nurse
Director / CEO
Karak International Oil

Randi Hagemann
Advisor SUS Env Tech FT SST ERO
Statoil ASA, Norway

Paul Quinn
International Consultant
Ambet LLC

Abdelhafid Foughalia
Engineer, **TTU Galoter, Russia**

Dr. Omar Salim Ayed Al-Ayed
Professor, Al-Balqa **Applied University Jordan**

Prof. Alvar Soesoo
Institute of Geology
University of Tartu Estonia

Eng. Munther Shakeeb Bsieso
Technical Manager
Al-Qamer for Energy & Infrastructure. Jordan

Dr. Peng GUO
Sinopec, China

Eyas Shuaibi
Partner
Trowers & Hamlins LLP

Birgit Maaten
Research Scientist
Department of Thermal Energy, **Tallinn University of Technology, Estonia**

Dr. Waleed Bawaneh
Income and Sales Tax Department Jordan

David Argyle
Managing Director
Global Oil Shale Holdings Inc (GOSH)

International Energy Agency
Secure Sustainable Together

World Energy Outlook 2016

Jordan International Oil Shale Symposium
21 November 2016

The global energy context today

World Energy Outlook 2016

- **Key points of orientation:**
  - *Middle East share in global oil production in 2016 at highest level for 40 years*
  - *Transformation in gas markets deepening with a 30% rise in LNG*
  - *Additions of renewable capacity in the power sector higher in 2015 than coal, gas, oil and nuclear combined*
  - *Paris Agreement enters into force, as global emissions stall*
  - *Billions remain without basic energy services*
- **There is no single story about the future of global energy; policies will determine where we go from here**

Total energy demand increases

World Energy Outlook 2016

Energy demand by region in the New Policies Scenario

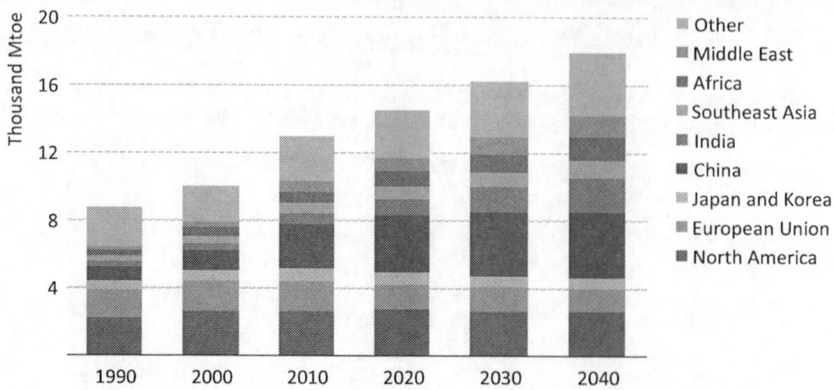

The geography of global energy demand continues to shift

A new 'fuel' in pole position

World Energy Outlook 2016

Change in total primary energy demand

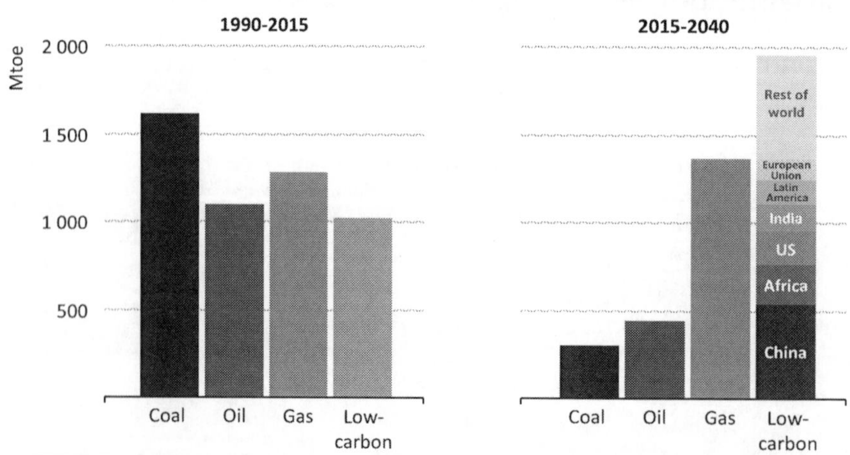

Low-carbon fuels & technologies, mostly renewables, supply nearly half of the increase in energy demand to 2040

A wave of LNG spurs a second natural gas revolution

World Energy Outlook 2016

Share of LNG in global long-distance gas trade

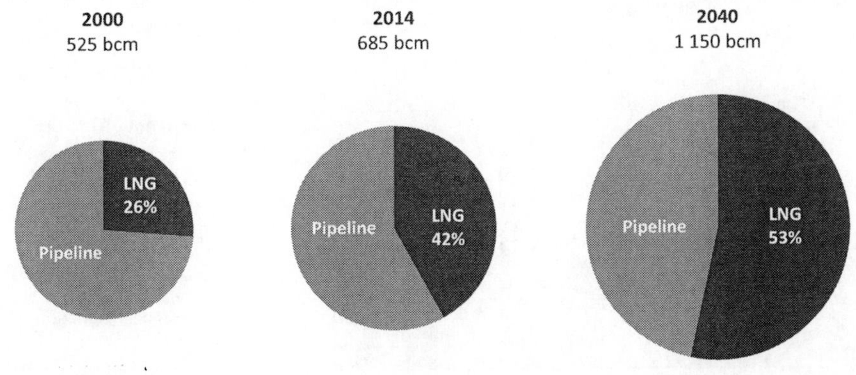

Contractual terms and pricing arrangements are all being tested as new LNG from Australia, the US & others collides into an already well-supplied market

Policy influence on oil demand

World Energy Outlook 2016

World oil demand and price by scenario

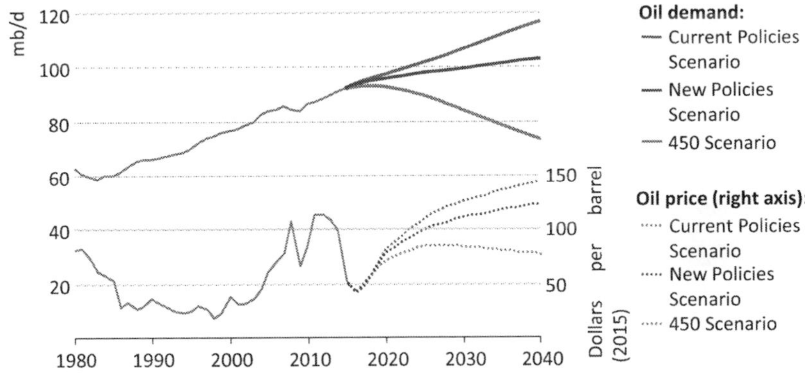

World oil demand exceeds 103 mb/d in 2040 in the New Policies Scenario, falls to mid-1990's level in 450 scenario

A suite of tools to address energy security

World Energy Outlook 2016

Net oil imports

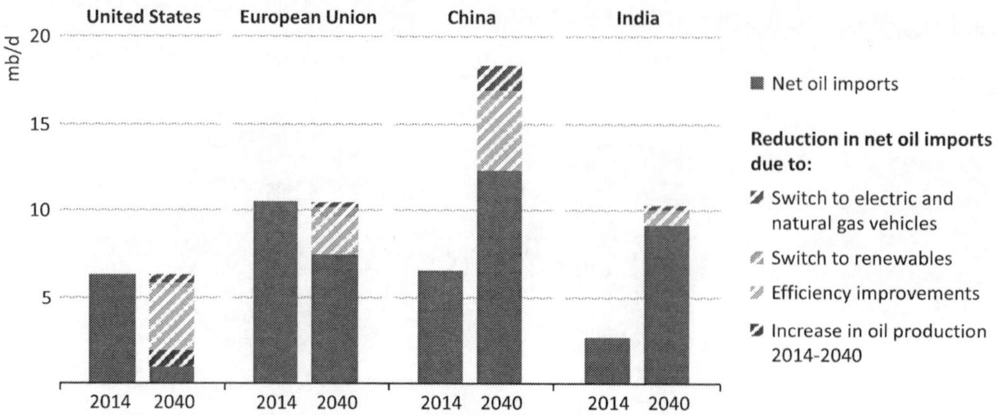

The energy transition provides instruments to address traditional energy security concerns, while shifting the spotlight onto electricity supply

No peak yet in sight, but a slowdown in growth for oil demand

World Energy Outlook 2016

Change in oil demand by sector, 2015-2040

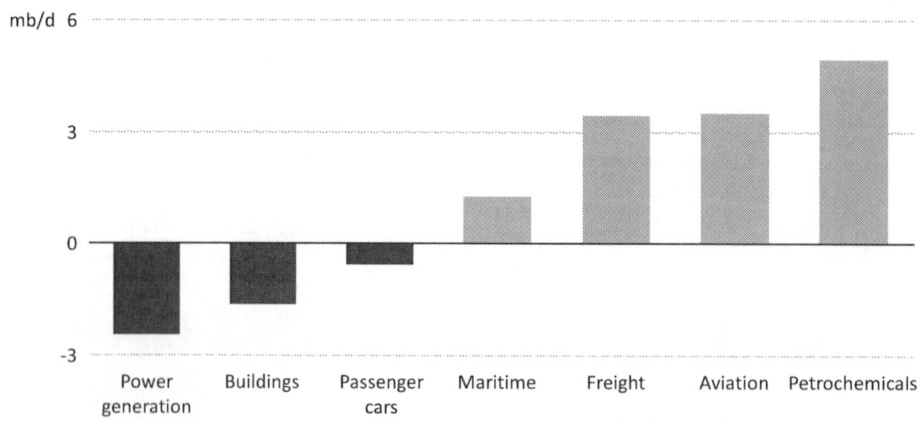

The global car fleet doubles, but efficiency gains, biofuels & electric cars reduce oil demand for passenger cars; growth elsewhere pushes total demand higher

Entering a period of greater oil market volatility

World Energy Outlook 2016

Conventional crude oil resources subject to new project approvals

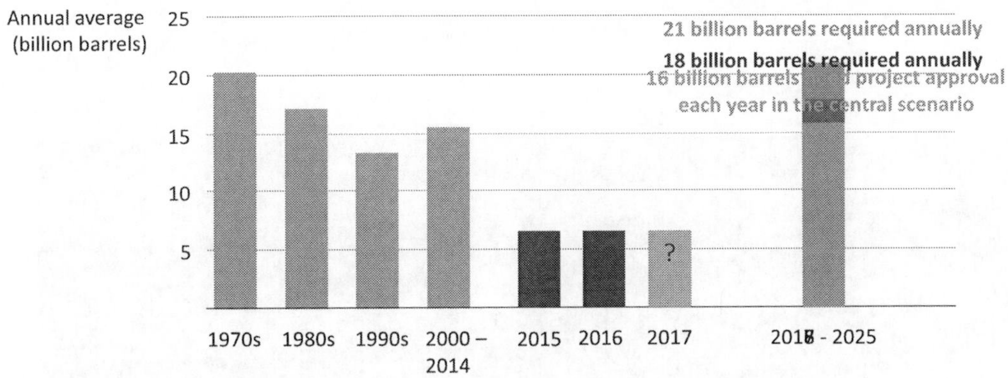

Without a pick-up in investment, or a rapid slowdown in demand growth, the stage is set for the next boom-and-bust cycle for oil

Conclusions

World Energy Outlook 2016

- **Energy security remains a major concern; potential vulnerabilities are growing, so too is the range of tools available to address them**
- **The Paris Agreement is a framework; its impact on energy depends on how its goals are translated into real government policy actions**
- **A wave of LNG is the catalyst for a second natural gas revolution, with far-reaching implications for gas pricing & contracts**
- **Transportation and industrial activity keep demand for oil high, even in a world regulated for climate change mitigation**
- **New oil market dynamics & subdued upstream investment are ushering in a new period of market volatility**

International Energy Agency
Secure Sustainable Together
World Energy Outlook 2016
www.worldenergyoutlook.org
@IEA @IEABirol
© OECD/IEA 2016

100 years in Estonia, 8 years in Jordan

Estonia
100 years of operation

Highest efficiency and lowest environmental impacts in the world

Jordan
8 years of development

554 MW oil shale fired power plant ready for construction

Enefit

Jordan 554 MW Oil Shale Fired Power Plant - Overview

- **2 x 277 MW** (gross) **mine-mouth oil shale fired Power Plant**
- **1st solid fuel** power project in Jordan using local oil shale
- Project comprises of:
  - Power Plant (2 CFB boilers)
  - Oil Shale Mine
  - Fuel & Mixing Yard & Ash Disposal Facilities
  - Water Wells
  - Housing Complex
  - Mine Infrastructure (mine camp, explosive storage, etc,)
- Project complies with World Bank & Jordanian regulations; EIA approved by Ministry of Environment in May 2013
- Total Project Cost: **USD 2.1 bn**
- Construction Period: **42 months from Financial Close**

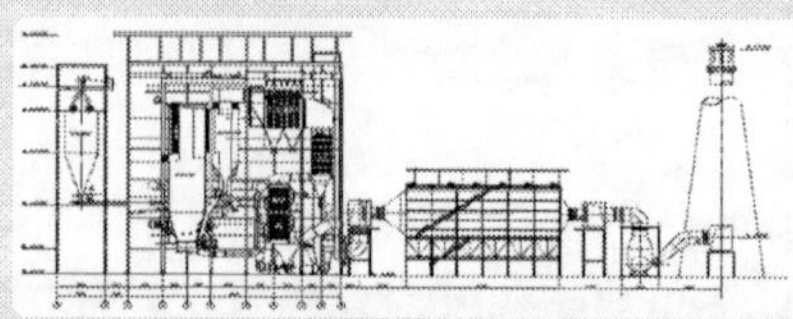

Jordan 554 MW Oil Shale Fired Power Plant – Financing Status

Financing agreements in place:

- Financing and Equity Agreements have been signed for the USD 2.1 bn Project
- Financial close expected in 2016, pending on the approval of Sinosure Export Credit Insurance Policy by Chinese Ministry of Finance
- 554 MW oil shale fired power station operational in 2020

Project sponsors:
Enefit, YTL, Yudean

Enefit Experience with Jordan Power Project

Opportunity Assessment | Conceptual Design | Pre-FEED, FEED | FINANCING | Construction | Operation and Maintenance

Opportunity Assessment

2008 **Power Project Development Agreement** between Enefit and GoJ

2010 **Concession agreement** awarded by GoJ

Conceptual Design / Pre-FEED, FEED

2010 - 2013 Geological & Hydrological Surveys, 3D modeling

2011 YTL from Malaysia joined the project

2011 Power Plant and Mine Conceptual Engineering

2012 Launch of EPC and mining tenders, EIA process, pilot testing with Attarat oil shale

2013 Legislative framework for solid fuels and EIA approval

2013 PPA negotiations with GoJ, JORC resource statement

2014 **EPC agreement** with GPEC

2014 **30-year Power Purchase Agreement** with GoJ

2014 Financing due diligence started

2015 JORC reserve status for oil shale, detailed design for mine and infrastructure completed

2015 Lenders Due Diligence completed, Equity Investor selected

FINANCING

2016 **Financing agreements** signed with Bank of China and Industrial and Commercial Bank of China

2016 **Yudean** joins as strategic investor

2016 Expected Financial Close

Construction

2016 Expected start of construction

Operation and Maintenance

2020 Power plant operational

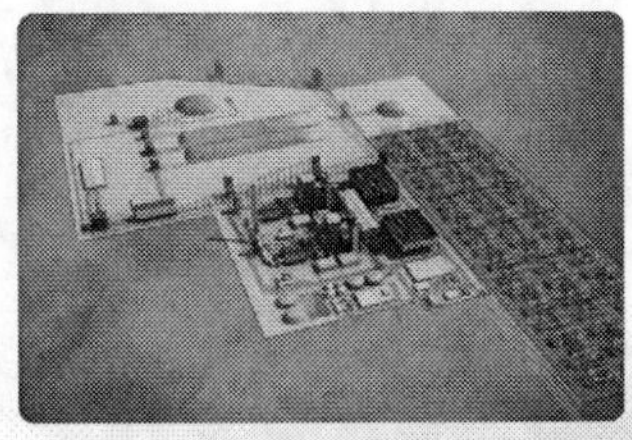

Enefit

One picture tells more than 1000 words...

Enefit

Geological
mapping
Enefit

Oil shale
drilling
during day…
Enefit

...and during
night
Enefit

Gammalogging
Enefit

Videologging
Enefit

Seismic studies
Enefit

Core samples
from the site…
Enefit

…were documented…
Enefit

...and analysed...
Enefit

...and double
checked.
Enefit

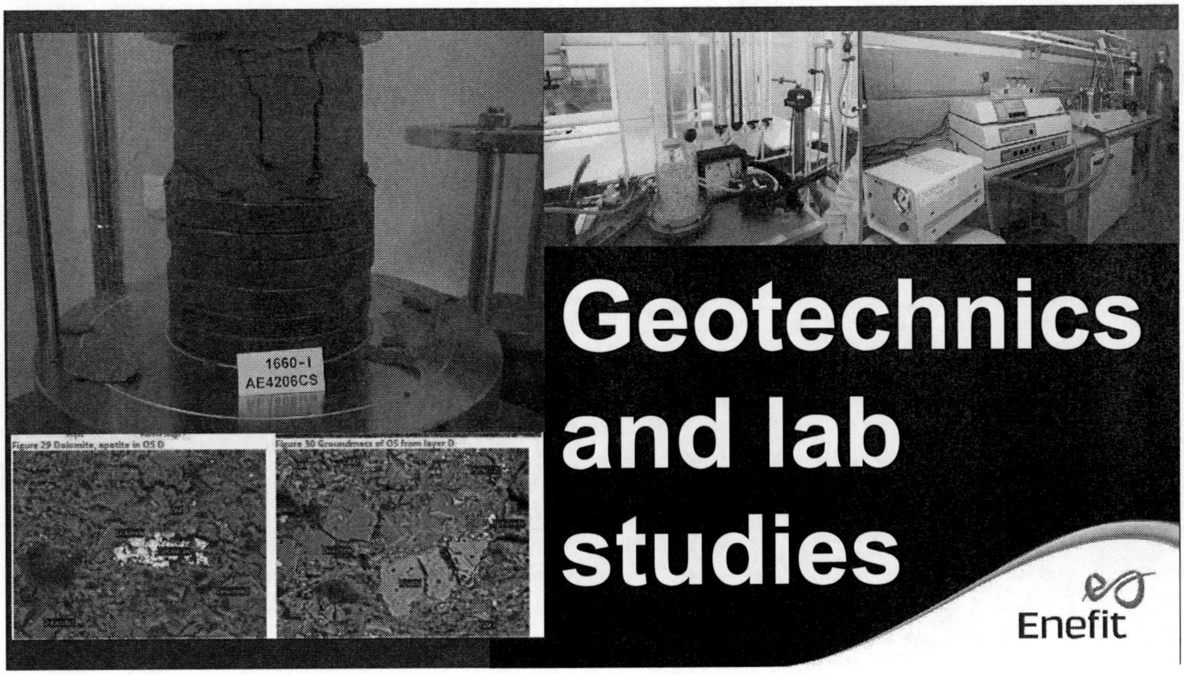
1660-I
AE4206CS
Figure 29 Dolomite, apatite in OS D
Figure 30 Groundmass of OS from layer D
Geotechnics and lab studies
Enefit

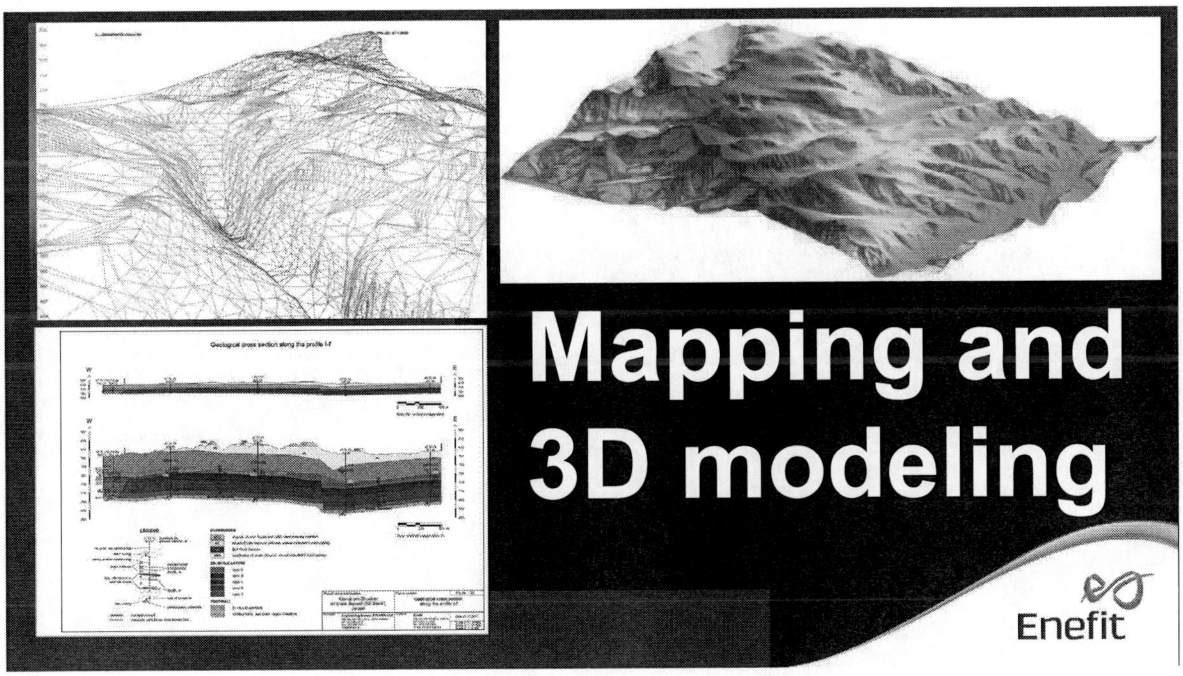
Geological cross section along the profile I-I'
Mapping and 3D modeling
Enefit

Hydrogeological
studies
Enefit

Short and long term
pumping tests
Enefit

Oil shale bulk sampling for pilot testing
Enefit

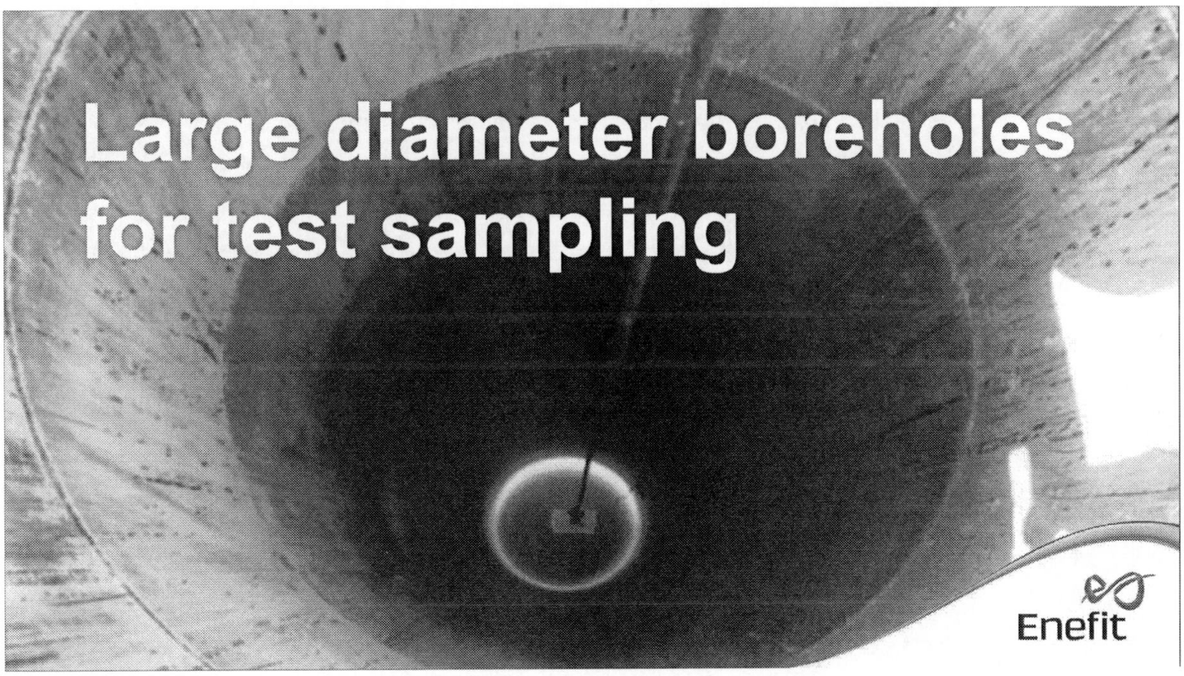
Large diameter boreholes for test sampling
Enefit

Oil shale sample prepared for the shipment
LD-4048
520-525
Enefit

Pilot testing
Enefit

Weather monitoring required
for min 12 months
Enefit

Pilot mining for
mine engineering
Enefit

H.E Dr. Mohammad Hamed
Mr. Sandor Liive
Mr. Joseph Tan
PPA signing in Jordan
Enefit

Enefit
EPC contract
signing in China
Enefit

Jordan Attarat Oil Shale Power Project Agreement Signing Ceremony
广东省粤电集团有限公司
GUANGDONG YUDEAN GROUP
Yudean
New investor signing in China
Enefit

Political support is critical
Enefit

Signing of strategic partnership pact between Jordan and China in 2015
Enefit

Oil Shale in Jordan – Energy Security and Economic Benefits
Long term energy security and stability
• Local energy resource
• Reduces fuel import bill
• Stable - power price not linked to oil price movements
• Kick-starts Jordan´s oil shale industry including oil production
Economic benefits
• 1,000 new jobs
• 5,500 people during construction
• Indirect employment
• Largest foreign investment to Jordan, 2.1 bn USD
• Multiplier effect to local GDP in various industry sectors

Key Elements of a Successful Oil Shale Project

	Oil Production	Power Production
Viable Resource	• Sufficient quality (oil yield, composition) and quantity • Certification: JORC / NI 51-101 • Acceptable mining cost (strip ratio)	• Sufficient quality (calorific value, composition) and quantity • Certification: JORC • Acceptable mining cost (strip ratio)
Proven Technology	• Many experimental technologies available, but only few proven commercial applications • Performance guarantees	• CFB boiler • Performance guarantees
Government Support	• Host Government interest to develop such industry and support the developers • Government guarantees	
Route to Market / Infrastructure	• Upgrader or raw shale oil • Pipeline vs truck vs rail • Available road, water, grid connections	• Local / Regional • PPA with single buyer / open market • Available road, water, grid connections
Environmentally Acceptable	• Existing regulation • Permits (EIA, water licenses, ash disposal)	
Favorable Economics	• All-in $/bbl break even • IRR • DSCR	• Tariff compared to alternatives (e.g. CCGT, coal) • IRR • DSCR

Enefit Offer for Oil Shale Developers

Enefit

Technology Choice (oil&power)

- First hand experience with Oil and Power production technologies
- Bench and pilot testing
- Choice between the best technology and concept for the specific resource.

Project Development Service

- Step-wise approach in reaching the feasibility status with the project
- Resource assessment
- Mine studies and engineering
- Technology choice and testing for oil and power production
- Project structuring for financing
- Legislation and contracting support
- Environmental studies (incl. EIA)
- Permitting support

Enefit

Step-Wise and Practical Approach

Our "stage-gated" approach provides for re-evaluation of the project feasibility and economics at intermediate stages or 'gates'.

Preparation

Opportunity Assessment

- Initial fuel quality assessment
- Checking suitability for combustion or retorting
- Market and infrastructure review
- Initial financial modelling

Conceptual Engineering

- Concept chosen for development
- Geological studies and 3D modeling
- Retort or combustion bench-scale and/or pilot tests
- Conceptual engineering for mine and power or oil plant
- Legislation overview
- Start EIA
- Project financial model and structuring for financing

Pre-FEED, FEED engineering

- Achieve resource and reserve statement for fuel
- Pre-FEED and FEED engineering for mine and power or oil plant
- Complete tendering and contracting
- Permitting and legislation (incl EIA)
- Tariff and offtake agreements
- Start financing
- Reach Feasibility level with the project

FINANCING, INVESTMENT DECISION

Execution

Construction

- Construction
- Testing
- Training
- Commissioning

Operation and Maintenance

- Operating and Maintenance of the plant

Enefit

Enefit´s Value Offer:

Enefit offers a stepwise and practical approach for developing the oil shale throughout the world. Enefit has successfully developed the cleanest and most energy efficient shale oil production technology and is able to design, build and operate large-scale projects in both the oil and power industries.

Enefit

Jordan Field Experiment

Yusuf Siddiqui
Shell Country Chairman & General Manager
Jordan Oil Shale Company (JOSCO)

November 2016

Definitions and Cautionary Note

The companies in which Royal Dutch Shell plc directly and indirectly owns investments are separate entities. In this presentation "Shell", "Shell group" and "Royal Dutch Shell" are sometimes used for convenience where references are made to Royal Dutch Shell plc and its subsidiaries in general. Likewise, the words "we", "us" and "our" are also used to refer to subsidiaries in general or to those who work for them. These expressions are also used where no useful purpose is served by identifying the particular company or companies. "Subsidiaries", "Shell subsidiaries" and "Shell companies" as used in this presentation refer to companies over which Royal Dutch Shell plc either directly or indirectly has control. Companies over which Shell has joint control are generally referred to "joint ventures" and companies over which Shell has significant influence but neither control nor joint control are referred to as "associates". In this presentation, joint ventures and associates may also be referred to as "equity-accounted investments". The term "Shell interest" is used for convenience to indicate the direct and/or indirect (for example, through our 23% shareholding in Woodside Petroleum Ltd.) ownership interest held by Shell in a venture, partnership or company, after exclusion of all third-party interest.

This presentation contains forward-looking statements concerning the financial condition, results of operations and businesses of Royal Dutch Shell. All statements other than statements of historical fact are, or may be deemed to be, forward-looking statements. Forward-looking statements are statements of future expectations that are based on management's current expectations and assumptions and involve known and unknown risks and uncertainties that could cause actual results, performance or events to differ materially from those expressed or implied in these statements. Forward-looking statements include, among other things, statements concerning the potential exposure of Royal Dutch Shell to market risks and statements expressing management's expectations, beliefs, estimates, forecasts, projections and assumptions. These forward-looking statements are identified by their use of terms and phrases such as "anticipate", "believe", "could", "estimate", "expect", "goals", "intend", "may", "objectives", "outlook", "plan", "probably", "project", "risks", "schedule", "seek", "should", "target", "will" and similar terms and phrases. There are a number of factors that could affect the future operations of Royal Dutch Shell and could cause those results to differ materially from those expressed in the forward-looking statements included in this presentation, including (without limitation): (a) price fluctuations in crude oil and natural gas; (b) changes in demand for Shell's products; (c) currency fluctuations; (d) drilling and production results; (e) reserves estimates; (f) loss of market share and industry competition; (g) environmental and physical risks; (h) risks associated with the identification of suitable potential acquisition properties and targets, and successful negotiation and completion of such transactions; (i) the risk of doing business in developing countries and countries subject to international sanctions; (j) legislative, fiscal and regulatory developments including regulatory measures addressing climate change; (k) economic and financial market conditions in various countries and regions; (l) political risks, including the risks of expropriation and renegotiation of the terms of contracts with governmental entities, delays or advancements in the approval of projects and delays in the reimbursement for shared costs; and (m) changes in trading conditions. All forward-looking statements contained in this presentation are expressly qualified in their entirety by the cautionary statements contained or referred to in this section. Readers should not place undue reliance on forward-looking statements. Additional risk factors that may affect future results are contained in Royal Dutch Shell's 20-F for the year ended December 31, 2015 (available at www.shell.com/investor and www.sec.gov). These risk factors also expressly qualify all forward looking statements contained in this presentation and should be considered by the reader. Each forward-looking statement speaks only as of the date of this presentation, November 21, 2016. Neither Royal Dutch Shell plc nor any of its subsidiaries undertake any obligation to publicly update or revise any forward-looking statement as a result of new information, future events or other information. In light of these risks, results could differ materially from those stated, implied or inferred from the forward-looking statements contained in this presentation.

We may have used certain terms, such as resources, in this presentation that United States Securities and Exchange Commission (SEC) strictly prohibits us from including in our filings with the SEC. U.S. Investors are urged to consider closely the disclosure in our Form 20-F, File No 1-32575, available on the SEC website www.sec.gov. You can also obtain these forms from the SEC by calling 1-800-SEC-0330.

Outline

- *Royal Dutch Shell Strategy*
- *In Situ* Conversion (ICP) Technology
- Jordan OSCA Review
- Jordan Field Experiment (JFE) Results

Our Strategy: **Shell manages its portfolio around three strategic themes: cash engines, growth priorities and future opportunities.**

Cash engines: today	Growth priorities: 2016+	Future opportunities: 2020+
Funds dividends + balance sheet	Cash engines 2020+	Material value + upside
Competitive + resilient	Affordable growth in advantaged positions	Path to profitability
Strong free cash flow and returns	FCF + ROACE pathway	Managed exposure
CONVENTIONAL OIL + GAS; INTEGRATED GAS; OIL SANDS MINING; OIL PRODUCTS	DEEP WATER; CHEMICALS	SHALES; NEW ENERGIES

Relentless portfolio high-grading

SHELL IN JORDAN PORTFOLIO OVERVIEW

UPSTREAM	MIDSTREAM	DOWNSTREAM	VALUE DELIVERY FOR JORDAN
JOSCO 100% Shell			→ Oil, Domestic & Export
	LNG SUPPLY 100% Shell		→ Power Generation
			→ Industrial Sector & Factories
		LUBRICANTS 100% Shell Local Distributor	→ Quality Performance Lubricants
		CHEMICALS 100% Shell B2B	→ Products
		FUEL 100% Shell B2B	→ Quality Refinery Products
		Shell Foundation / GroFin — Shell Foundation/ Grofin	→ SME Financing

In-Situ Conversion Process (ICP)

- ICP enhances the natural maturation of kerogen by *in situ* heating resulting in:
  - Thermal cracking
  - *In situ* hydrogenation
- High Sweep (thermal conduction not impacted by heterogeneity)
- High Recovery (vapor-phase in reservoir)
- High Value Products (heavy ends left behind)

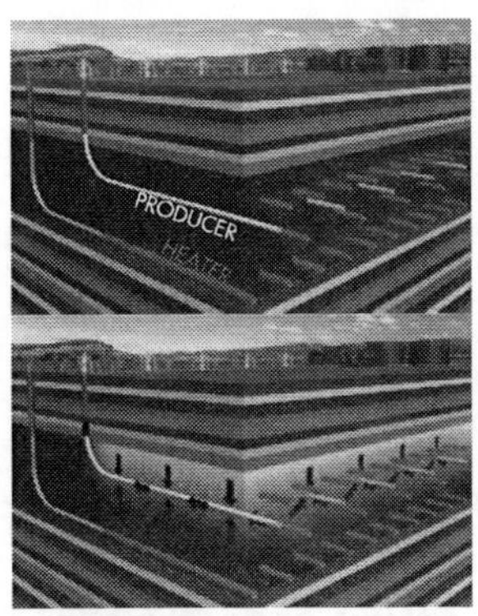

- *Heaters inserted into drilled holes*
- *Resource heated to ~330°C.*
- *Kerogen thermally cracked to lighter components*
- *Coke left behind*

JOSCO – Oil Shale Concession Agreement (OSCA)

- OSCA ratified by Jordan Parliament Aug 2009:
  - Initial acreage: 22,270 km²
- OSCA frames the path to a Commercial Project
- Assess Phase drilling began in 2010; 340 wells drilled
- Appraise Phase began May 2013; reduced holding to 15,000 km². Focus: Define final 1,000 km² and execute JFE
- Excellent Governmental support

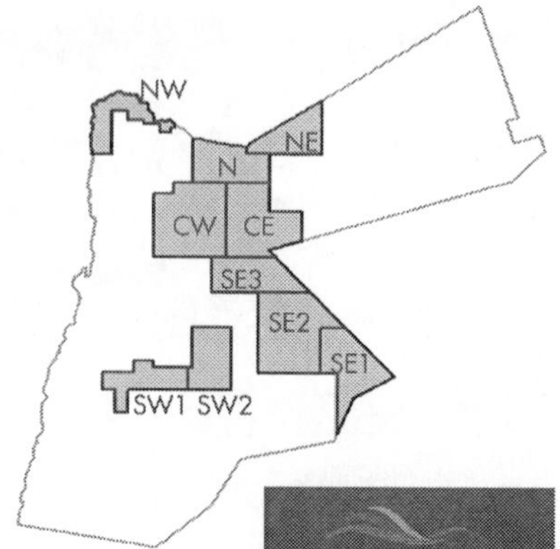

JOSCO
JORDAN OIL SHALE COMPANY

Oil Shale Concession Agreement – Project Phases

Identify	Assess	Appraise	Pilot	Design	FID
Is Jordan of interest and can we secure acreage?	Does the play exist/work?	Is the materiality and productivity present?	Can we produce commercially?	How is the development defined?	Final Investment Decision
Regional resource evaluation: • Data integration: • Oil Shale wells • Petroleum wells • Water wells • Oil Shale Richness **Negotiate Concession Agreement:** • Investment Climate • Commercial Terms	**Explore for sweet spots:** • 9 months Venture Set Up • Drill/Learn/Drill campaign • Assess Richness, Thickness and Overburden • Regional G&G Evaluation • Define key risks	**Focused maturation:** • Define sweet spots - infill drilling • Define energy balance - kerogen conversion kinetics • Show heaters work & integrity can be maintained – rock mechanics • Show we have containment – hydrogeology • Demonstrate ICP works in Jordan – Jordan Field Experiment (JFE) • Characterize resultant products	**Field Trials to demonstrate:** • Key ICP parameters: • Well design & performance • Well spacing • Surface components • Product volumes • Waste products • Market requirements • Cost reduction	**FEED:** • Scope: • Plateau oil rate • Power generation • Oil & Gas treating • Refining strategy • Water • Cost • Efficiency • Learning Curve Application • Schedule • Drilling pace • Capital phasing	
✓ OSCA ratified	✓ Sweet spots identified	✓ Sweet spots appraised ✓ Successful JFE	○ ICP commerciality ○ Successful pilot	○ Commercial project defined	

Jordan Field Experiment (JFE)

- Overview
- Wells/Heaters
- Facilities
- Operational Excellence
- Milestones
- Production
- Results

JFE – Drilling Operations

JFE – Water Treaters

JFE Overview

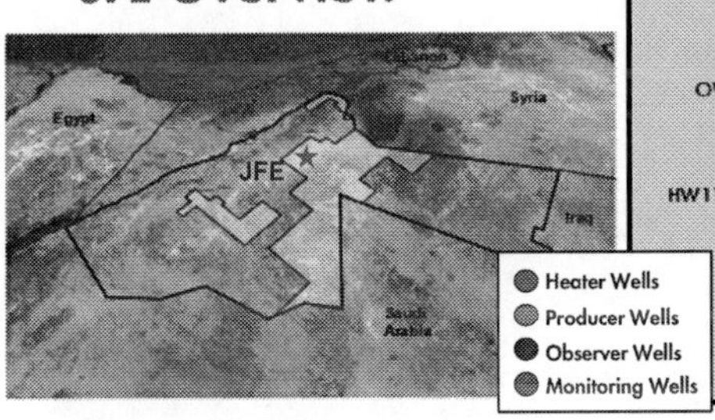

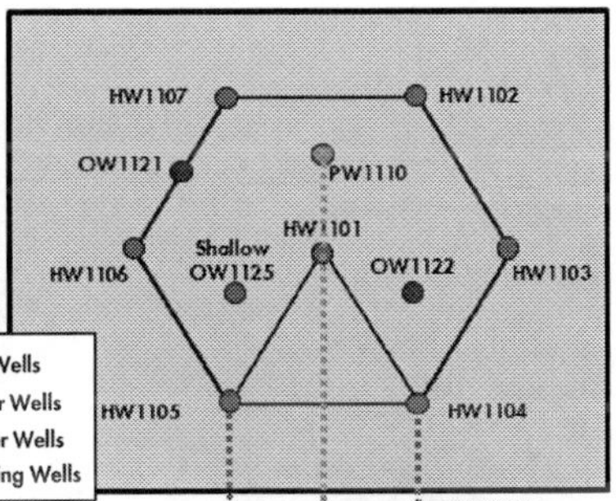

Timeline:
- Skid Fabrication – 2014/15
- Onsite Drilling and Construction – Apr 2014 – Sep 2015
- Heaters On – Sep 26, 2015
- First Gas – Nov 24, 2015
- First Oil – Mar 16, 2016
- Heaters Off – Aug 10, 2016
- Cool Down through mid 2017

Objectives:
- Produce light, high value hydrocarbons via ICP
- Calibrate subsurface models
- Assess in-country capabilities

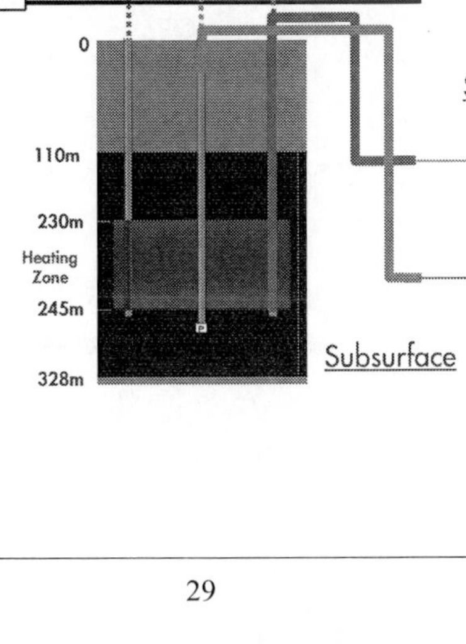

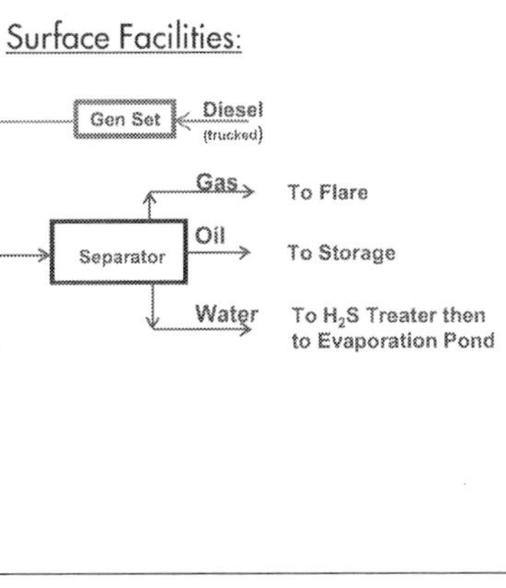

JFE - Wells/Heaters

- Drilling/Completion
  - ***7 heaters, 17 total wells***
  - Drilled to within **2 ft radial tolerance**
  - Instrumented for pressure & temperature measurement.
- Wells/Heaters
  - Three phase, Mineral Insulated (MI) electrical heaters
  - Vertical deployment, 15 m heated interval for pilot application
  - Heater powered by diesel generators

Note: A full length, long life, high power heater has been designed, built, and tested and is available for commercial deployment.

JFE – Well Completions

JFE – Heater Deployment

JFE - Facilities Construction

- Simple Oil, Gas & Water separation, sampling, and measurement. **Focus on high reliability data collection**
- Extensively instrumented. **Real time data** available in Amman and Houston.
- Skid fabricated in USA, Europe, and Jordan. Engineering contractor: W. S. Nelson (Houston)
- Drilling conducted concurrently with Construction. SIMOPS effectively managed.
- Construction contractor: CCC- Jordan.
- Excellent safety record during construction (over 600 day "Goal Zero" attainment during JFE construction).

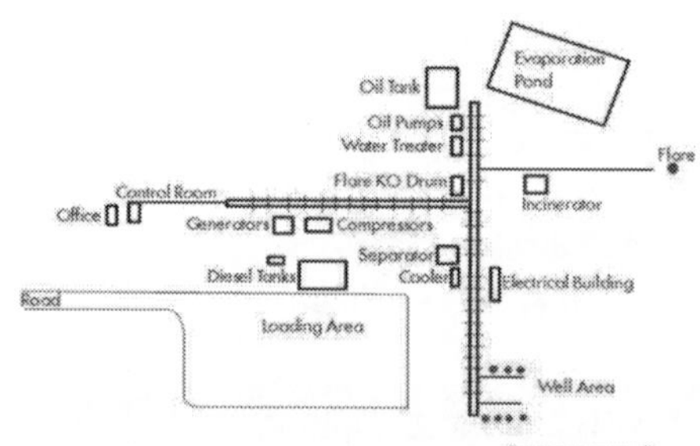

JFE – Overhead Layout

JFE – Separator Skid

JFE – Aerial View

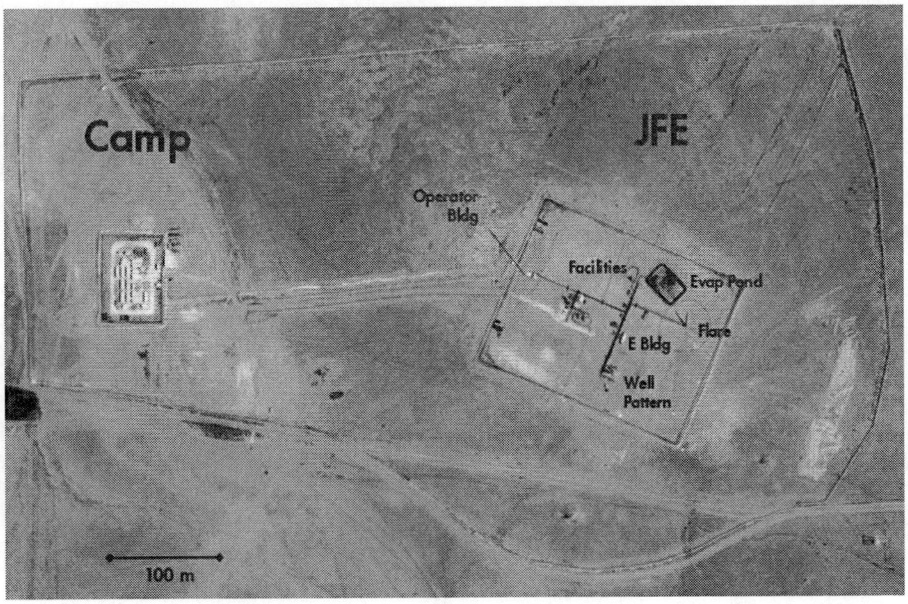

JFE Operational Excellence

- Challenges:

Sour Process: With very high H_2S composition 50+% in the production stream, and unexpected operational challenges

Novel ICP Technology: Unconventional development applied for the first time with Jordanian oil shale

- Actions:

Staff Development & Training: Raised the qualifications and capabilities of JOSCo employees to operate a sour facility with trainings in other Shell operations.

Proactive O&M Strategy: Developed a reliable M&I schedule, and involvement with P&T Houston for critical operational challenges

- Results:

Achieved 1000+ GZD: no LTI's, no process safety incidents, no leaks and proved full containment of oil shale reservoir

100 % Uptime Facility: 0% downtime as a result of full compliance to JFE M&I system, and robust monitoring and protection systems

ICP Technology Proves To Be Successful: Production of oil from Jordanian oil shale

JFE - Milestones

- Heaters Energized: Sep 26, 2015
- Full Power: Oct. 1, 2015
- First Gas: Nov 24, 2015
- First Oil: Mar 16, 2016
- Target Heat Injected: Aug 10, 2016 (Heaters off)

Official opening of JFE by H.E. Minister of MEMR – Sep 29, 2015

First Gas

Heaters Energized

First Oil – 16 March, 2016

JFE Production

- First Gas – Nov 24, 2015
- First Oil – Mar 16, 2016
- Oil – 20 – 30 API
- Gas – Similar to Colorado tests except higher H_2S (due to more sulfur in rock). Treated gas* will be rich in NGLS (~25% C_{2+}) and have high hydrogen content (>40%).
- Production dropped rapidly after turning off heaters (Aug 10, 2016)

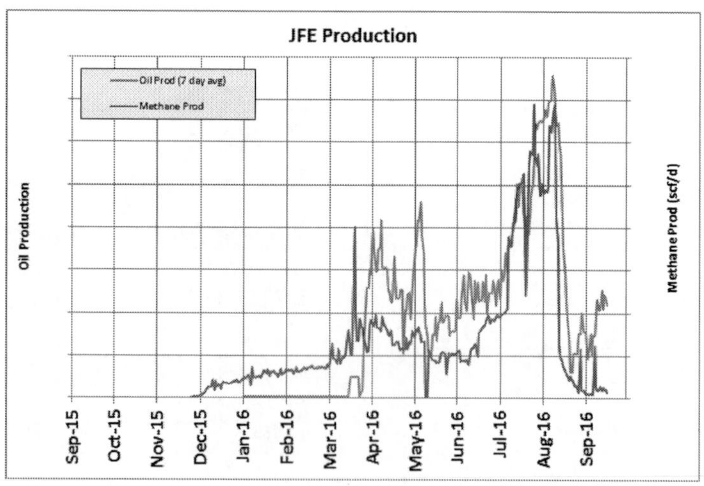

*after CO_2 and H_2S removal

JFE Results vs. Key Objectives

✓ ■ Produce light, high value hydrocarbons via ICP – Significant volume of oil produced.

✓ ■ Assess in-country capabilities – Achieved excellent performance as exemplified by 1000+"Goal Zero" days and 100% heater uptime

Ongoing ■ Calibrate subsurface models – Significant data collected mostly within expectations. Reservoir history match and project report out are ongoing.

<u>In addition:</u>

✓ ■ Demonstrated control of water ingress and excellent containment of ICP pyrolysis products

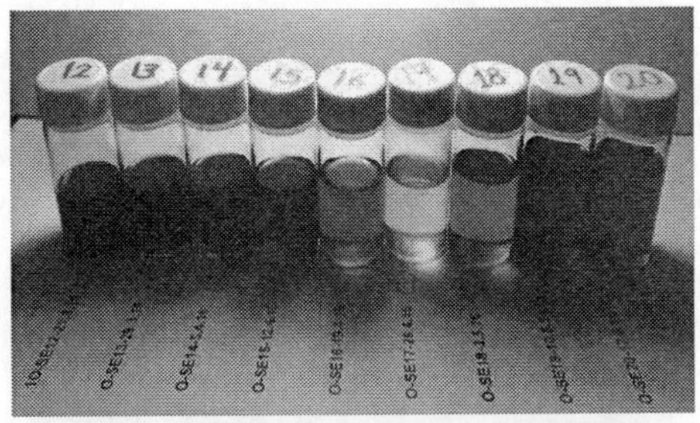

Weekly Oil Samples

Operational Excellence

Al Lajjun Oil Shale Project – Jordan

Jordan International Oil Shale Symposium Presentation

Karak International Oil

21st November 2016

KIO Oil Shale Project Progress

- KIO Oil Shale Project Progress
- Positive Local impact of KIO's project
- Challenges of oil shale development

2

KIO Project Progress
Al Lajjun Shale into barrels of Oil

KARAK INTERNATIONAL OIL

- Infill drilling to delineate the CA resource to NI43-101 standard
- Trial mining of c. 50 kt of material, with a representative sample of 500 tonnes of oil shale sent to Calgary, Canada for pilot plant runs
- Multiple pilot plant runs to confirm the yields and quality of low-sulphur liquid products after hydro-treating – confirmed the recovery of 117litres/tonne and potential for finished products
- Assessing the market needs for finished products now and in future

Mining Oil Shale → Processing Oil Shale using ATP → ATP trial Jordanian Shale Oil → Upgrading the raw shale oil → High quality finished products

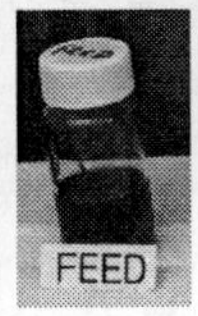

3

ATP oil shale processor
How does the technology work?

KARAK INTERNATIONAL OIL

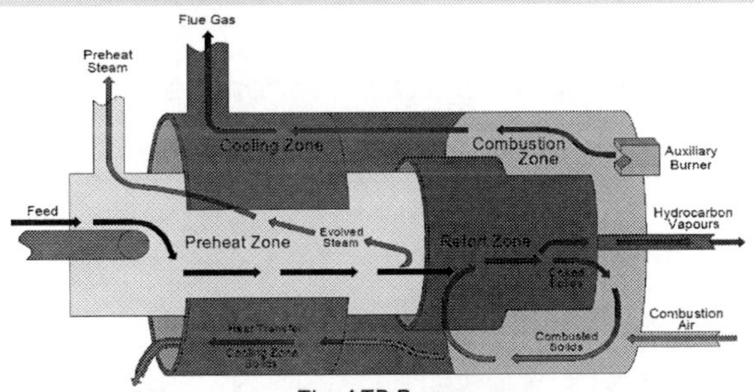

The ATP Processor

1. Preheat Zone
- Feed enters ATP Processor
- Heat transfer cooling zone heats feed to ~250°C–300°C
- Feed dried generating steam
- Steam-rich vapours removed
- Solids moved to the retort zone

2. Retort Zone
- Heated to ~500°C by spent solids from combustion zone
- Pyrolysis in anaerobic atmosphere generates vapours and coke-coated solids.
- Vapours generate liquid and gas products
- Coked solids to combustion zone

3. Combustion Zone
- Air added to burn coked solids
- Temperature ~740°C
- Coke combustion usually supplies all process heat
- Portion of solids recycled to retort zone as a heat source

4. Cooling Zone
- Heat from hot waste material used to heat preheat zone
- Waste material cooled to ~350°C–450°C
- Solids sent to spent shale moistening
- Gases sent to flue gas handling system

4

ATP oil shale processor
Technology currently exceeding expectations

KARAK INTERNATIONAL OIL

Plant achieved 91% availability for first six months of 2016, 87% for 12 month period June 2015-June 2016.

5

Engineering progress

KARAK INTERNATIONAL OIL

- **Continually aiming to optimising the project economics by enhancing the engineering – size of plant, arrangement of plant, potential sourcing of plant equipment and resource utilisation**
- **Enhanced crushing and transportation studies**
- **ATP retort sizing optimisation - target 25,000 bpsd capacity**
- **Front End Engineering on Upgrading and Utilities**
- **Product slate alignment with local market needs**

6

Environmental & Social Impact Assessment Study (ESIA) – Approved by Ministry of Environment

KARAK INTERNATIONAL OIL

1. Support to Environmental Clearance Process (Scoping Sessions)
2. Technical Scopes & Baseline Reporting
3. Impact Assessment & Mitigation
4. Impact Assessment Workshops
5. Environmental & Social Management Plan (ESMP)
6. Stakeholder Engagement & Stakeholder Engagement Plan (SEP)
7. ESIA Report

❖ **ESIA conforms to:**

- **Concession Agreement requirements**
- **IFC Performance Standards**
- **Equator Principles III**

7

ESIA Data Collection

KARAK INTERNATIONAL OIL

- Climate
- Air Quality
- Surface & Sub-Surface Water Resources
- Archaeology and Cultural Heritage
- Soil Conditions & Flora and Fauna
- Uranium/Radon
- Socio-Economics; Gender and human rights; Community health, safety and security
- Traffic and Transport
- Noise and Vibration

8

ESIA
Air Quality Devices

Osiris – Wind Speed, Direction

AQMESH – Air Quality - Sox. Nox

Tubes to collect dust particles

- Monitoring data sent by SMS for KIO to access via online website
- 12 months baseline data collected for ESIA

9

Social Investment Projects

KIO has shown commitment to the best interest of the local community in Karak – the local neighbourhood of the project - through its Corporate Social Investment initiatives

KIO's Corporate Social Investments, which provide the social license to operate focus on the following:

➢ **Training**

➢ **Education**

➢ **Health Care**

10

KIO's Sustainable Development Projects
Mutah University Lab

- **Location:** Mutah, Al-Karak
- **The Project:** Establishing a computer lab
- **Project Objectives:** The lab offers the following services:
  - College testing system
  - Supporting e-learning examination system
  - Theses discussions
  - TOEFL examinations
  - University Entrance Exam

11

KIO's Sustainable Development Projects
Lajjun Primary School

- **Location:** Lajjun, Al-Karak
- **The Project:** Adding three classrooms to Al-Lajjun Primary School
- **Project Objectives:**
  - Building new classes to reduce the number of students in the already overcrowded classes
  - Providing space to add higher classes
  - Providing space to separate the mixed gender classes
  - Allowing the school to accommodate new students

12

KIO's Sustainable Development Projects
Maternity Care Facility

KARAK INTERNATIONAL OIL

- **Location:** Al-Ghwair
- **The Project:** The maternity facility in Al-Ghwair area, 12 km to the south of Lajjun area, was built & furnished by KIO
- **Project Objectives:** The center provides the following services:
  - Maternity care
  - Vaccination/Immunization to new born babies and toddlers

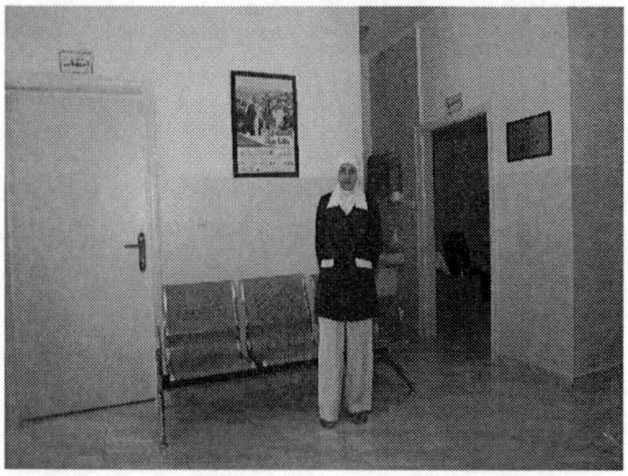

13

Project Challenges

KARAK INTERNATIONAL OIL

- **Project Financing**
- ✓ **Raising $2 billion of project financing**
- ✓ **Natural resources versus renewables – existing oil shale projects initiated with public support**

- **Economics**
- ✓ **The decline of the global oil prices**
- ✓ **Long term off-take contract to underpin revenues**

- **Regional Stability**
- ✓ **Syrian Conflict created refugee crisis adding strains on Jordan's economy, infrastructure, and resources - underpins the need for projects creating long term skilled jobs in Jordan.**

14

Questions ?

KIO team available to discuss further with interested parties at our stand.

15

SACOS

الشركة السعودية العربية للصخر الزيتي ذ.م.م.

Saudi Arabian Corporation for Oil Shale

IOST

Inv. Oily Shale Technology

SACOS

الشركة السعودية العربية للصخر الزيتي ذ.م.م.

Saudi Arabian Corporation for Oil Shale

SAUDI ARABIAN CORPORATION FOR OIL SHALE

SACOS

SINCE 2008

SACOS OBJECTIVES

- Extraction and processing of oil shale.
- Extraction of energy generation related products.
- Hydrocarbon production from oil shale resources.

SACOS BACKGROUND

- On October 13, 2008 a Memorandum of Understanding (MOU).
- On March 13, 2014 The Concession Agreement.

Investment of Oil Shale Technology

IOST

Since 2007

IOST Objectives

- IOST is a technology provider company in the field of oil shale.
- IOST is a specialized company in the research of oil shale industry.
- IOST provides a comprehensive study for the investment of oil shale industry.

IOST Background

- IOST established in 2007 AD as a research company.
- IOST built the first proto type in 2008 AD that treats 3Kg of oil shale.
- IOST built more advanced proto type in 2009 AD that treats 15 Kg of oil shale.
- IOST built its first pilot plant in 2011 that treats 1700 Kg/hour of oil shale.
- IOST granted four patents of oil shale processing by 2015 AD.

  http://goo.gl/g3MVoZ
- IOST registered the oil shale treatment technology world widely in 2016 AD by the validation of JACOBS.

الشركة السعودية العربية للصخر الزيتي ذ.م.م.

Saudi Arabian Corporation for Oil Shale

SACOS and IOST Alliance

ALLIANCE OBJECTIVES

- SACOS & IOST Alliance have the required financial and technical capability to conduct and develop oil shale enterprises in the Hashemite Kingdom of Jordan.
- IOST provides the oil shale treatment technology.
- IOST provides comprehensive and definitive study for SACOS to be engaged in the actual investment stage.
- In the investment stage; SACOS is planning to build a processing unit to obtain a minimum of 300 shale oil barrels per day, accordingly; IOST is taking the responsibility to do so.

IOST REGISTERED TREATMENT TECHNOLOGY

Technology overview.

- Extraction metallurgy.
- The principle of pyrometallurgy (Thermal Mining).
- The principle of thermal dismantling (Calcinations).
- Combustion gases purification unit.

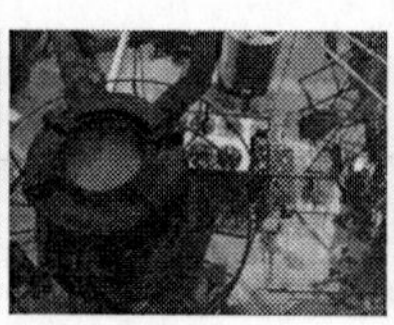

Investment side

- The relation between the oil shale treatment and the price of the crude oil.
- IOST takes advantage from the oil shale ash.
- The proposed 300 shale oil unit is still lucrative even with the 36$ for the crude oil barrel.

Environmental impact

- No enrichment operation needed.
- No water or any other gages are needed to separate the organic materials.
- No solvent materials are needed to separate the organic materials.
- The treatment processes are performed in a closed thermo-dynamic cycle.
- Combustion gases purification unit.

SACOS

الشركة السعودية العربية للصخر الزيتي ذ.م.م.

Saudi Arabian Corporation for Oil Shale

IOST

Inv. Oily Shale Technology

We would like to express our thanks and appreciation to the **Hashemite Kingdome of Jordan** headed by **His Majesty King Abdullah II** for providing the appropriate environment for the scientific research and investment. Our thank extend to **JACOBS** for there efforts.

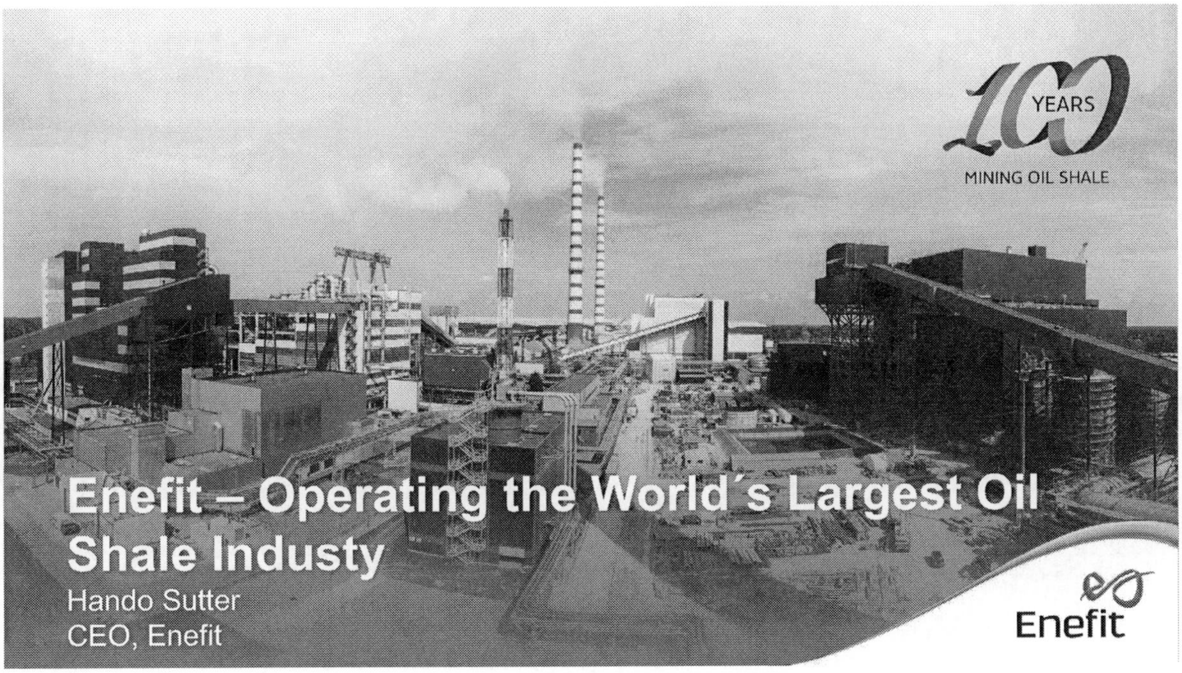
100 YEARS
MINING OIL SHALE
Enefit – Operating the World´s Largest Oil Shale Industy
Hando Sutter
CEO, Enefit
Enefit

1.1
Billion tons of oil shale has been mined in Estonia during the past 100 years
650
TWh of electricity has been produced in 100 years
200
Million barrels of shale oil has been produced during 100 years
Whole towns have been built in Estonia because of oil shale industry

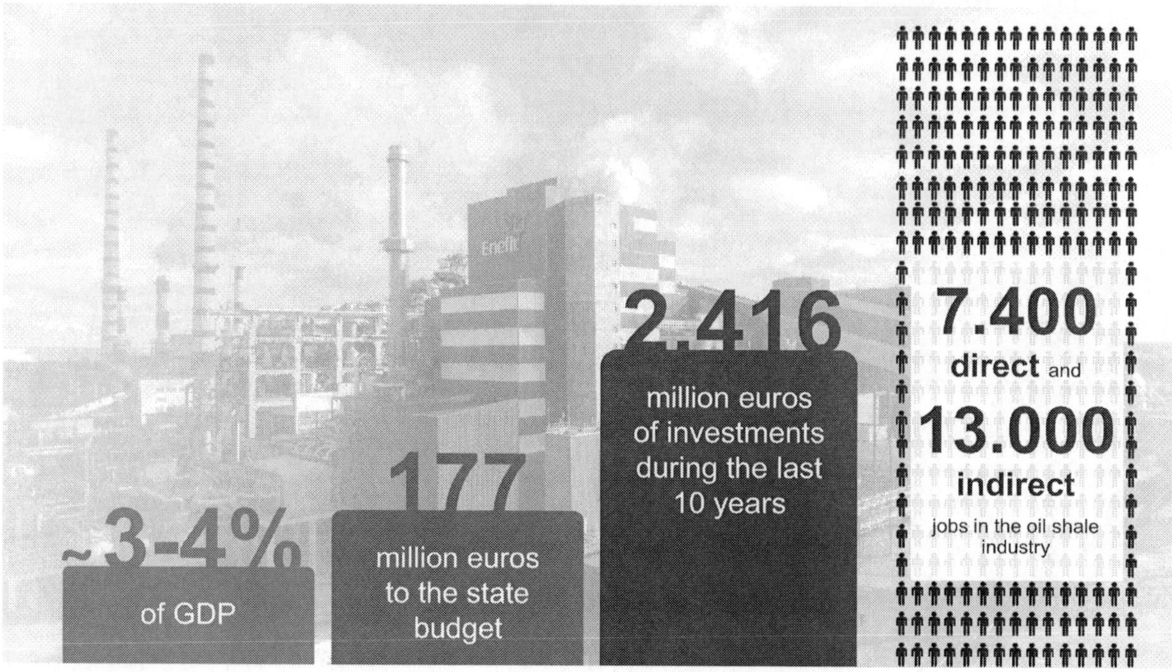

~3-4%
of GDP
177
million euros
to the state
budget
2.416
million euros
of investments
during the last
10 years
7.400
direct and
13.000
indirect
jobs in the oil shale
industry

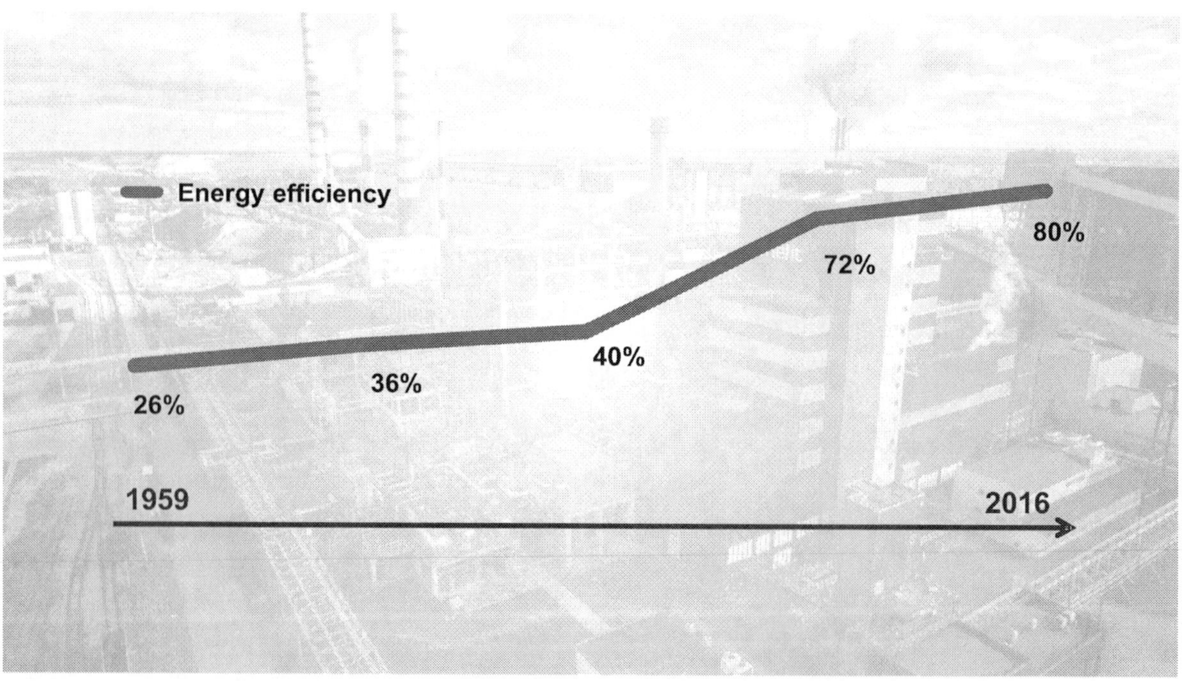

Energy efficiency
26%
36%
40%
72%
80%
1959
2016

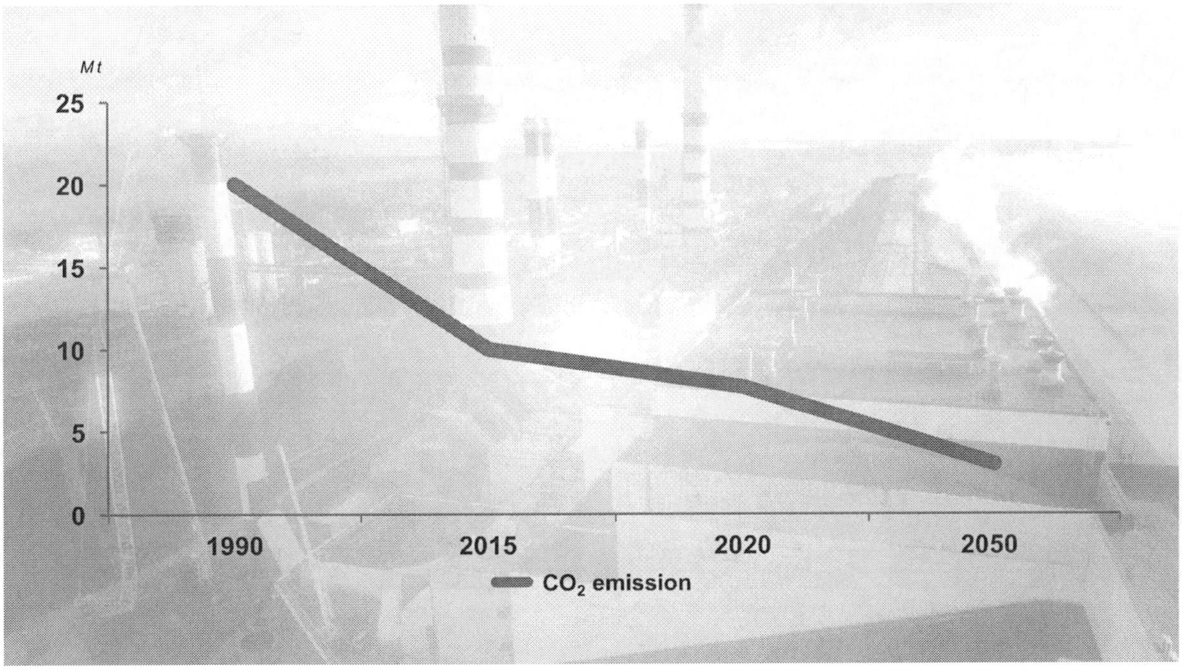
Mt
25
20
15
10
5
0
1990
2015
2020
2050
CO_2 emission

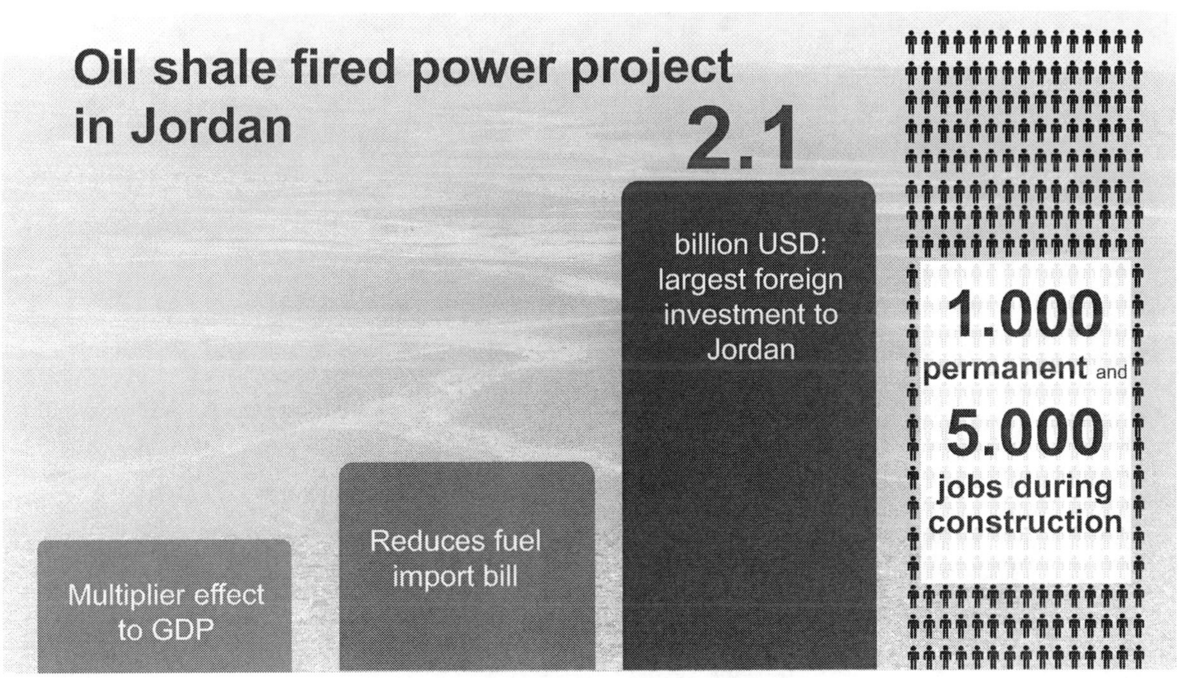
Oil shale fired power project in Jordan
2.1
billion USD: largest foreign investment to Jordan
Reduces fuel import bill
Multiplier effect to GDP
1.000
permanent and
5.000
jobs during construction

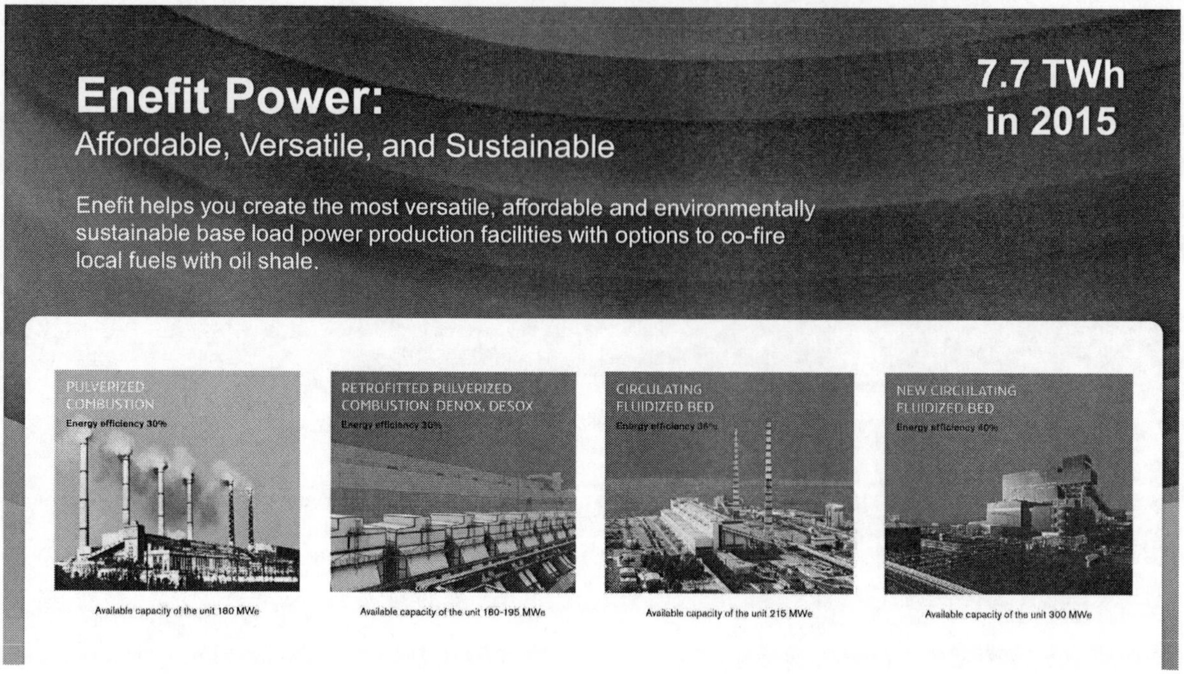

Enefit Power:
Affordable, Versatile, and Sustainable
7.7 TWh
in 2015
Enefit helps you create the most versatile, affordable and environmentally sustainable base load power production facilities with options to co-fire local fuels with oil shale.
PULVERIZED COMBUSTION
Energy efficiency 30%
Available capacity of the unit 180 MWe
RETROFITTED PULVERIZED COMBUSTION: DENOX, DESOX
Energy efficiency 30%
Available capacity of the unit 180-195 MWe
CIRCULATING FLUIDIZED BED
Energy efficiency 36%
Available capacity of the unit 215 MWe
NEW CIRCULATING FLUIDIZED BED
Energy efficiency 40%
Available capacity of the unit 300 MWe

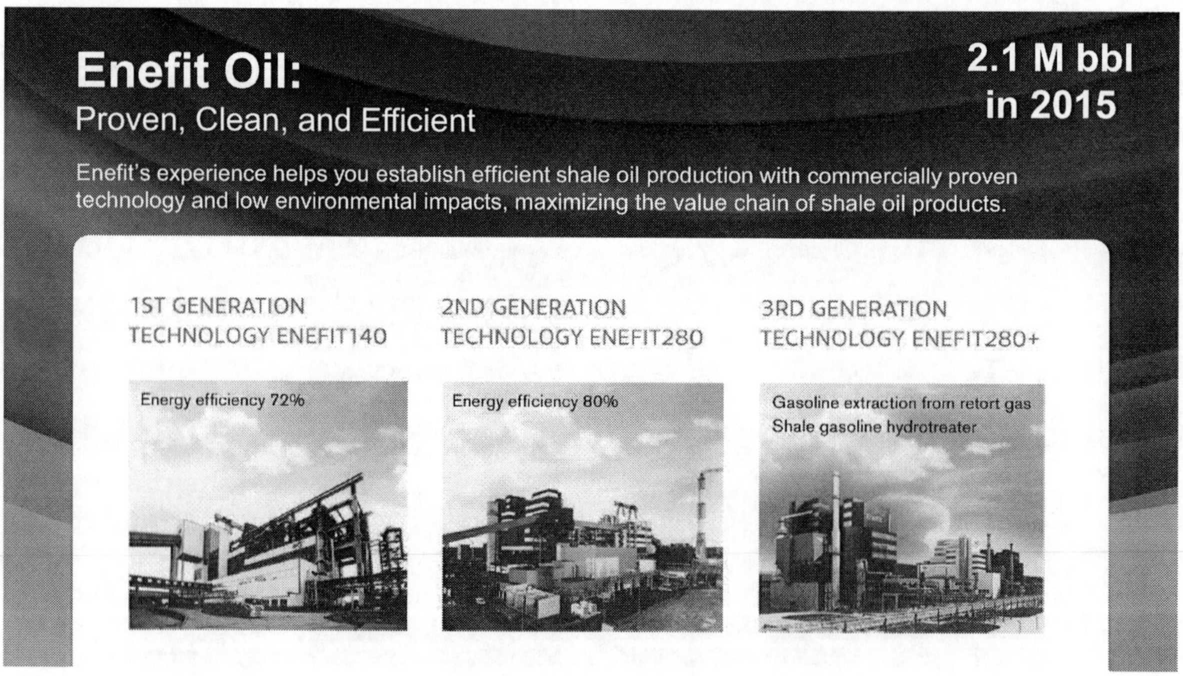

Enefit Oil:
Proven, Clean, and Efficient
2.1 M bbl
in 2015
Enefit's experience helps you establish efficient shale oil production with commercially proven technology and low environmental impacts, maximizing the value chain of shale oil products.
1ST GENERATION TECHNOLOGY ENEFIT140
Energy efficiency 72%
2ND GENERATION TECHNOLOGY ENEFIT280
Energy efficiency 80%
3RD GENERATION TECHNOLOGY ENEFIT280+
Gasoline extraction from retort gas
Shale gasoline hydrotreater

By-products:
Value-added Revenue Streams
Burnt shale as construction material
Burnt shale and waste rock for road and port construction
Burnt shale as soil fertilizer
Burnt shale as foundation for new energy projects
Municipal waste for electricity and heat
Producing oil from old tyres

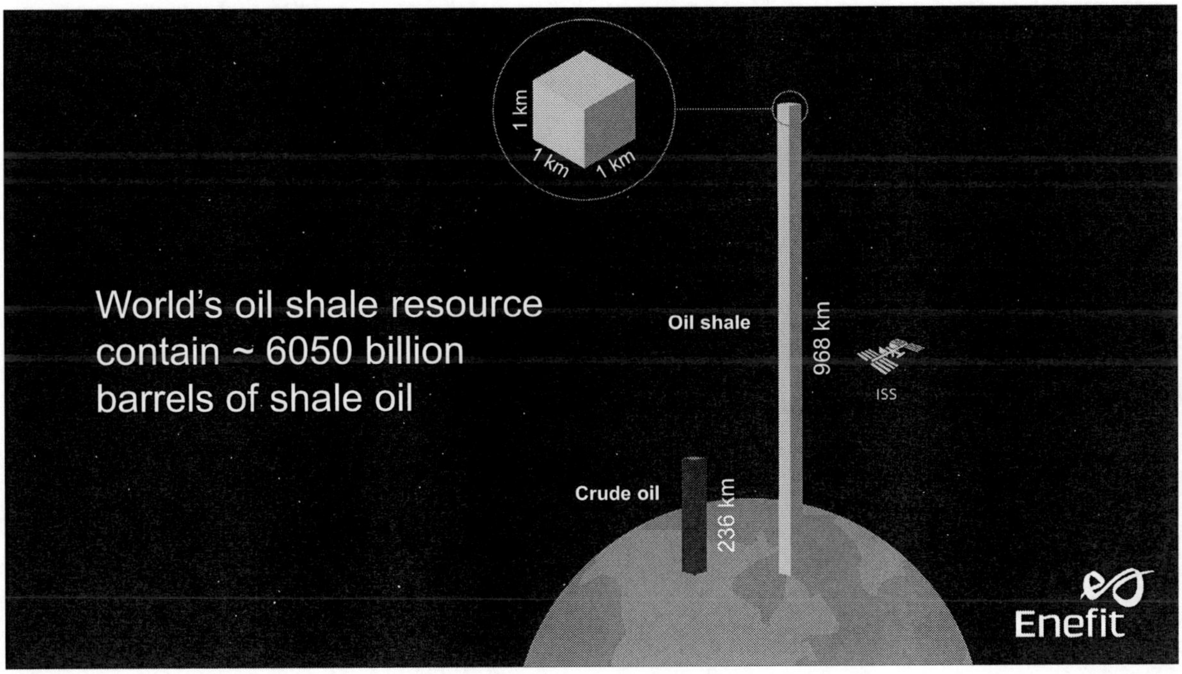
1 km
1 km
1 km
World's oil shale resource contain ~ 6050 billion barrels of shale oil
Oil shale
968 km
ISS
Crude oil
236 km
Enefit

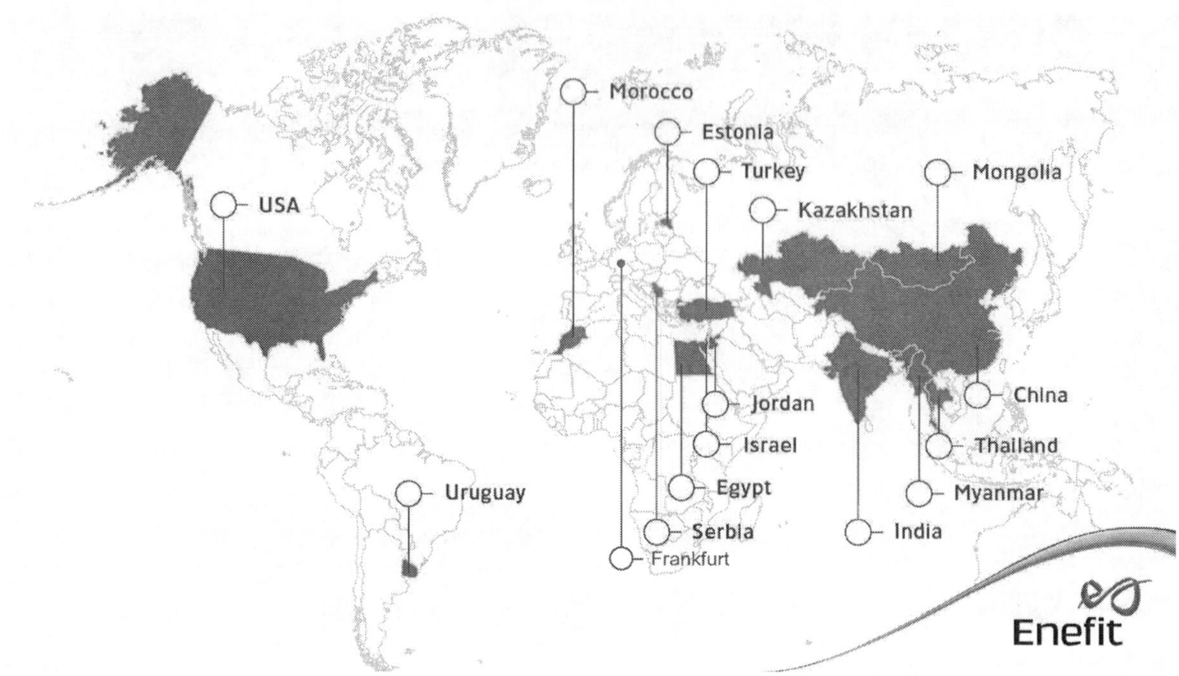
Morocco
Estonia
Turkey
Mongolia
USA
Kazakhstan
Jordan
China
Israel
Thailand
Egypt
Uruguay
Myanmar
Serbia
India
Frankfurt
Enefit

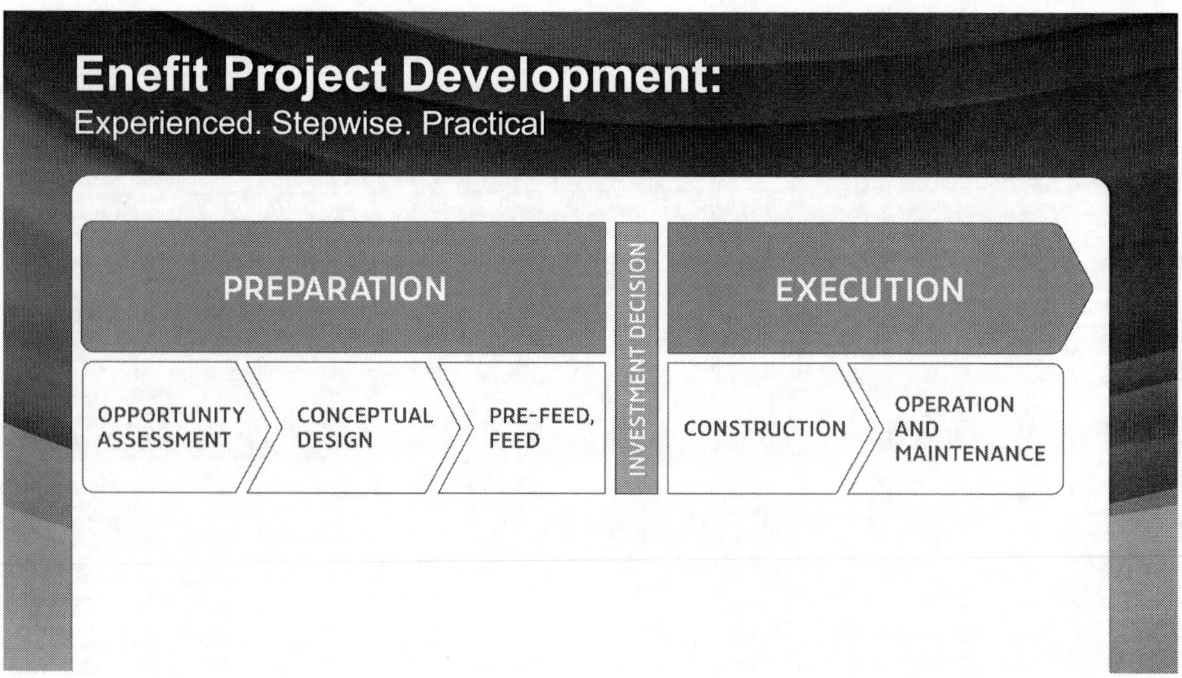
Enefit Project Development:
Experienced. Stepwise. Practical
PREPARATION
INVESTMENT DECISION
EXECUTION
OPPORTUNITY ASSESSMENT
CONCEPTUAL DESIGN
PRE-FEED, FEED
CONSTRUCTION
OPERATION AND MAINTENANCE

Enefit Project Development:

Experienced. Stepwise. Practical

- Identifying the oil shale resource
- Resource optimization studies
- Engineering throughout the entire value chain, from mining to end product
- Commercially proven technology for shale oil production – the cleanest and most energy efficient in the world
- Maximizing the value chain of oil shale products
- Attracting potential investors and financing
- Designing, building and operating large-scale projects in both the oil and power industries

The Resource Assessment Of The Isfir-Jafr Oil Shale In Southern Jordan

John Brodylo, Vice President Exploration – Questerre Energy Corporation
Mohammad Abweny, Senior Geologist – Questerre Energy Jordan

Isfir-Jafr MOU

- MOU is 182 km south Amman, 25 km east of Ma'an city.
- Isfir-Jafr MOU covers 38,790 ha
- Route 5 passes between Blocks A and B
- MOU area is located in an uninhabited and undeveloped area within El Jafr Depression.
- Dominated by broad desert plains covered by chert gravel pavement, wide wadis, and mud flats.

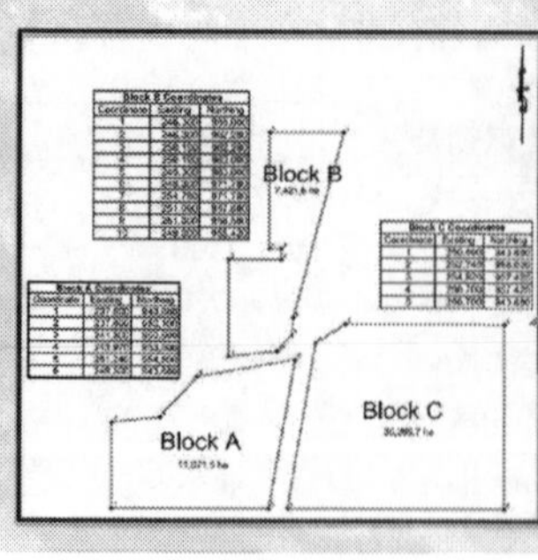

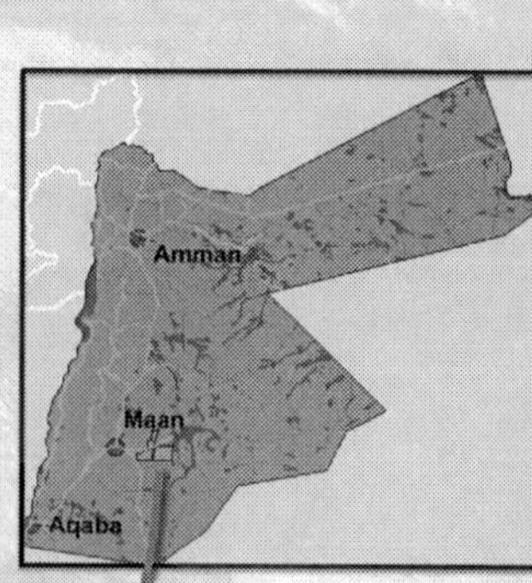

2

Stratigraphy

- Oil Shale occurs in MCM which is divided into ULH and LRH and rests on the Al Hisa Phosphate
- Overburden of MCM is the Umm Rijam Chert-Limestone (URC) and a thin veneer of Pleistocene fluvial sediments, Holocene desert pavements, and fluvial-lacustrine sediments of the Al Jafr

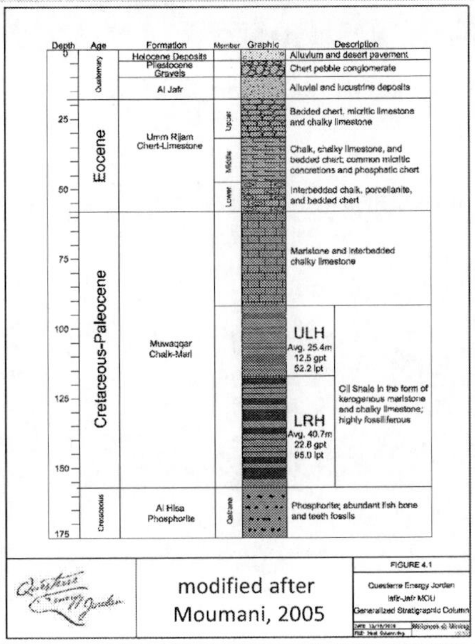

3

Surface Geology

- Surface composed of Pleistocene fluvial sediments, Holocene desert pavement, and fluvial-lacustrine sediments of Al Jafr Fm
- Generally, bedding dips at < 1 degree towards the northeast.
- "Relatively" structurally undisturbed
- local block faulting and folding exhibit a NE-SW trend
- The largest vertical fault displacement is 15 m across Kubayda fault

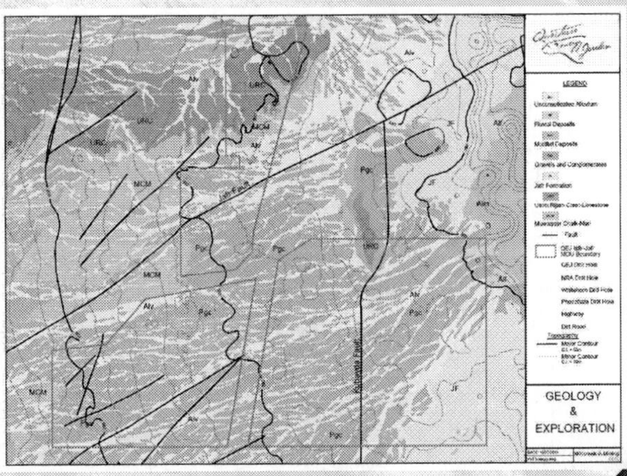

modified after Moumani, 2005

4

MOU Exploration

- NRA conducted a VES survey and drilled and assayed 35 core holes
- QEJ also conducted a detailed VES survey and drilled and analyzed five core holes
- Tested refraction seismic
- Commissioned a ground water study
- Conducted field mapping of surface geology and surface topography
- Completed independent Resource estimate
- Engineering Studies currently underway

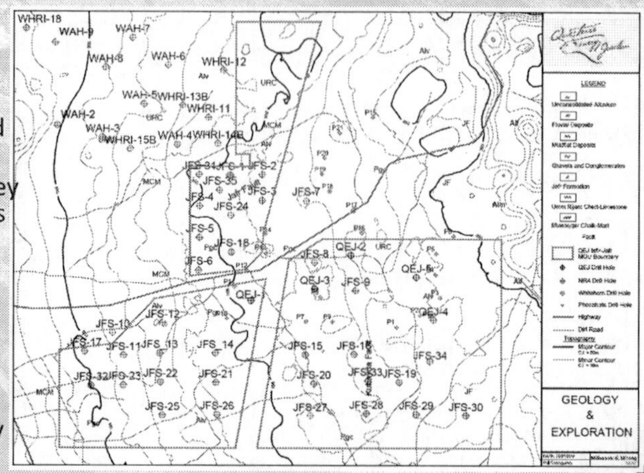

5

MCM Oil Shale

- ULH vs LRH determined by 15 GPT cutoff
- ULH has average thickness on the MOU of 25.4 m with weighted average oil yield of 12.5 GPT within the drill cores
- LRH has average thickness of 40.7 m with weighted average oil yield of 22.78 GPT
- Both the ULH and LRH sub crop to the south

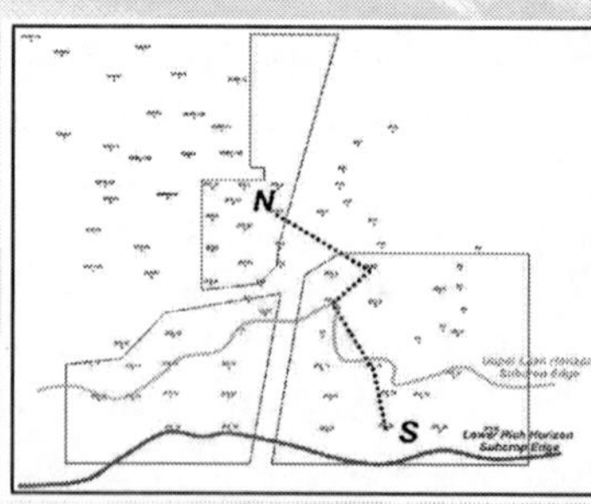

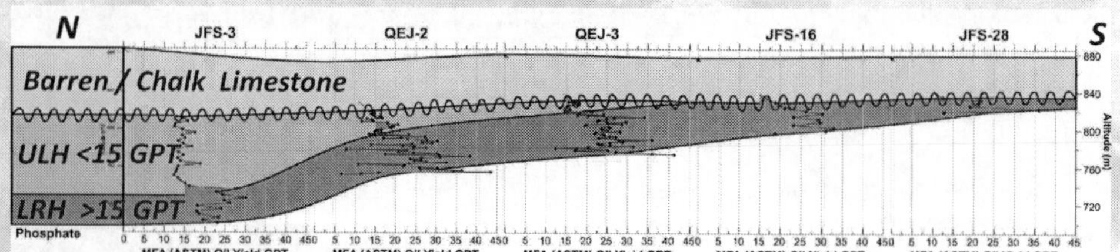

Total MCM Oil Shale Isopach

- Total MCM oil shale thickness ranges from 220 m in the North to 0 m in the South at the sub crop
- Total MCM includes both ULH and LRH

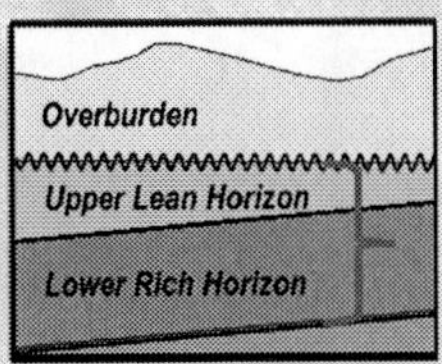

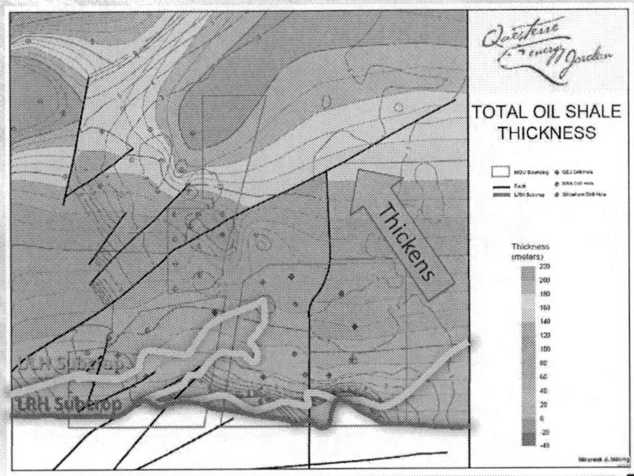

7

Lower Rich Horizon Overburden Map

- Overburden thickness for the LRH ranges from 180 m in the North to 57 m in the South
- LRH overburden includes both barren overburden and ULH
- Northern portion of block too deep

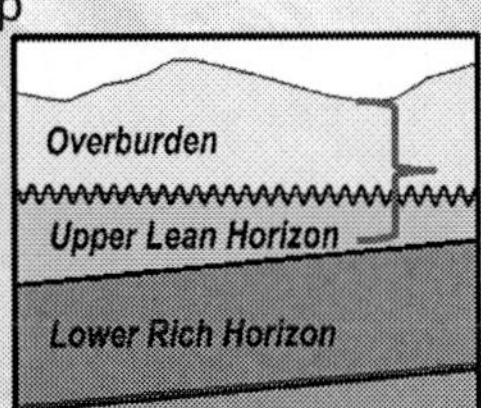

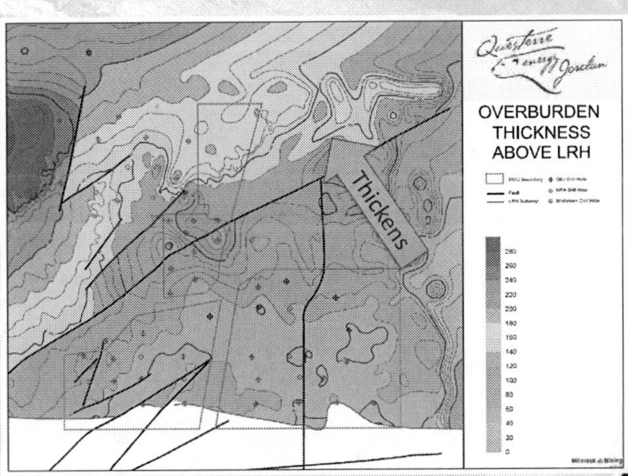

8

Rich Horizon Strip Ratio Map

- 2:1 Strip ratio cutoff
- 15 gpt cutoff
- MFA calibrated to ASTM
- Mean strip Ratio 1:1.21
- Area 213 km2

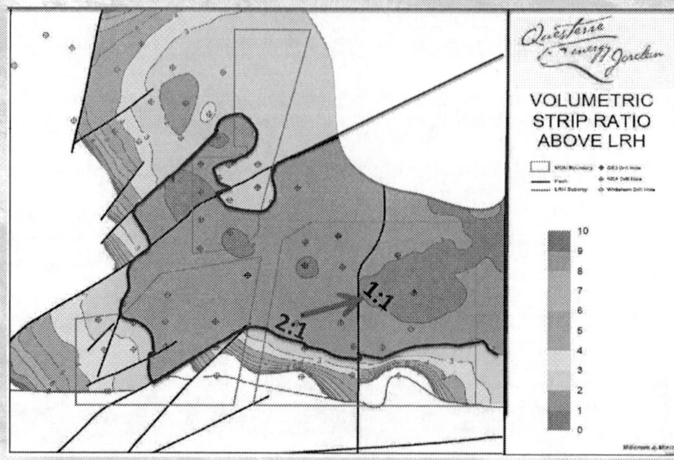

Total Petroleum Initially In place Resources

- Independent Engineering firm Millcreek Mining Group estimate compliant with National Instrument 51-101
- Mean strip Ratio 1: 0.88
- Area 375 km2
- Mean thickness 40.7 m
- Mean oil Yield 20.12 gpt
- Low = 1 km ,Best = 2 km, High = 4 km

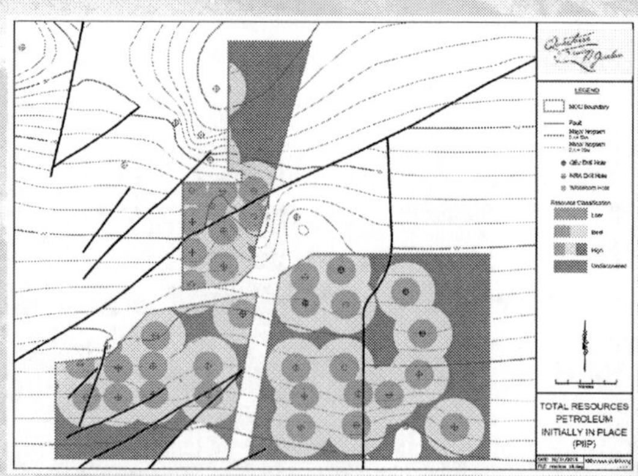

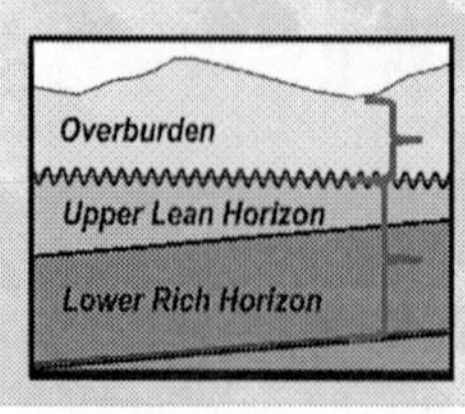

Billion Barrels	Low	Best	High	Undiscovered
Total	5.6	12.2	17.4	1.7

Mineable Petroleum Initially In place Resources

- 2:1 Strip ratio cutoff
- 15 gpt cutoff
- Mean strip Ratio 1:1.21
- Area 213 km2
- Mean thickness 50.7 m
- Mean oil Yield 22.6 gpt

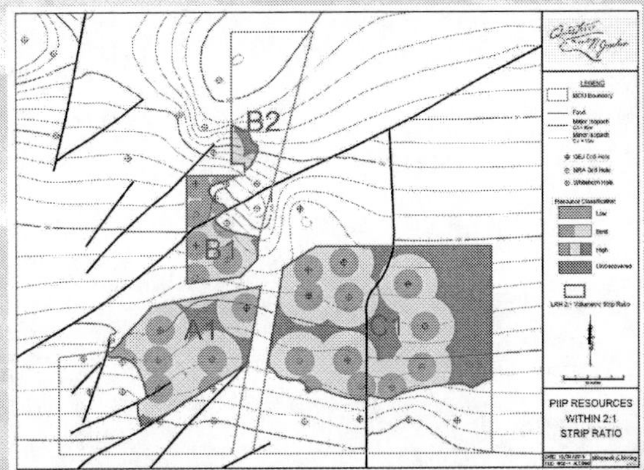

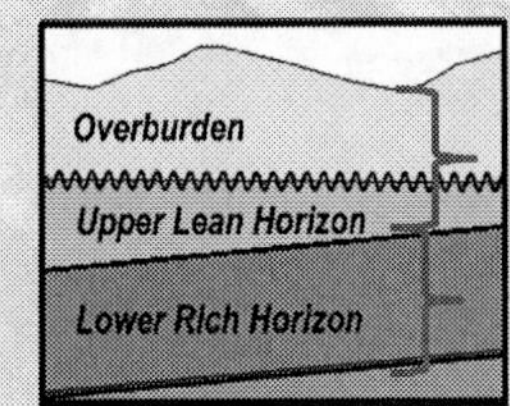

Billion Barrels	Low	Best	High	Undiscovered
Total	**3.7**	8.6	**11.0**	**0.4**

11

Conclusion

- Questerre Isfir-Jafr MOU contains a world class resource with Total PIIP of 12.2 billion barrels of oil at an average grade of 20.12 gpt.
- An independent evaluator has identified a surface mineable resource of 8.6 billion barrels with a 2:1 strip ratio at an average grade of 22.66 gpt.
- Millcreek Mining Group is of the opinion that this resource has a good potential of becoming a viable petroleum reserve that relies on mining and surface processing to produce synthetic crude oil.
- Technology review ongoing

12

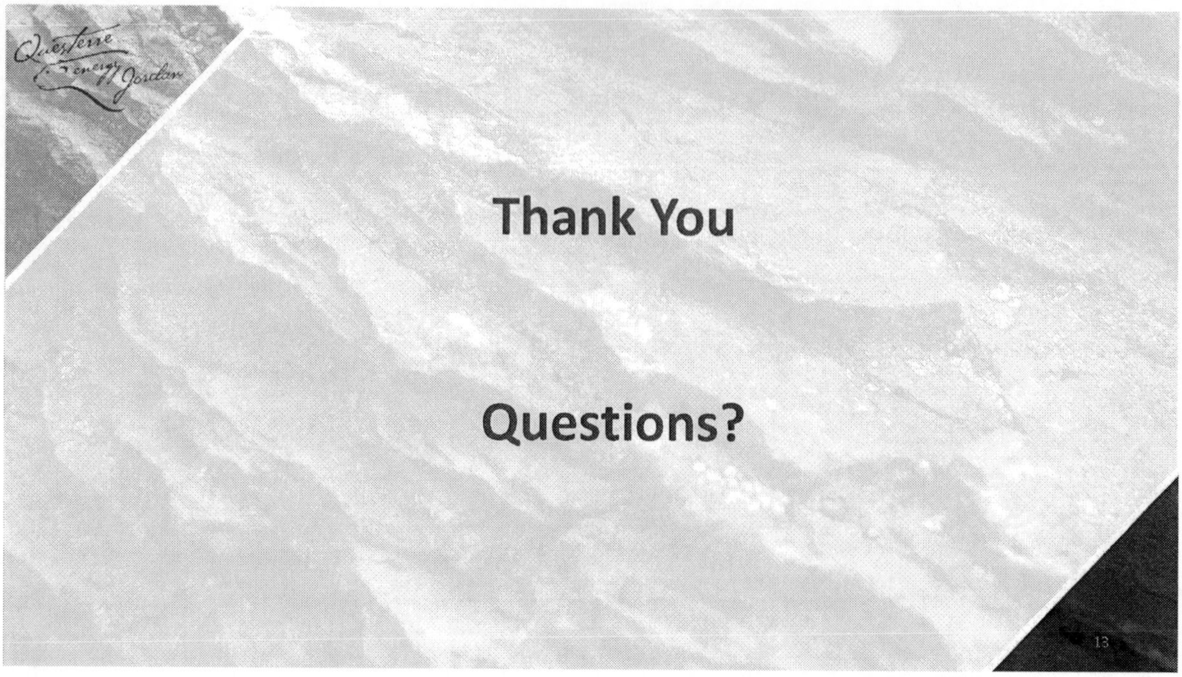
Thank You
Questions?
13

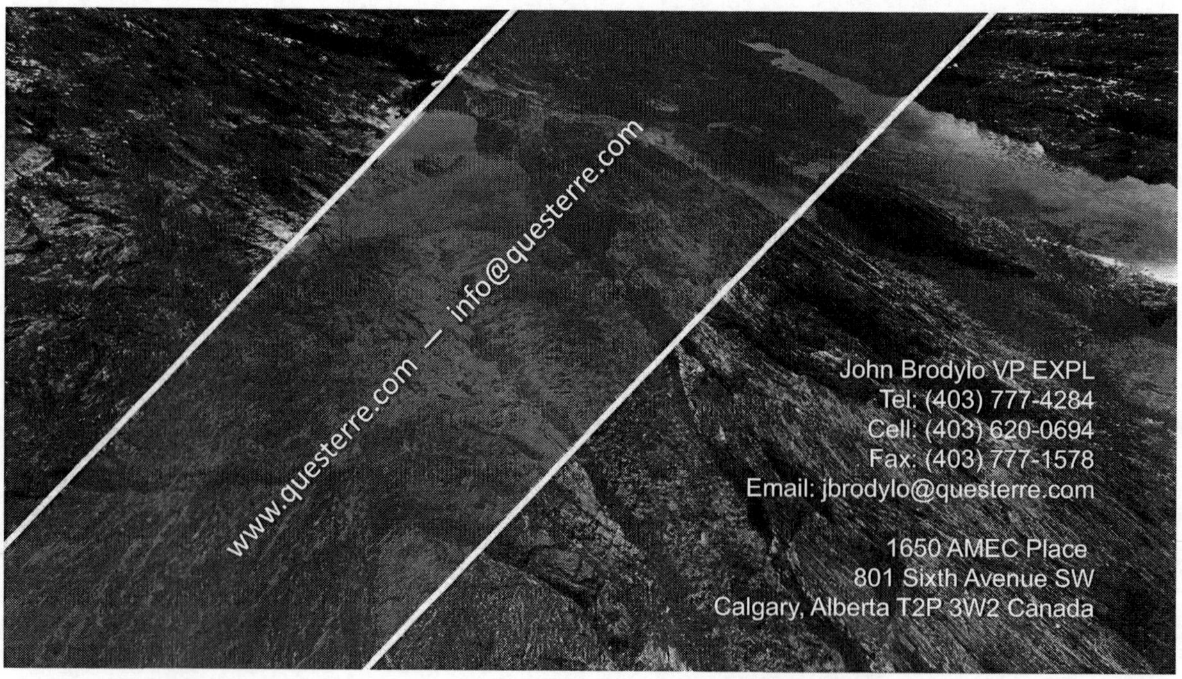
www.questerre.com — info@questerre.com
John Brodylo VP EXPL
Tel: (403) 777-4284
Cell: (403) 620-0694
Fax: (403) 777-1578
Email: jbrodylo@questerre.com
1650 AMEC Place
801 Sixth Avenue SW
Calgary, Alberta T2P 3W2 Canada

Oil Shale Sulphur Content Variations with Different Analytical Methods

Birgit Maaten*, MSc
Alar Konist, PhD

Tallinn University of Technololgy

Jordan International Oil Shale Symposium 2016

Estonian oil shale [1]

- Low-calorific fuel, LHV = 7 - 11 MJ/kg, typically 8.5 MJ/kg
- Moisture: 10 – 12 %
- High alkali and chlorine content
- Mineral matrix consists of mainly calcite, but also of dolomite and terrigenous phases
- High ash content (~50 %)
- Molar ratio of Ca/S = 8-10
- Sulphur content up to 3 %, sulphur both in the organic and inorganic phase
- Pyrite (FeS_2) as clumps in the structure

[1] Konist, A., Pihu, T., Neshumayev, D. & Külaots, I. Low Grade Fuel Oil Shale and Biomass Co-Combustion in CFB Boiler. *Oil Shale* **30,** 294 (2013)

Jordan International Oil Shale Symposium 2016

2/13

Measuring sulphur content - why?

- In case of firing oil shale, sulphur is the source of SOx emissions – limitations 200 (CFB) or 400 (PF) mg/Nm[3]
- In case of pyrolysis, sulphur ends up in the oil

→ Necessity to characterize materials and products well – analyze the chemical composition

- WD-XRF, elemental analysis and quantitative chemical analysis were used to analyze Estonian oil shale
- TGA-MS was used to characterize the evolving of SO_2

Characterization of the used sample

- Organic part content: 28.5 %
- Carbonate minerals: 18.5 %
- $Q_b^d = 10.0$ MJ/kg
- A = 47 %

Oil shale mineral part

mineral	amount, %
Dolomite	18.6
Calcite	41.8
Quartz	12.1
K-feldspar	5.87
Illite	16.4
Chlorite	1.77
Kaolinite	tr.
Pyrite	3.27

Laboratory ash

compound	amount, %
SiO_2	30,7
FeO_3	4,8
Al_2O_3	6,1
CaO	39
MgO	8,7
K_2O	1,8
Na_2O	0,1

Quantitative chemical analysis

- Eschka method of total sulphur determination
- ISO 334:2013
- Gravimetric analysis – sulphur is converted to $BaSO_4$ and the total sulphur content is obtained
- Considered the most accurate of the listed methods – the repeatability should be less than 0.05 %
- Very time-consuming

- S^{total} = 1.52 %

Jordan International Oil Shale Symposium 2016

5/13

Wavelength dispersive X-ray fluorescence spectroscopy (WD-XRF)

- One of the two types of XRF instrumentation – analyzing the elemental content by exciting electrons and measuring the energy difference between the expelled and replacement electrons
- Non-destructive, high resolution, minimal spectral overlaps
- Amorphous compounds interfere with obtaining correct results

- S^{total} = 1.2 %

Jordan International Oil Shale Symposium 2016

6/13

Elemental analysis

- Measuring the content of C, H, N, S with vario MACRO
- Can analyze complex matrices and different consistencies
- The sample is burned at ~1150 °C in an oxygen-rich atmosphere with helium as the carrier gas
- WO_3 is used as catalyst
- Easy and fast analysis, small amounts of sample

Jordan International Oil Shale Symposium 2016

7/13

Elemental analysis results

- Varying the sample to catalyst ratio and the sample weight

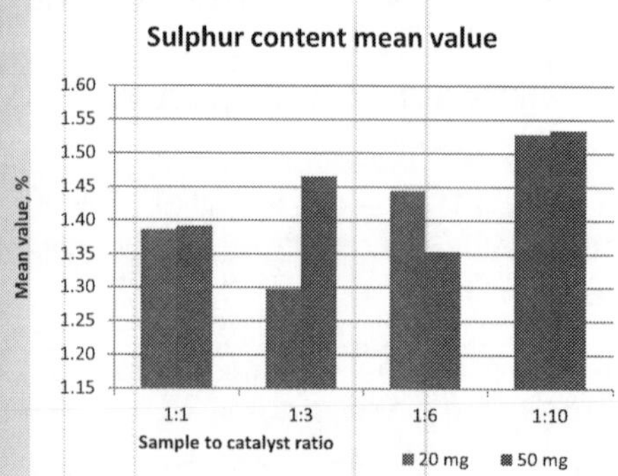

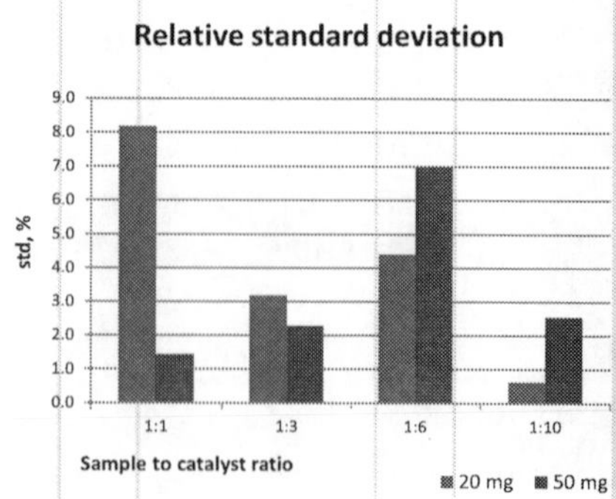

Jordan International Oil Shale Symposium 2016

8/13

Comparison of the methods

- Quantitative chemical analysis: S^{total} = 1.52 %
- Elemental analysis:
  20 mg, 1:10 S^{total} = 1.53 ± 0.01 %
  50 mg, 1:10 S^{total} = 1.53 ± 0.01 %
- XRF: S = 1.2 %

Jordan International Oil Shale Symposium 2016

9/13

Thermogravimetric analysis coupled with mass spectrometry (TGA-MS)

- Combustion in 80 % N_2 and 20 % O_2
- MS was used to analyze the evolving of SO_2
- CO_2 was analyzed to confirm the degradation
- Set heating rate: 35 °C/min
- Organic part content: 28.5 %
- Carbonate minerals: 18.5 %
- Ash content: 47 %

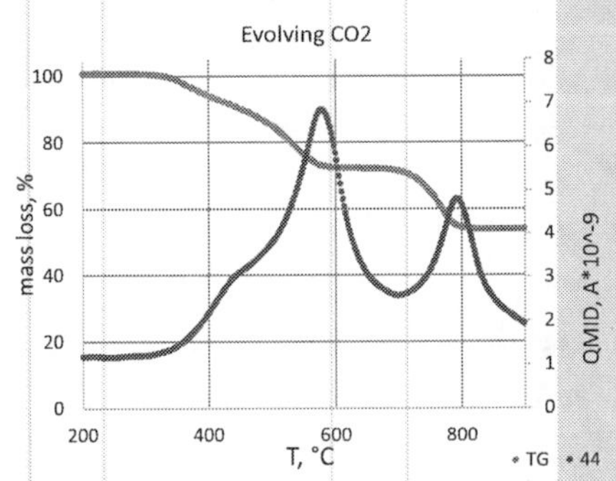

Jordan International Oil Shale Symposium 2016

10/13

TGA-MS

- SO_2 identified via m/z 64 and 48
- Main peak: 460 to 640 °C
- Pyrite decomposition: 400 to 500 °C
- No SO_2 forming at temperatures above 640 °C
- Small delay between signals due to connection of equipment

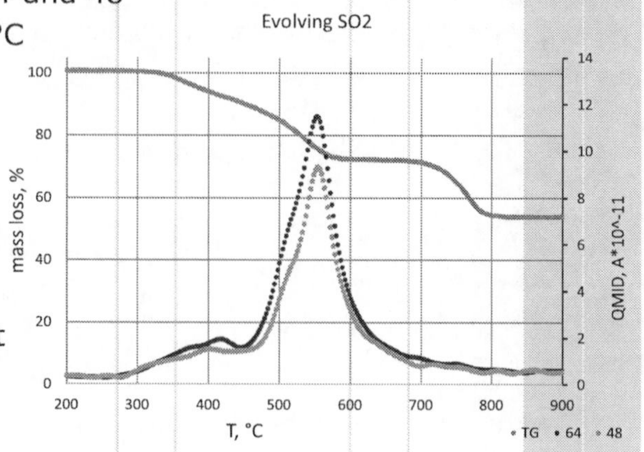

Jordan International Oil Shale Symposium 2016

11/13

Future plans

- Further investigations of sulphur behaviour during oil shale firing
- e.g. using oxy-fuel combustion
- 60 kW Circulating Fluidized Bed Combustion Test Facility

Jordan International Oil Shale Symposium 2016

12/13

Thank you for your attention!
Questions?
birgit.maaten@ttu.ee
Jordan International Oil Shale Symposium 2016
13/13

Characterization of Shale Oil Produced by Retorting Sultani Oil Shale

- Prof. Dr. Omar Al-Ayed[1] and Dr. Mohammad Amer[2]
- [1]Al-Balqa Applied University, Department of Chemical Engineering, Salt, Jordan.

Omar.alayed@bau.edu.jo

- [2]Jordan University, Department of Chemistry, Amman, Jordan

Jordan International Oil Shale Symposium and 36th CSM Oil Shale [illegible]
21-22 November 2016
Dead Sea, Jordan

The pyrolysis temperature varied from 360 °C to 540 °C with particle size less than 12mm. The generated shale oil is characterized by determining the heating value, sulfur content, API degree, distillation range, heat capacity as function of temperature, and oil yield.

- Experimental setup

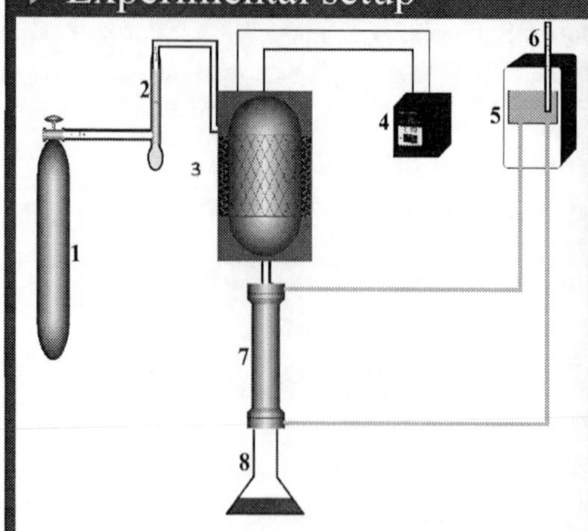

- Sultani oil shale sample, Fischer Assay

Sample No.	Water	Oil	Spent shale	Gas
1	3.0	14.37	78.12	4.51
2	2.5	14.2	79.2	4.1
3	2.0	13.56	80.15	4.29

Pyrolysis Results, Weight loss, Yield

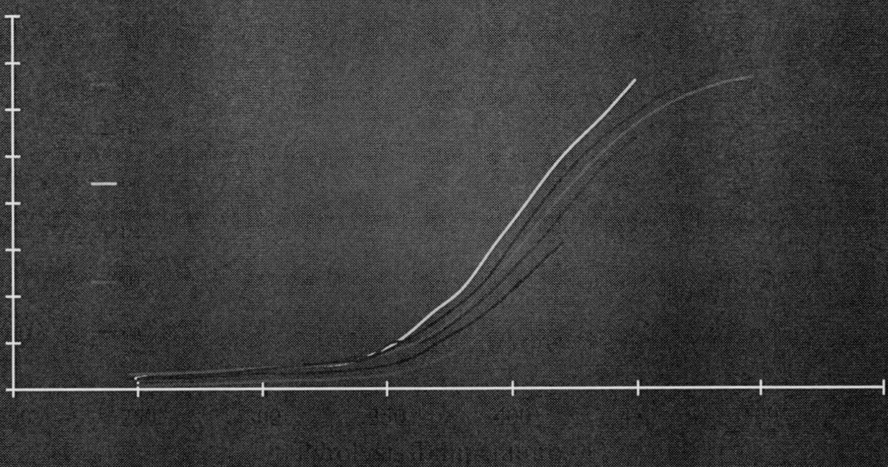

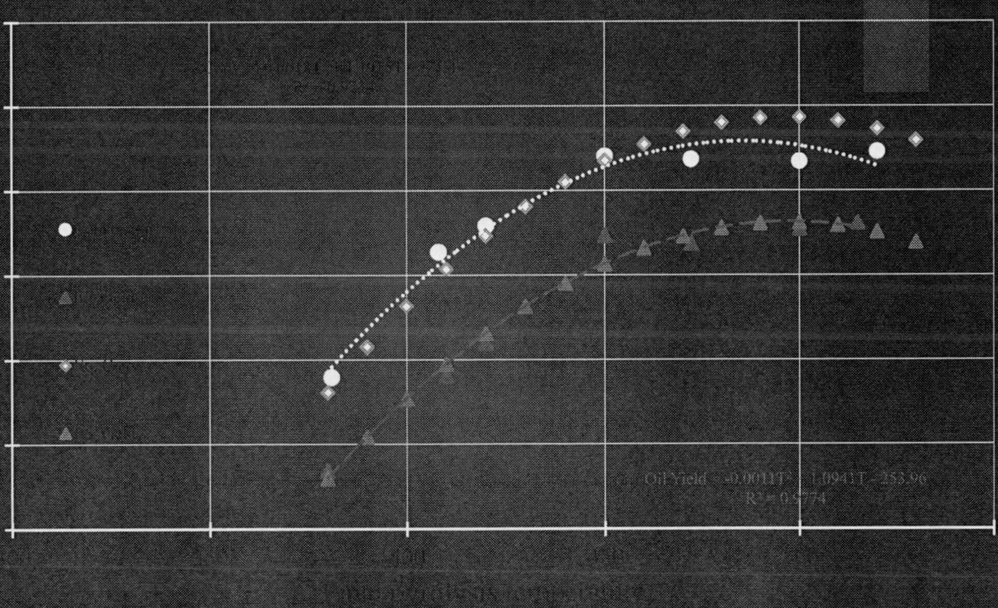

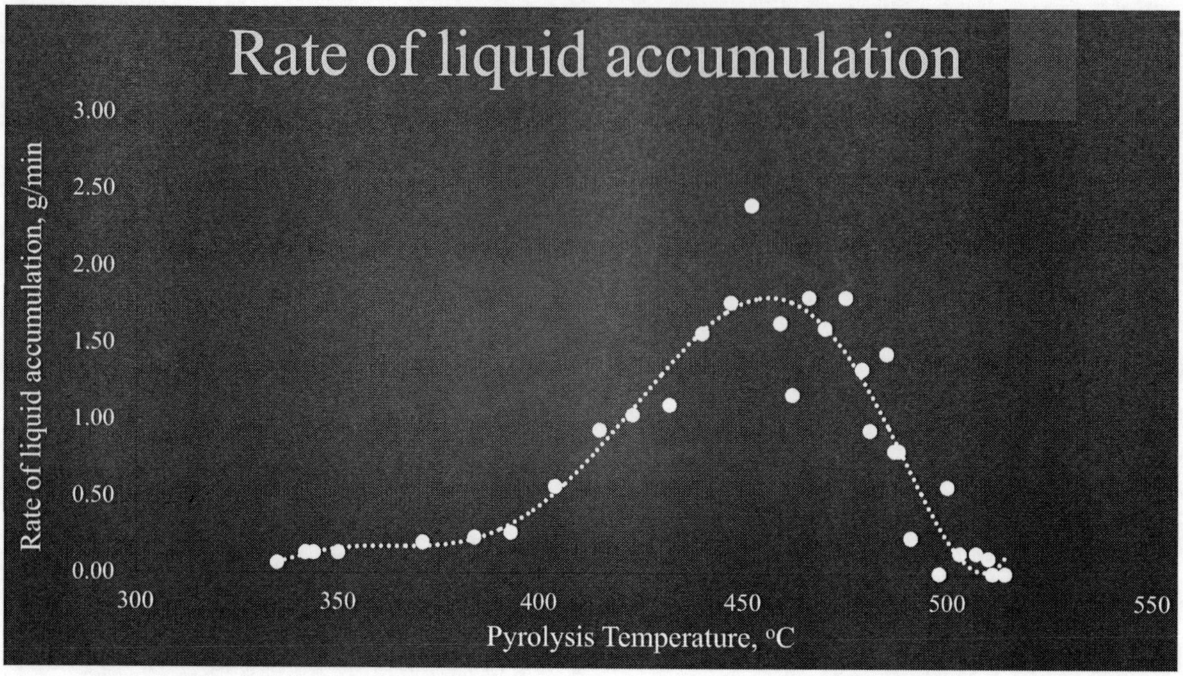
Rate of liquid accumulation
Rate of liquid accumulation, g/min
3.00
2.50
2.00
1.50
1.00
0.50
0.00
300
350
400
450
500
550
Pyrolysis Temperature, °C

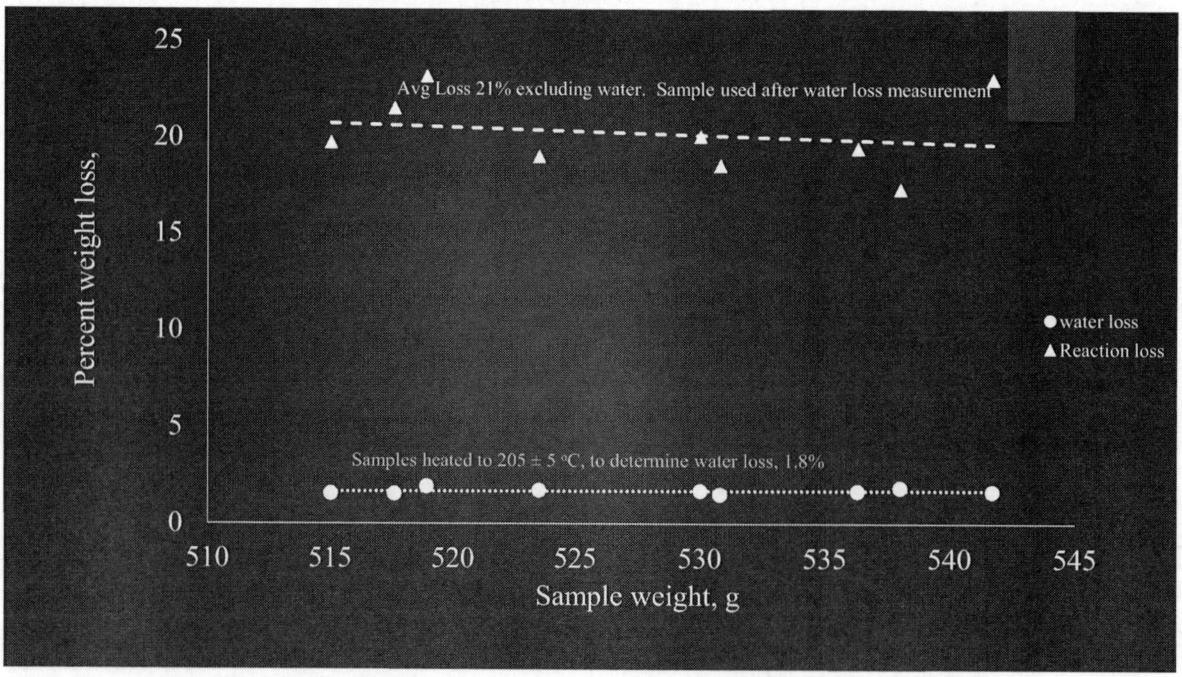
Avg Loss 21% excluding water. Sample used after water loss measurement
Samples heated to 205 ± 5 °C, to determine water loss, 1.8%
Percent weight loss,
25
20
15
10
5
0
510
515
520
525
530
535
540
545
Sample weight, g
water loss
Reaction loss

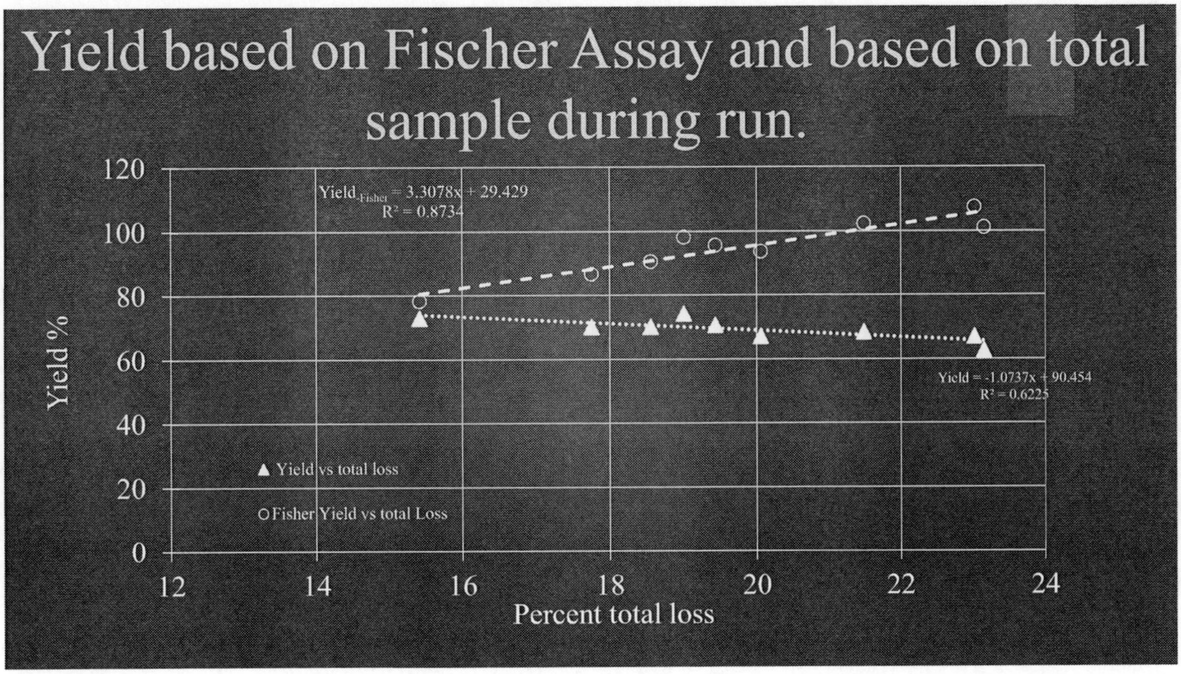
Yield based on Fischer Assay and based on total sample during run.
Yield-Fisher = 3.3078x + 29.429
R² = 0.8734
Yield = -1.0737x + 90.454
R² = 0.6225
Yield %
120
100
80
60
40
20
0
12
14
16
18
20
22
24
Percent total loss
Yield vs total loss
Fisher Yield vs total Loss

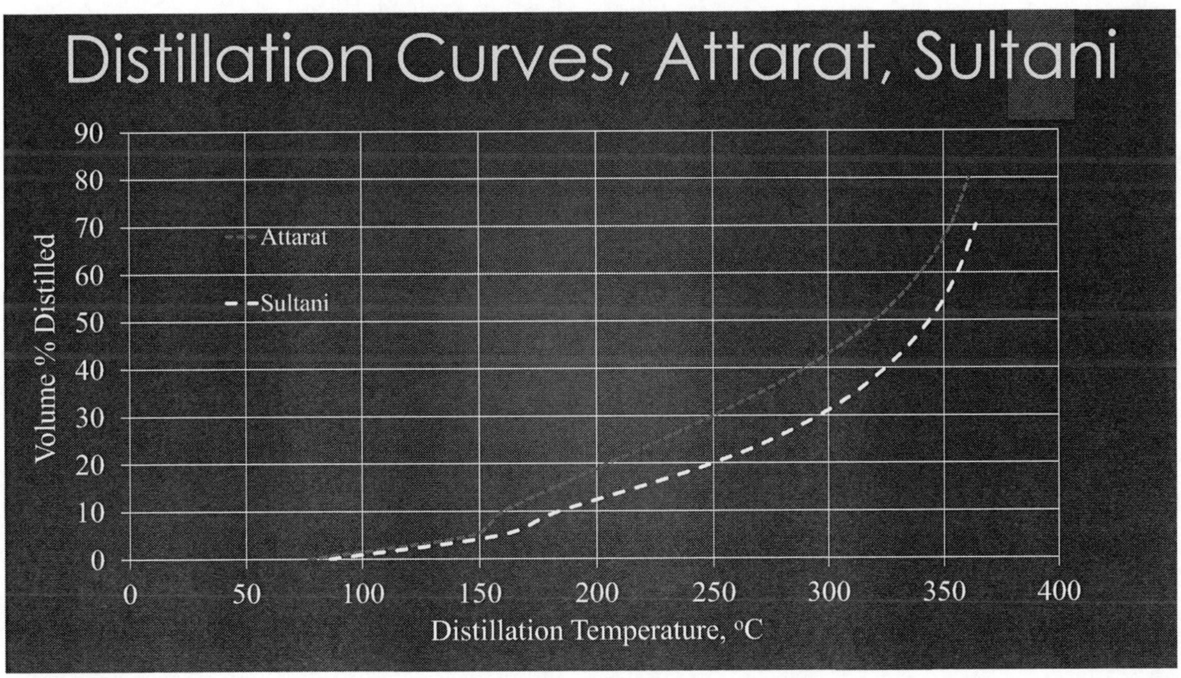
Distillation Curves, Attarat, Sultani
Volume % Distilled
90
80
70
60
50
40
30
20
10
0
Attarat
Sultani
0
50
100
150
200
250
300
350
400
Distillation Temperature, °C

Shale oil properties

Property	Attarat		Sultani
Density, @ 15 °C.	0.9774		0.9568
Residue, vol%	15		23
Sulfur, wt. %	8.954		8.294
API,° (ASTM D4052)	13.3		16.4

Experimental results at different pyrolysis temperatures, Sultani

Sample ID	Calorific Value (cal/g) ASTM D240	Density @ 15°C ASTM D4052	API @ 15°C	Sulfur Content wt% ASTM D4294
			ASTM D4052	
360	9517	0.965	14.99	6.48
380	9521	0.9587	15.94	6.38
400	9527	0.9684	14.47	6.45
420	9505	ND	ND	-
440	9528	0.9974	10.22	6.5
520	9530	0.9892	11.42	6.41
540	9492	ND	ND	-

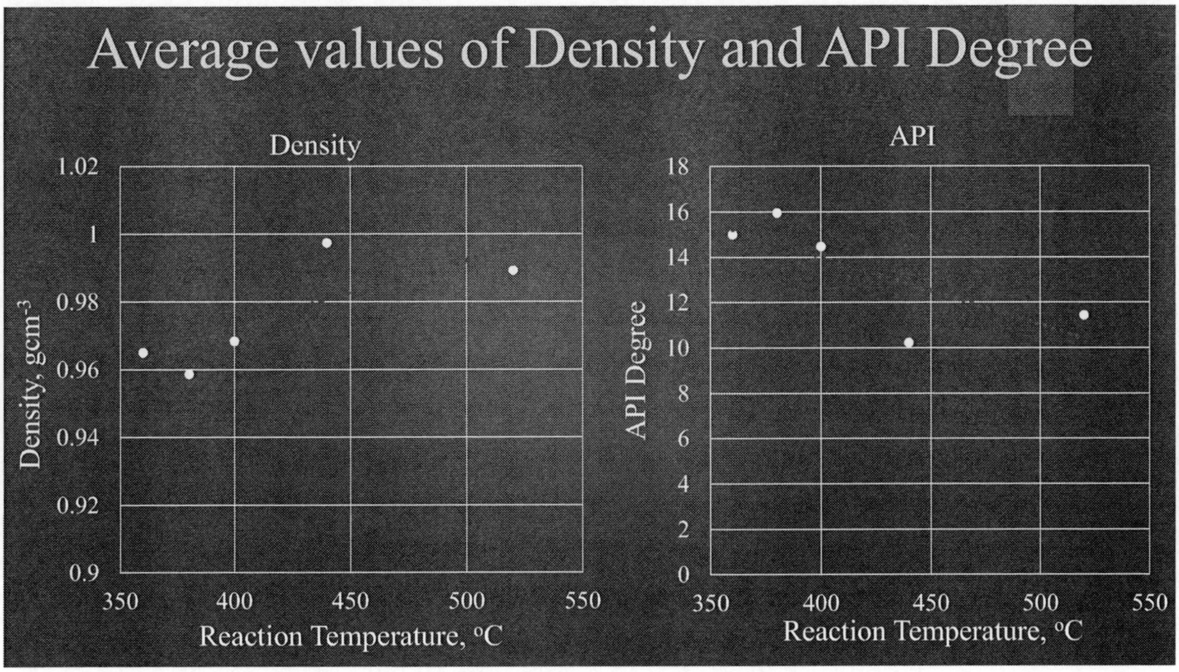
Average values of Density and API Degree
Density
Density, gcm-3
1.02
1
0.98
0.96
0.94
0.92
0.9
350
400
450
500
550
Reaction Temperature, °C
API
API Degree
18
16
14
12
10
8
6
4
2
0
350
400
450
500
550
Reaction Temperature, °C

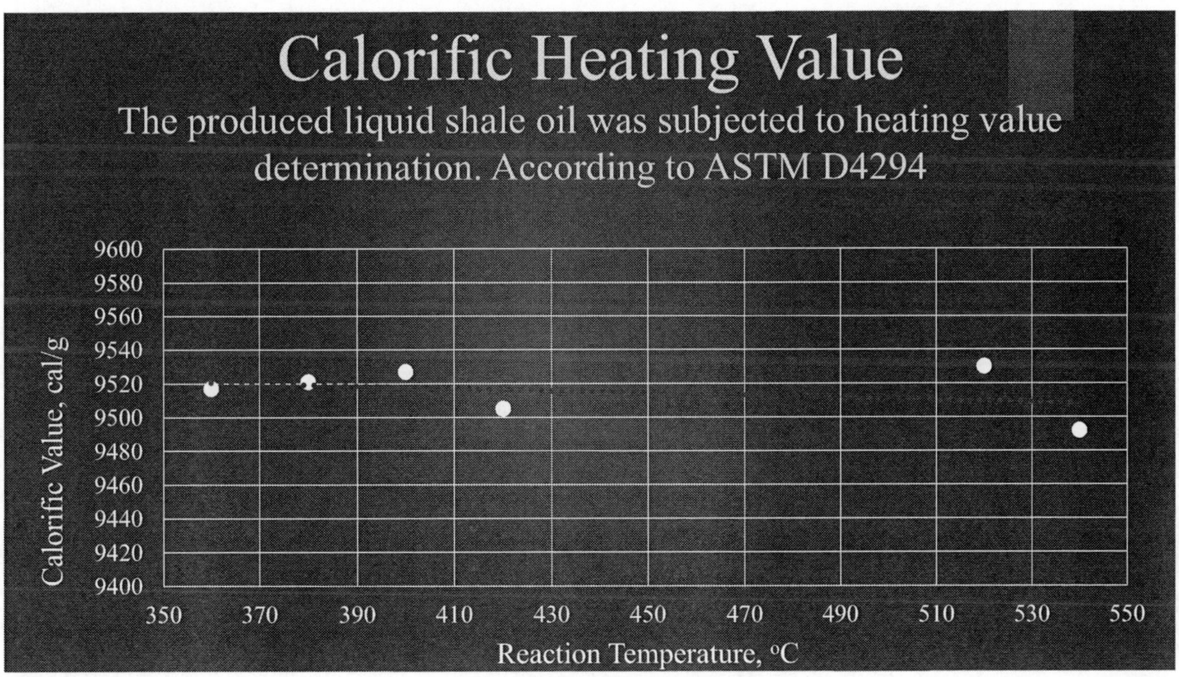
Calorific Heating Value
The produced liquid shale oil was subjected to heating value determination. According to ASTM D4294
Calorific Value, cal/g
9600
9580
9560
9540
9520
9500
9480
9460
9440
9420
9400
350
370
390
410
430
450
470
490
510
530
550
Reaction Temperature, °C

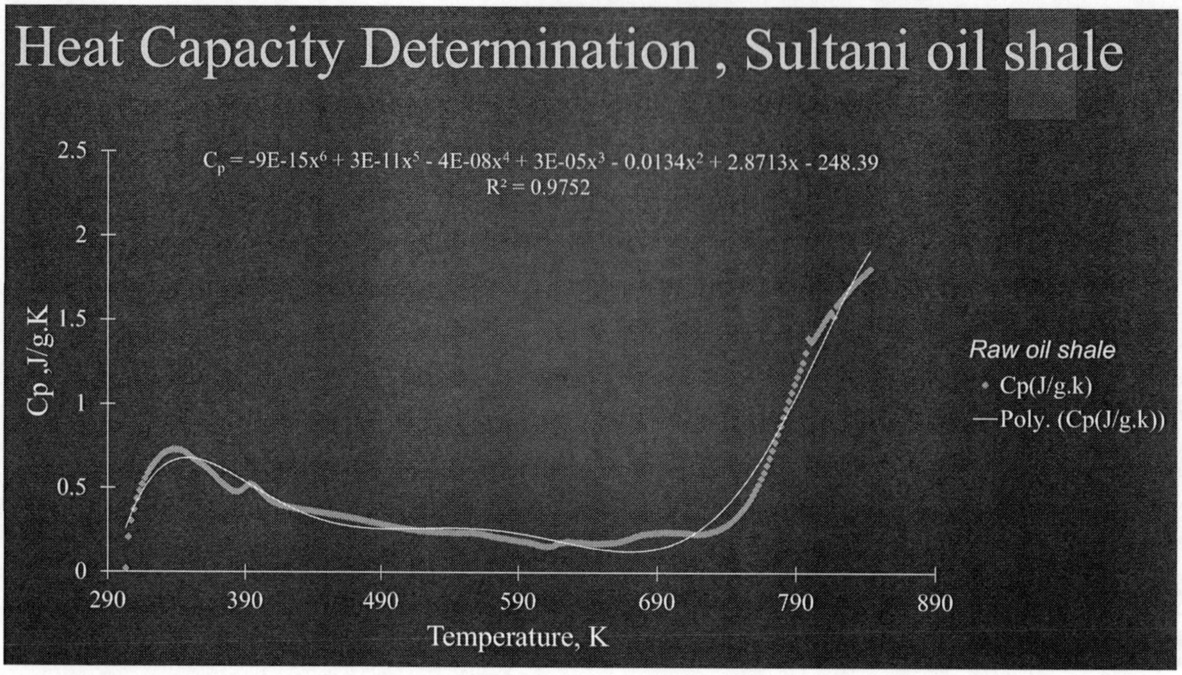
Heat Capacity Determination , Sultani oil shale
Cp = -9E-15x6 + 3E-11x5 - 4E-08x4 + 3E-05x3 - 0.0134x2 + 2.8713x - 248.39
R² = 0.9752
Cp ,J/g.K
2.5
2
1.5
1
0.5
0
290
390
490
590
690
790
890
Temperature, K
Raw oil shale
Cp(J/g.k)
Poly. (Cp(J/g.k))

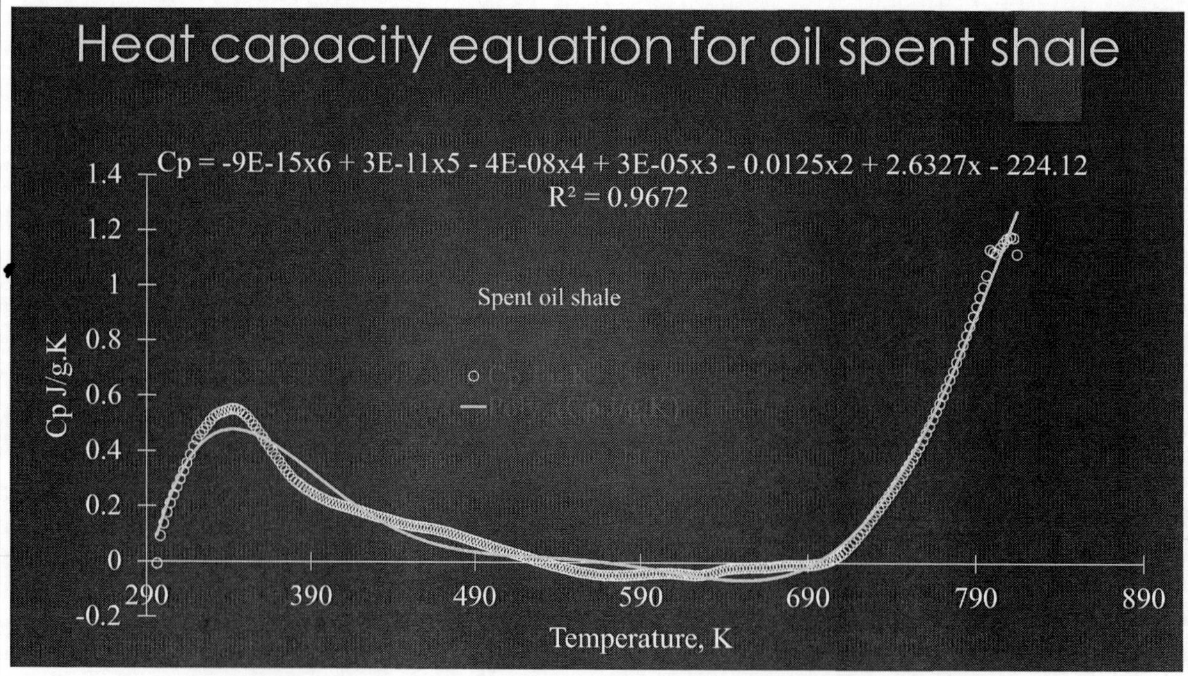
Heat capacity equation for oil spent shale
Cp = -9E-15x6 + 3E-11x5 - 4E-08x4 + 3E-05x3 - 0.0125x2 + 2.6327x - 224.12
R² = 0.9672
Spent oil shale
Cp J/g.K
Poly. (Cp J/g.K)
Cp J/g.K
1.4
1.2
1
0.8
0.6
0.4
0.2
0
-0.2
290
390
490
590
690
790
890
Temperature, K

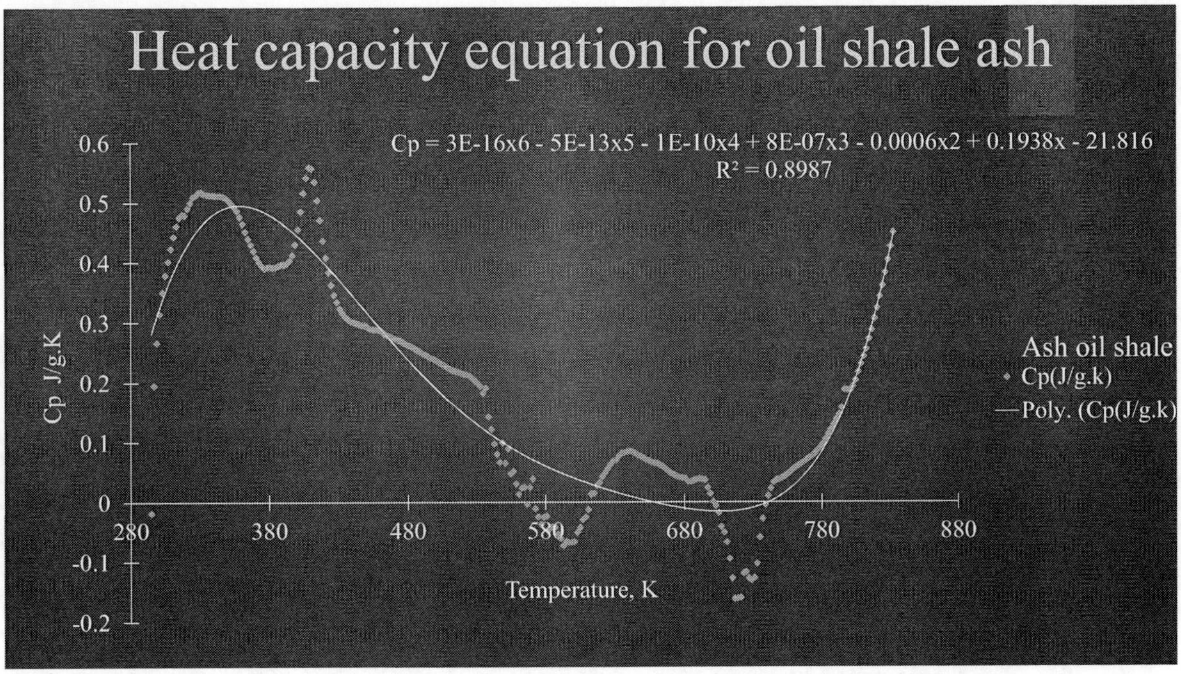
Heat capacity equation for oil shale ash
Cp = 3E-16x6 - 5E-13x5 - 1E-10x4 + 8E-07x3 - 0.0006x2 + 0.1938x - 21.816
R² = 0.8987
0.6
0.5
0.4
0.3
0.2
0.1
0
-0.1
-0.2
280
380
480
580
680
780
880
Cp J/g.K
Temperature, K
Ash oil shale
Cp(J/g.k)
Poly. (Cp(J/g.k)

Petra, the "rose city", the moment of entrance from Sieq

Thank you

- **Acknowledgement**
- This project is funded by the 'Support to Research, Technological Development & Innovation in Jordan' (SRTD – II), an EU funded project managed by the Higher Council for Science & Technology. Jordan.
- The Authors would like to express their thanks to The Higher Council for Science and Technology, SRTD II project implementation office for their help, facilities and follow up during the project period.

Assessing the geochemical variability of Jordan oil shale: laboratory geochemistry vs. multivariate statistics

Alvar Soesoo, Väino Puura, Margus Voolma
University of Tartu, Estonia

alvar.soesoo@gmail.com

In the start:

- The Cretaceous - Palaeogene oil shale (OS) of Jordan is mostly a calcareous mudstone with intervals of siliceous minerals, quartz and cristobalite–tridymite, dolomite. Oil shale is commonly rich in organic sulphur and trace metals.
- Presently, in complex studies of OS units, the variation in chemical composition is of primary interest, especially in the context of genetic, oil production and mining aspects.

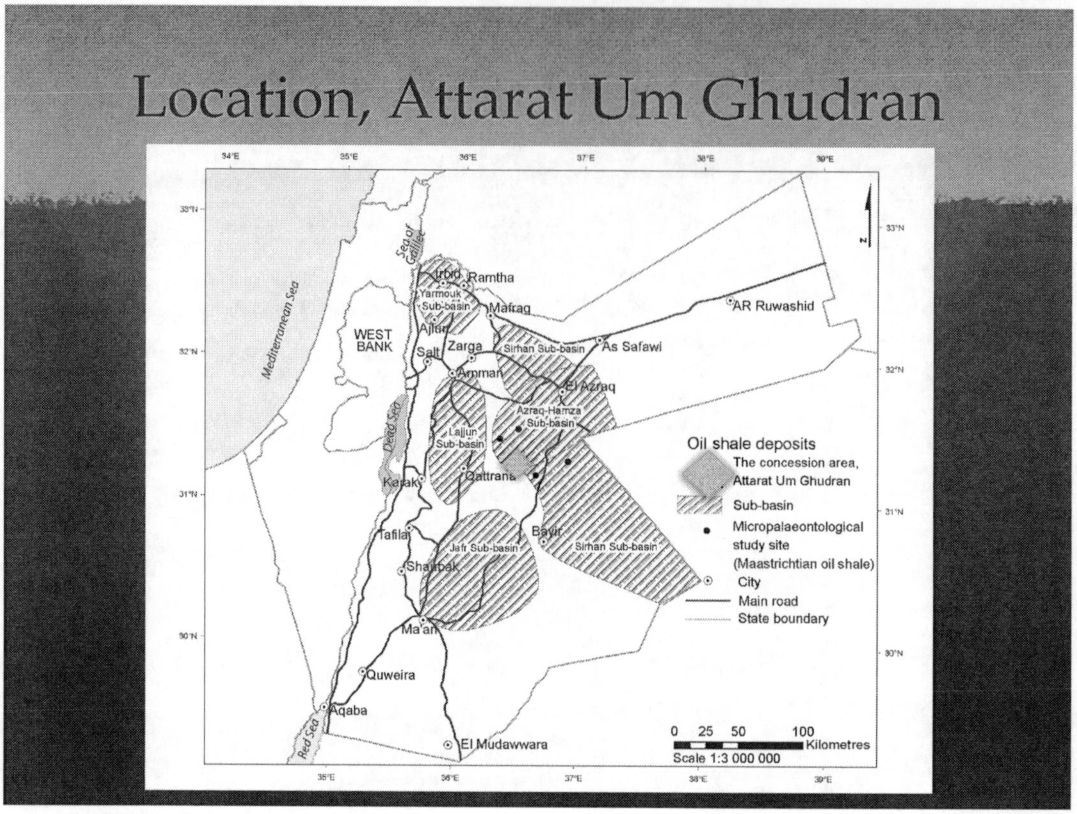
Location, Attarat Um Ghudran
Mediterranean Sea
Sea of Galilee
WEST BANK
Dead Sea
Red Sea
Irbid
Ramtha
Yarmouk Sub-basin
Ajlun
Mafraq
Salt
Zarqa
Amman
AR Ruwashid
As Safawi
Sirhan Sub-basin
El Azraq
Azraq-Hamza Sub-basin
Lajjun Sub-basin
Qattrana
Karak
Tafila
Jafr Sub-basin
Shawbak
Bayir
Ma'an
Quweira
Aqaba
El Mudawwara
Oil shale deposits
The concession area, Attarat Um Ghudran
Sub-basin
Micropalaeontological study site (Maastrichtian oil shale)
City
Main road
State boundary
0 25 50 100 Kilometres
Scale 1:3 000 000

OS – usually considered as massive, homogeneous unit, however....

- Oil shales are different in visual appearance
- Types and variations of composition of both organic (OM) and mineral matter (MM) can be determined in laboratory etc. tests

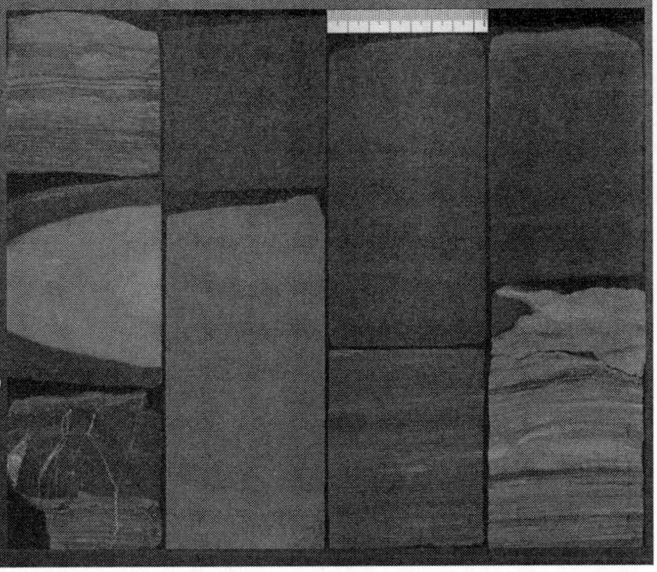

Finely laminated mudstone OS (richest)

Grain-bearing laminated OS

Barren carbonate and chert inclusions

Barren carbonate and chert interlayers

Geochemical structure of oil shale layers: our approach

- Detailed internal lithology
- Very detailed geochemistry, physical properties
- By these detailed studies we have established, that the simple – looking Jordan oil shale is not such a simple geochemical structure at all!

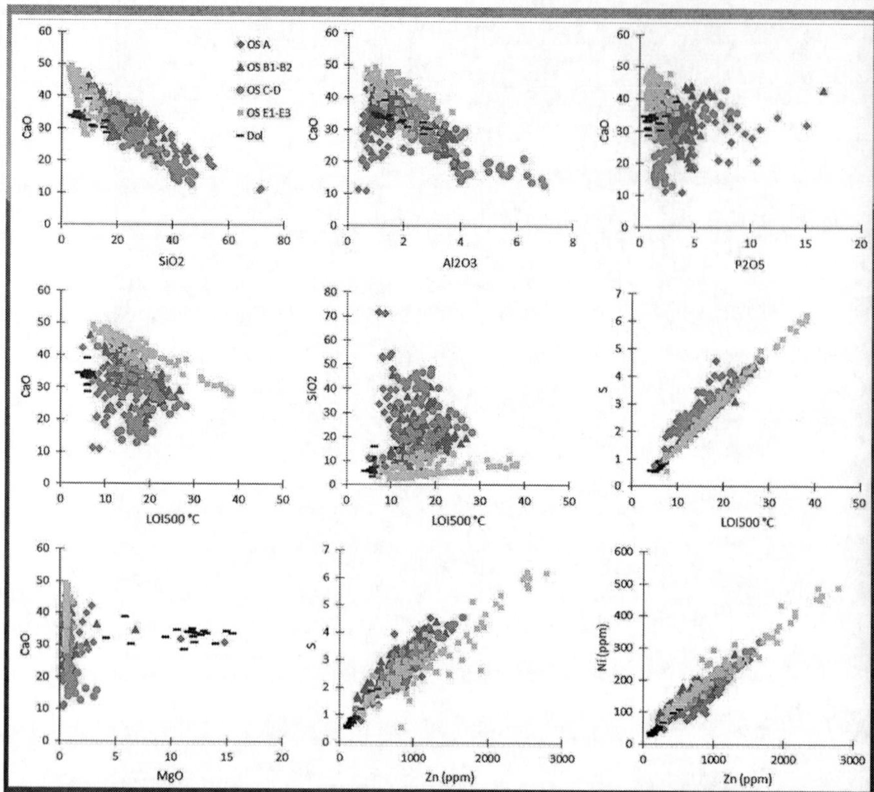

A selection of X-Y plots between of selected elements.

Chemically distinctive oil shale sub-layers

	Total data			Average by distinguished layers										
	Mean	Min	Max	A	A/B	B1	B2	B/C	C	D	D/E	E1	E2	E3
Number of samples	392 (398)			66	7 (9)	38	84	7 (9)	44	28	7 (9)	45	22	44
Thickness (m)	1.59	0.50	2.00	1.51	0.78	1.70	1.73	0.55	1.68	1.54	1.15	1.63	1.57	1.69
SiO_2	18.90	2.59	71.70	29.23	5.97	19.75	19.88	8.79	22.21	40.13	9.21	8.84	7.59	5.13
TiO_2	0.08	0.02	0.29	0.05	0.04	0.05	0.09	0.05	0.10	0.20	0.07	0.08	0.06	0.05
Al_2O_3	2.02	0.38	6.97	1.28	1.15	1.18	2.29	1.67	2.56	4.84	1.99	2.20	1.64	1.20
Fe_2O_3	0.63	0.21	1.77	0.49	0.37	0.46	0.65	0.43	0.76	1.27	0.49	0.65	0.66	0.44
MgO	1.37	0.22	15.50	1.13	11.99	0.46	0.84	9.88	0.77	1.28	12.29	0.63	0.54	0.59
CaO	33.91	10.79	49.26	30.34	34.73	36.20	32.98	33.13	30.82	18.87	31.97	41.48	34.75	43.84
K_2O	0.25	0.06	0.62	0.22	0.11	0.18	0.34	0.21	0.35	0.41	0.13	0.22	0.19	0.12
P_2O_5	2.82	0.47	16.58	5.03	1.38	3.85	3.01	1.63	3.20	1.80	0.72	1.91	0.97	1.12
S	2.68	0.54	6.22	2.65	0.80	2.66	2.82	1.18	2.98	2.81	0.72	2.24	4.73	2.37
OM	16.02	4.13	38.28	13.17	6.13	14.85	17.30	8.03	18.65	16.37	5.77	14.53	29.13	15.43
Zn	862	100	2784	876	176	764	769	288	946	856	149	657	2035	954
Ni	173	28	488	148	43	158	167	66	193	163	35	151	378	203
Cr	292	83	626	230	108	241	312	152	331	328	105	284	488	320
Cu	60	<10	131	61	16	62	61	25	65	53	12	53	106	60
Mo	143	12	614	136	17	96	133	36	170	180	19	103	348	157
V	436	78	2040	508	113	341	292	114	388	427	98	347	1395	508
Zr	38	<10	85	28	16	33	43	21	43	48	29	39	44	36
Sr	680	229	1855	585	375	729	697	432	710	568	565	759	769	770

A principal profile of layered and partially weathered oil shale unit, Attarat (AUG)

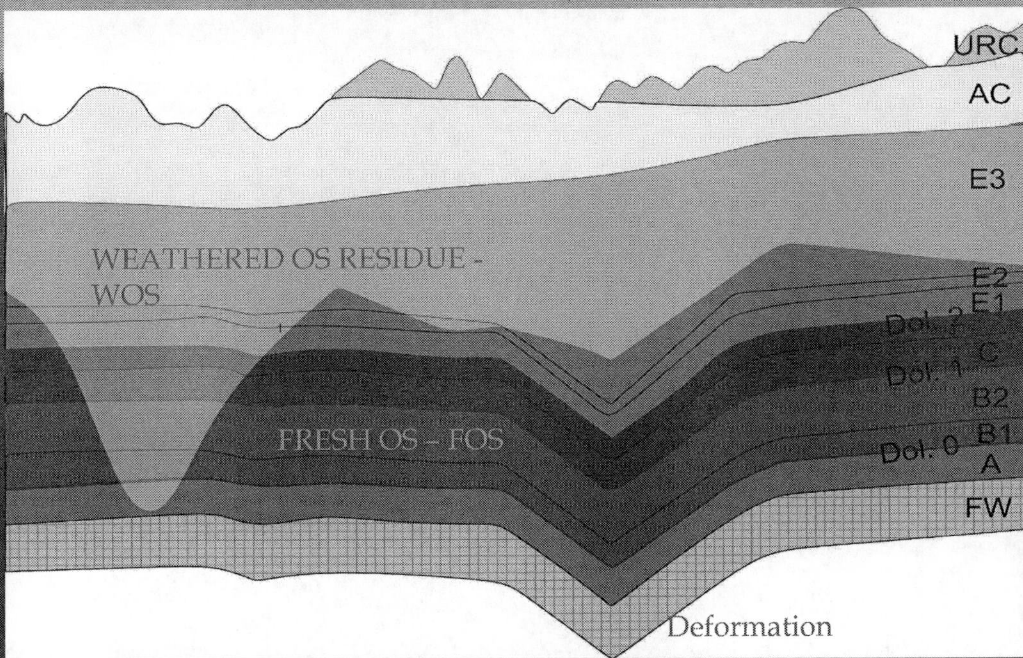

Appearance of Weathered Oil Shale

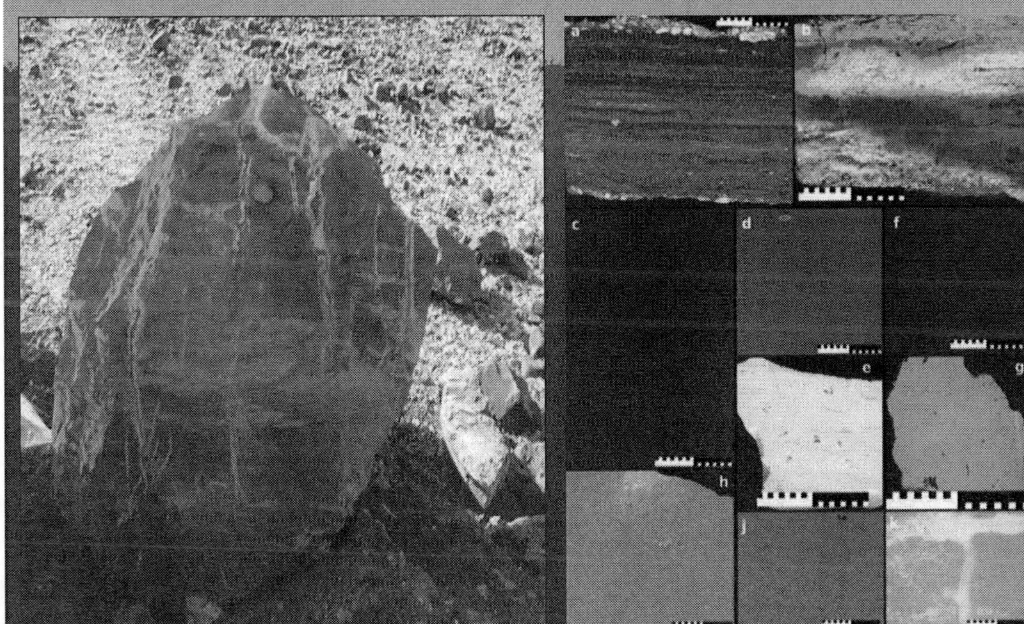

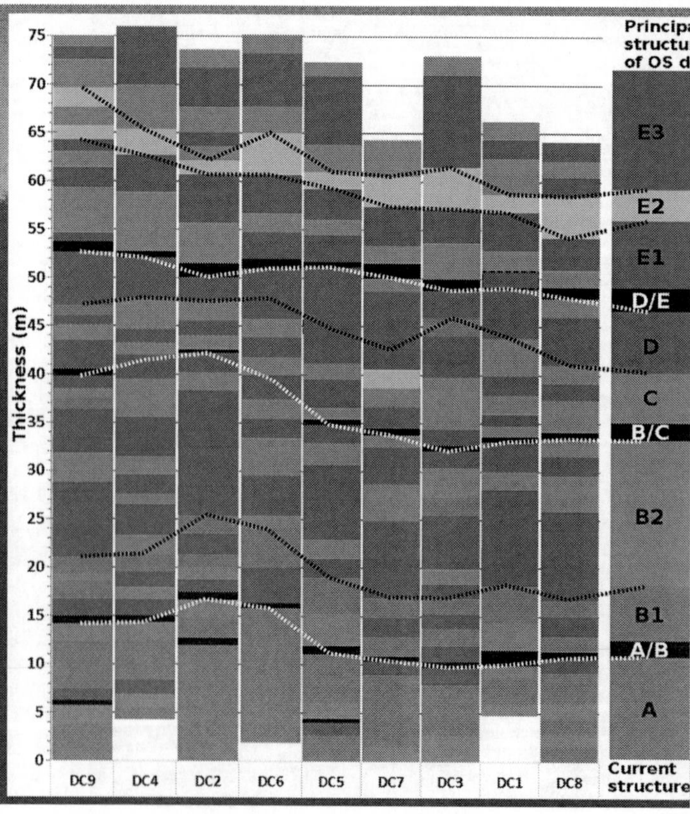

Internal lithological-geochemical structure of Attarat Um Ghudrun oil shale units

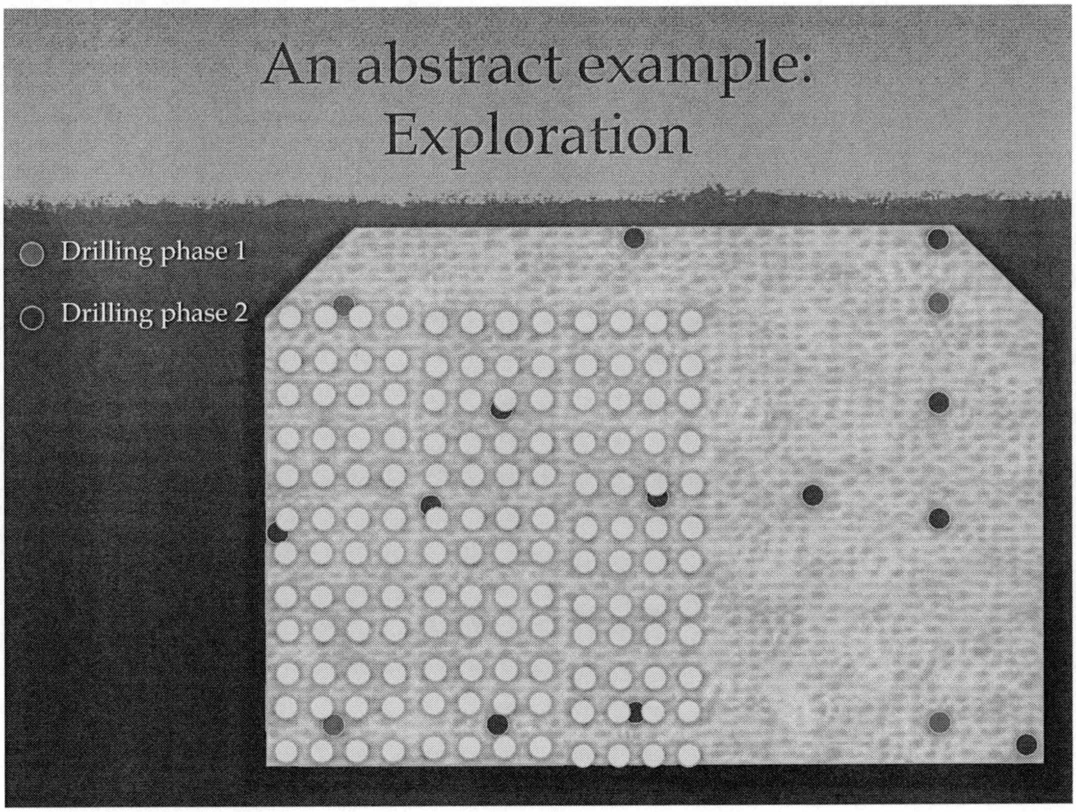
An abstract example:
Exploration
Drilling phase 1
Drilling phase 2

An abstract example:
Exploration
Drilling phase 1
Drilling phase 2
Detailed geochemical-mathemathical work
Now we do carefully analyse the data!!!

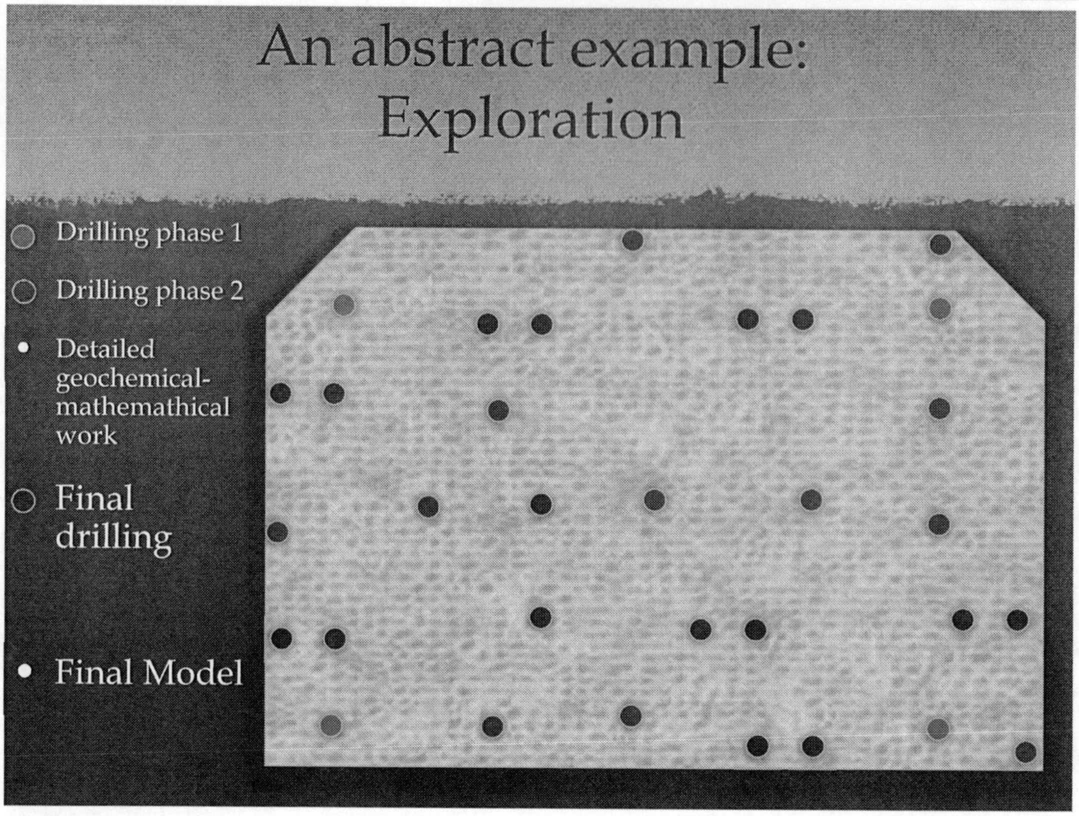
An abstract example:
Exploration
Drilling phase 1
Drilling phase 2
Detailed geochemical-mathemathical work
Final drilling
Final Model

A statistical analysis

- A representative collection of Attarat Um Ghudran (AUG) deposit samples, has been geochemically analysed. As an example, 392 samples from 9 drill cores reliably characterises the sequence of the OS seam, on average 70 m thick.
- The composition of AUG OS varies significantly. The contents of CaO and SiO_2 range within 3–70 wt% and 10–50 wt%, respectively, and the contents of organic matter, MgO, P_2O_5, Al_2O_3 and S vary considerably.
- The concentrations of metals (e.g. Zn, V, Cr, Ni and Mo) vary many dozens of times in the cross section.

- Two multivariate statistical analysis methods were selected: principal components analysis (PCA) and hierarchical clustering of PCA groups. The dataset is composed of 392 data rows as chemical analyses and 18 columns as elements (9 drill cores by 18 elements; analyses carried out by software 'R').
- The four principal components account for 88.6% of OS geochemical variability.
- The first component PC1 accounts for 47% of the variance, showing the highest correlation with **organic matter, S, Cr, Cu, Ni, Zn, Mo (these metals related to organic matter),**
- PC2 accounts for 22.82% and shows positive correlation with **TiO_2, Al_2O_3, Fe_2O_3, SiO_2 (terrigenous) and K_2O** and negative correlation with **CaO** (in opposite to carbonate formation).
- PC3&4 -- covariance with CaO and MgO. This statistical analysis provides detailed information for the deposit compositional heterogeneity and allows better exploration targeting.

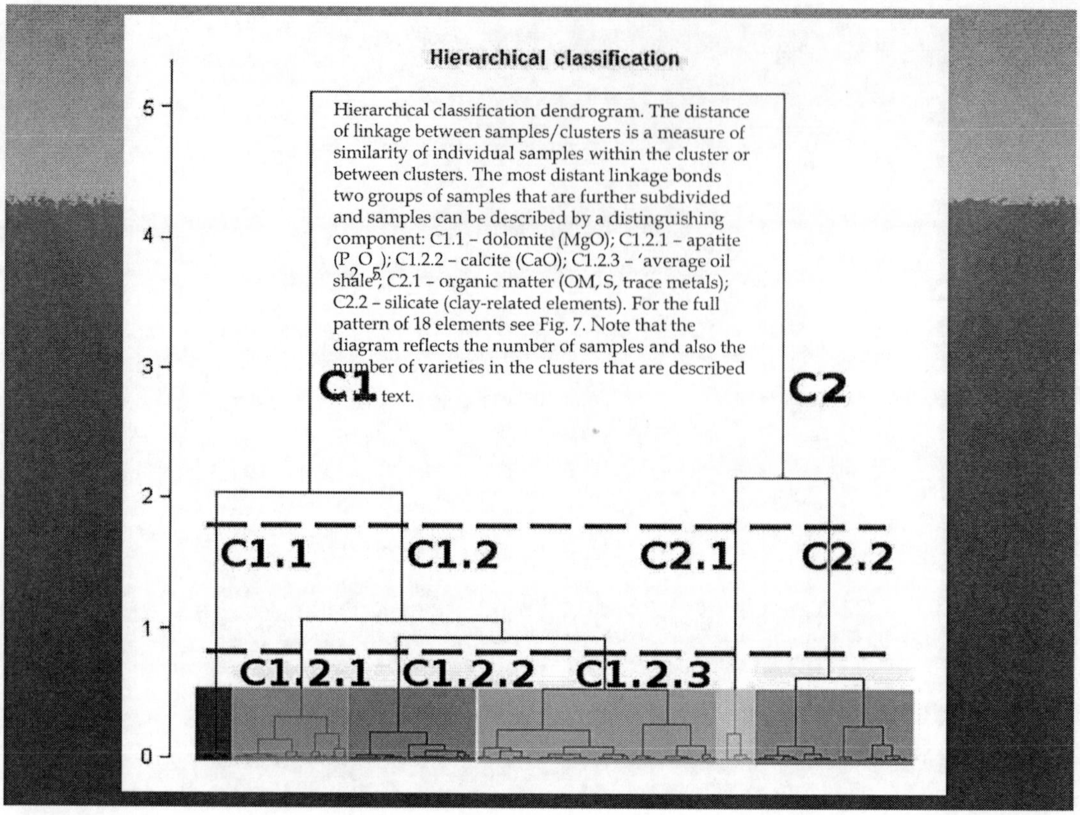

Hierarchical classification dendrogram. The distance of linkage between samples/clusters is a measure of similarity of individual samples within the cluster or between clusters. The most distant linkage bonds two groups of samples that are further subdivided and samples can be described by a distinguishing component: C1.1 – dolomite (MgO); C1.2.1 – apatite (P_2O_5); C1.2.2 – calcite (CaO); C1.2.3 – 'average oil shale'; C2.1 – organic matter (OM, S, trace metals); C2.2 – silicate (clay-related elements). For the full pattern of 18 elements see Fig. 7. Note that the diagram reflects the number of samples and also the number of varieties in the clusters that are described in the text.

On principal components planes constructed by PC1–PC2 and PC1–PC3 (and others not presented here), the cloud of the analysed samples grouped into clusters (C1.1, C1.2.1, etc.) demonstrates the correlation between samples and principal components. Note the different positioning of clusters on different planes, showing which component (and elements forming the corresponding component) best describe the samples in the cluster. On the plane PC1 (S, OM, trace metals) – PC2 ('clay components') clusters C1.2.1 (P_2O_5 with SiO_2, typical for the layers A and B1), C1.2.2 (Sr with CaO) and C1.2.3 (nearly corresponding to average oil shale) partially overlap, whereas on the plane PC1–PC3 (Sr and CaO with SiO_2) these clusters are located separately.

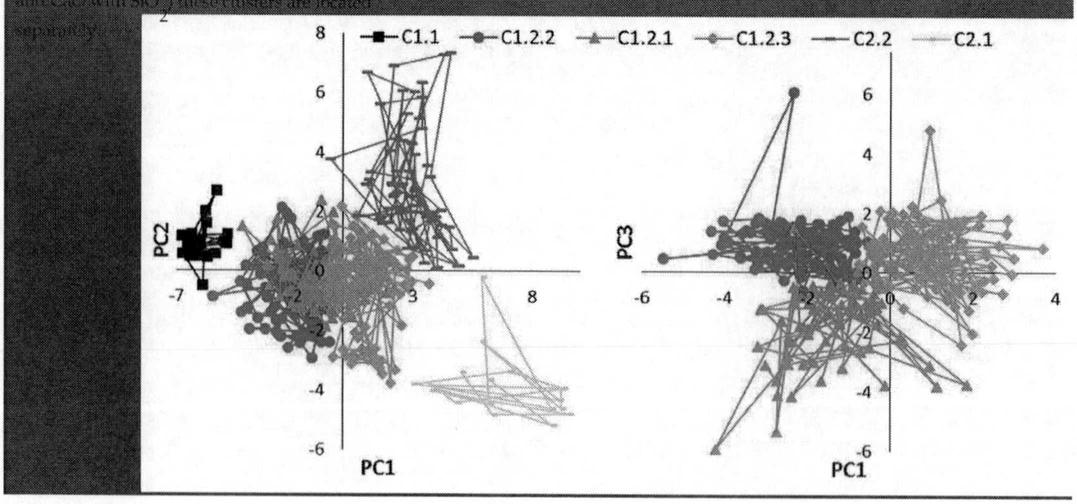

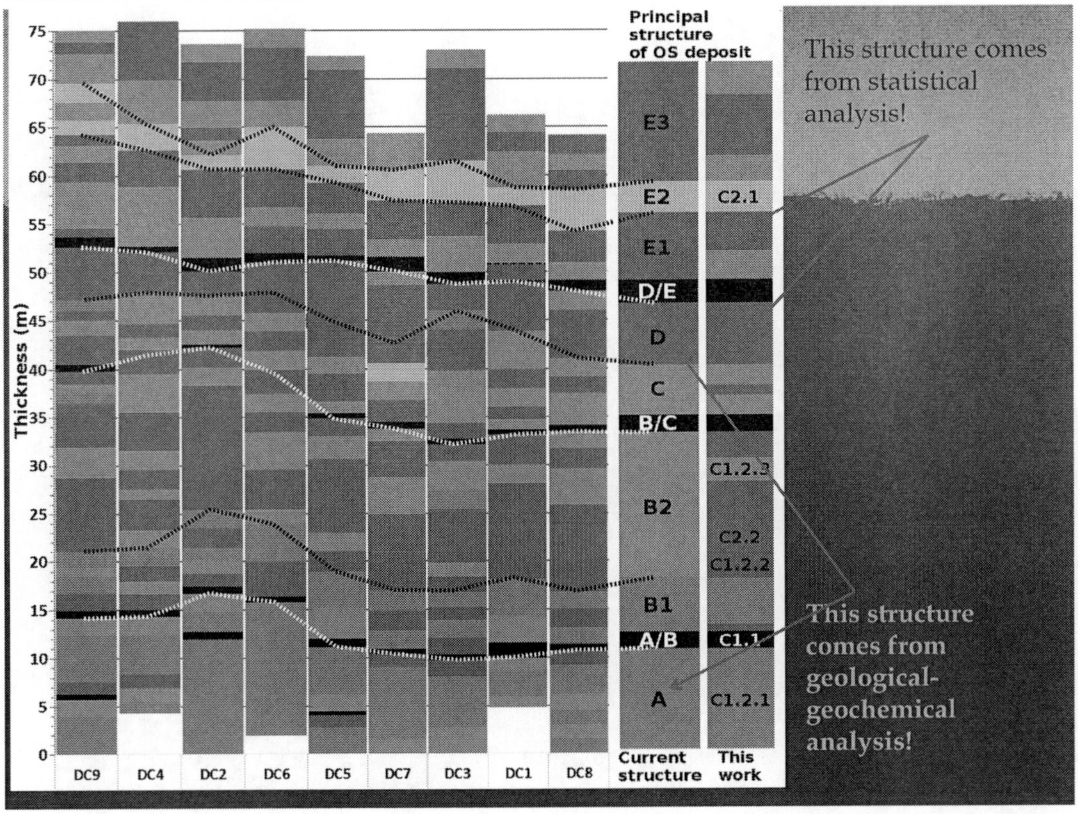

Conclusions

- Even Jordan oil shale geology looks simple, the chemical heterogeneity is large & OS has suffered in places remarkable alteration
- integrated geochemical investigations combined with multivariate statistical analysis of well-constrained chemical data provides a base for further exploration strategies, and likely reduces exploration costs
- apart to energy and oil source, this OS can also be considered as a metal reserve in the future
- the applied geochemical-statistical methods are powerful tools for interpretation of complex and variable structure of OS units in Jordan and elsewhere in similar settings.

Thanks!
… and lets keep high-level research on... in order to reduce exploration 6 mining costs
Contact us:
alvar.soesoo@gmail.com

Shale oil desulfurization by thiophenes extraction.

By the extraction of thiophenes and the recycling of heavy fraction in Galoter oil shale retorting process the content of sulfur in shale oil can be decreased up to 70%

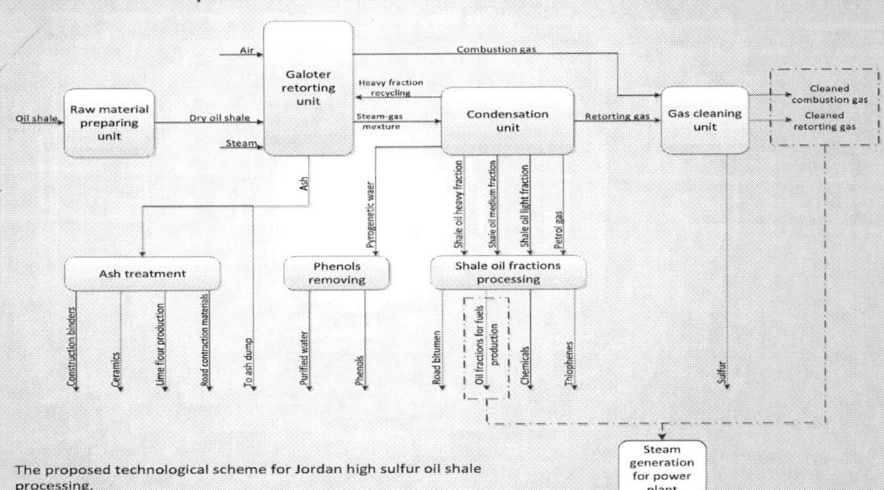

The proposed technological scheme for Jordan high sulfur oil shale processing.

The lab retorting installation simulates the industrial oil shale retorting technology Galoter.

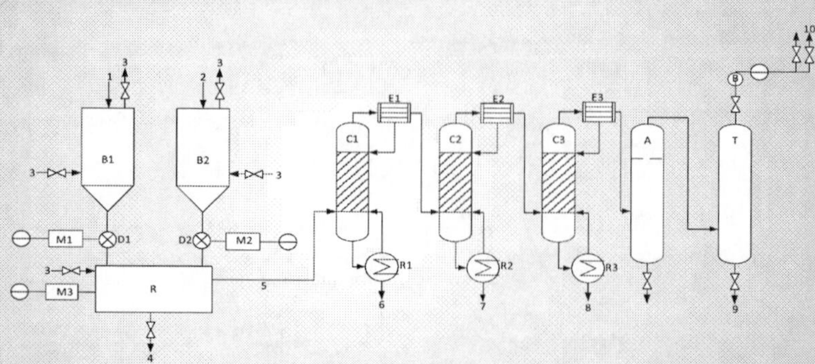

Schema: Galoter oil shale lab retorting installation.

B1: Oil shale bunker; B2: Heat carrier bunker; D1,D2: Dosers, M1,M2,M3: Gear motors, R: Reactor; C1,C2,C3: Rectification columns; E1,E2,E3: Condensers; R1,R2,R3: Reboilers; A: H_2S absorber; T: Drip pan; B: Blower. 1: Oil shale, 2: Ash as heat carrier; 3: N_2; 4: Semi-coke; 5: Steam-gas mixture, 6: Shale oil heavy fraction; 7: Shale oil medium fraction; 8: Shale oil light fraction; 9: Water; 10: Pyrolysis gas.

The most applicated thiophene derivatives are P3HT in photovoltaic and PEDOT in organic conductors.

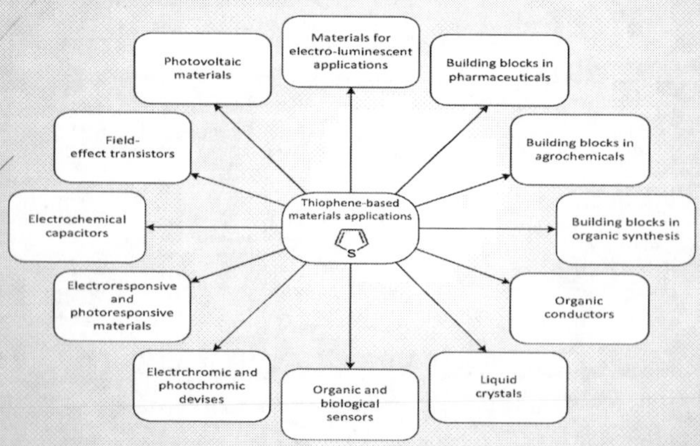

Some thiophene-based materials applications.

Poly(3,4-ethylenedioxythiophene), PEDOT, is the most industrially important organic conductor with a wide range of applications:

Antistatic coatings;

Electrically conducting coatings in organic solar cells (OSCs);

PEDOT:PSS as a transparent conductor in electroluminescent devices; conducting layer in Capacitors;

Conducting layers for printed wiring board manufacture

PEDOT layers with 'electronic' functions:

- -As a hole injection layer in OLEDs
- -Layers for photovoltaics and sensors
- -Conductive layer in all-organic thin-film transistors
- -Utilizing the redox behavior of PEDOT-electrochromic windows based on PEDOT:PSS

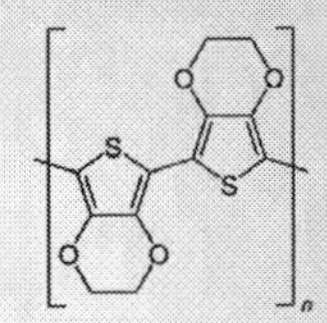

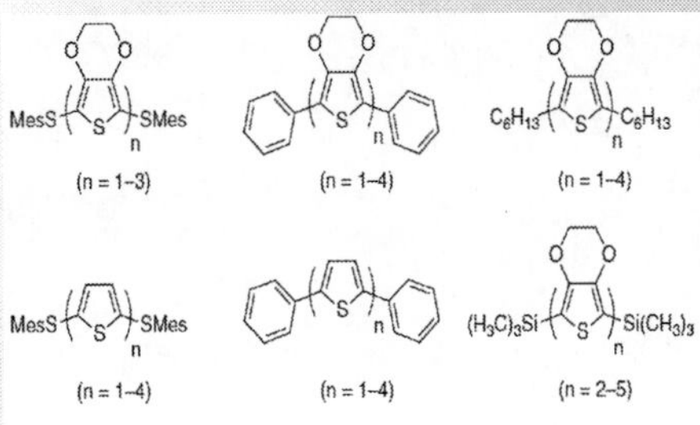

Oligomers for Conductors

Polythiophenes as chemical sensors for cations and Anions detection

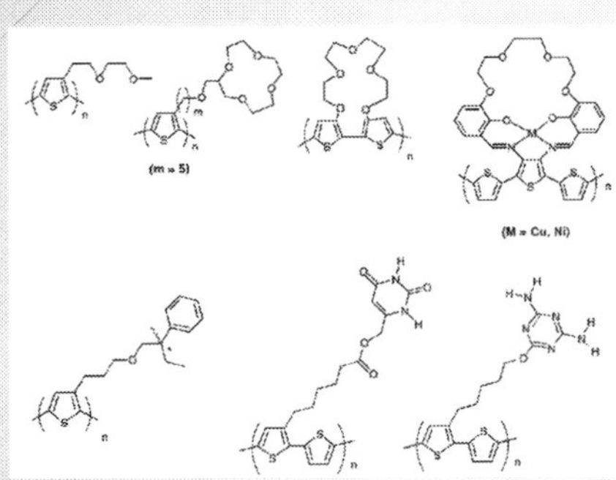

Polythiophenes as biological sensors for detection of low molecular weight biological molecules, DNA and proteins

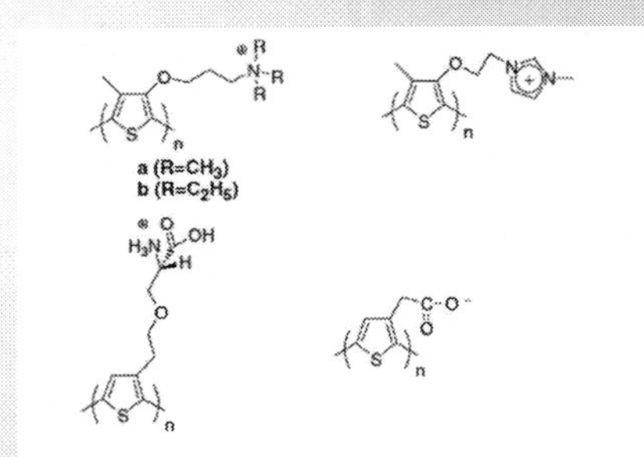

Liquid crystals polythiophenes derivatives

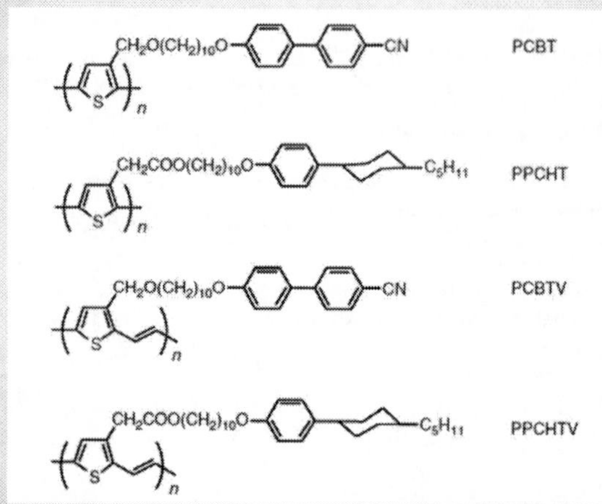

Thiophenes are prepared by the reaction of 1,4-diketones, diesters or dicarboxylates with sulfidizing reagents as in the Paal-Knorr's synthesis method.

Thiophene extraction method from coal tar

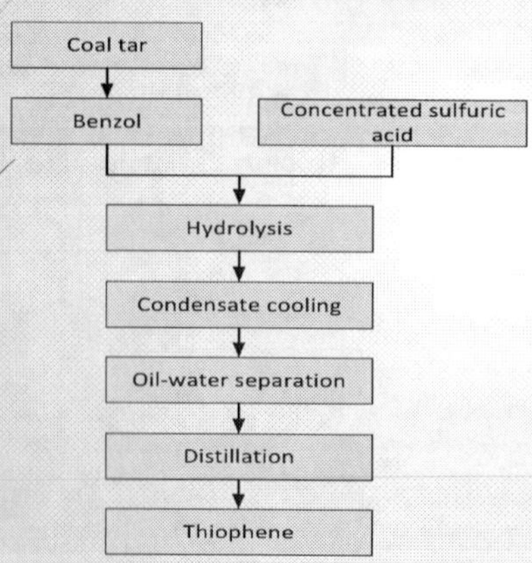

Paal-Knorr's thiophenes synthesis method

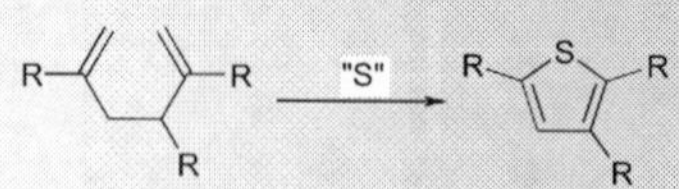

"S": $H_2S/HCl/H_2O$; P_4S_{10}
Lawessons reagent

Global major thiophene manufacturers

Manufacturers	Website
PCAS	http://www.pcas.com/
SKC	http://www.skcinc.com/
Fisher-Scientific	http://www.fichersci.com/
Minstar Industries	http://www.minstargroup.com/
Dongsheng Chemical	http://www.xysdshg.com/
Sincere Chemicals	http://www.sjzchx.com/
Yuanly Science and technology	http://www.yuanlichem.com.cn/
Jinan Fufang Chemical	http://www.fufangchem.com/

Thiophene preparing involves the vapor phase reaction of carbon disulfide and butanol with an oxide catalyst at 500–550 °C, Several reactions conditions and catalysts requirement make the thiophenes synthesis process slightly profitable.

Table Asia Thiophene Capacity Production Price Cost Production Value

Year	2010	2011	2012	2013	2014	2015
Capacity (T)	4626.80	4645.95	4729.09	4749.92	4782.85	4820.34
Production (T)	3044.60	3125.10	3218.93	3273.83	3347.48	3421.65
CapacityUtilization	65.80%	67.27%	68.07%	68.92%	69.99%	70.98%
Price (USD/T)	5495.04	5391.20	5321.33	5172.68	5064.85	4953.13
Production value (K USD)	16730.22	16848.03	17128.97	16934.49	16954.49	16947.84
Cost (USD/T)	4754.86	4700.05	4698.20	4595.41	4519.36	4431.56
Gross (USD/T)	740.18	691.15	623.13	577.27	545.48	521.56
Gross Margin	13.47%	12.82%	11.71%	11.16%	10.77%	10.53%

Source: QYR Thiophene Research Center, Sep 2015

Frequently detected components in light fraction of high sulfur shale oils by GCMS.

№	Component	CAS	Formula	BP, °C	Density
1	1-Hexene	592-41-6	C6H12	60-66	0.678
2	Benzene	71-43-2	C6H6	80	0.874
3	Thiophene	110-02-1	C4H4S	84	1.051
4	1-Heptene	592-76-7	C7H14	94	0.697
5	1-Butanethiol	109-79-5	C4H10S	98	0.842
6	Hexane	110-54-3	C6H14	98	0.692
7	Heptane	142-82-5	C7H16	98	0.71
8	Pyridine	110-86-1	C5H5N	97	0.983
9	Toluene	108-88-3	C7H8	111	0.866
10	Thiophene, 2-methyl-	554-14-3	C5H6S	113	1.014
11	Thiophene, 3-methyl-	616-44-4	C5H6S	114	1.016
12	1-Octene	111-66-0	C8H16	122-123	0.715
13	Octane	111-65-9	C8H18	125-127	0.703
14	Thiophene, tetrahydro-2-methyl-	1795-09-1	C5H10S	132	0.96
15	Thiophene, 2-ethyl-	872-55-9	C6H8S	132-134	0.99
16	Thiophene, 2,5-dimethyl-	638-02-8	C6H8S	134	0.985
17	Ethylbenzene	100-41-4	C8H10	134.6	0.867
18	P-Xylene	106-42-3	C8H10	138	0.861
19	Thiophene, 2,4-dimethyl-	638-00-6	C6H8S	139-141	0.994
20	Thiophene, 3-ethyl-	1795-01-3	C6H8S	140	1
21	Thiophene, 2,3-dimethyl-	632-16-6	C6H8S	140-141	1.002
22	o-Xylene	95-47-6	C8H10	143-145	0.879
23	Nonane	111-84-2	C9H20	151	0.72
24	Thiophene, 2-(1-methylethyl)-	4095-22-1	C7H10S	-	-
25	1-Butanone, 1-(2-thienyl)-	5333-83-5	C8H10OS	159-160	1.075
26	Thiophene, 2-ethyl-5-methyl-	40323-88-4	C7H10S	160	0.986
27	Benzene, 1,3,5-trimethyl-	108-67-8	C9H12	163-166	0.8675
28	Benzene, 1-ethyl-2-methyl-	611-14-3	C9H12	164-165	0.887
29	Decane	124-18-5	C10H22	174	0.735
30	Indane	496-11-7	C9H10	176	0.894
31	Thiophene, 2-methyl-5-propyl-	33933-73-2	C8H12S	179.1	0.97
32	Thiophene, 2,5-diethyl-	5069-23-8	C8H12S	181	0.97
33	Phenol	108-95-2	C6H6O	182	1.071
34	Thiophene, 3,4-diethyl-	35686-14-7	C8H12S	-	-
35	Benzene, 4-ethyl-2,3-dimethyl-	874-41-9	C10H14	186	0.88
36	Phenol, 2-methyl-	95-48-7	C7H8O	191	1.048
37	4-Undecene, (E)-	693-62-9	C11H22	191.2±7	0.757
38	1-Undecene	821-95-4	C11H22	192-193	0.75
39	Undecane	1120-21-4	C11H24	196	0.74
40	Thiophene, 5-propyl-2-ethyl	54244-74-5	C9H14S	196-197	-
41	1H-Indene, 1-methyl	767-59-9	C10H10	199	0.987
42	Ethanone, 1-(2-thienyl)-	1468-83-3	C6H6OS	208-210	1.142

The presence of non-aromatic hydrocarbon azeotropes with thiophenes and aromatics makes the separation of thiophenes by a simple rectification or by an azeotropic distillation not possible.

Azeotropes of thiophene, boiling point= 84˚C

Component	Boiling point of component, ˚C	Boiling point of mixture, ˚C	% by weight
Hexane	69.0	68.7	96.4
Methylcyclopentane	71.9	71.7	84.0
2,2-Dimethylpentane	79.1	75.9	53.7
2,4-Dimethylpentane	80.6	77.6	78.0
Cyclohexane	80.9	77.9	41.2
Cyclohexene	82.7	78.9	35.3
2,3-Dimethylpentane	89.9	79.4	21.2
Heptane	98.4	80.1	0.7

Azeotropes of benzene, boiling point= 80˚C

Component	Boiling point of component, ˚C	Boiling point of mixture, ˚C	% by weight
Hexane	69.0	68.5	16.8
Methylcyclopentane	71.9	71.5	86.0
2,4-Dimethylpentane	80.6	76.6	57.3
Cyclohexane	80.9	77.9	58.8
Cyclohexene	82.7	82.5	85.0
2,3-Dimethylpentane	89.9	80.9	36.0
tranc-1,3-Dimethyl-cyclopentane	90.8	82	32.3
Heptane	98.4	83.1	16.8

Azeotropes of 3-methylthiophene, boiling point=114˚C

Component	Boiling point of component, ˚C	Boiling point of mixture, ˚C	% by weight
Ethylcyclopentane	103.5	102.8	95.1
2,5-Dimetylhexane	109.2	107.1	68.3
1,1,2-Trimetylcyclo-pentane	113.8	110.5	56.7
2-Metylheptane	117.7	111.9	41.2
Trans-1,3-Dimetyl-cyclohexane	120.3	113.2	34.0
Oktane	125.7	114.2	18.0

Azeotropes of 2-methylthiophene, boiling point= 113˚C

Component	Boiling point of component, ˚C	Boiling point of mixture, ˚C	% by weight
heptane	98.4	97.8	97.8
2,2-Dimethylhexane	106.9	104.6	32.2
2,4-Dimethylhexane	109.2	106.1	60.4
2-Metylheptane	117.7	110.0	61.2

Extractive rectification allows separating up to 80% of thiophenes and also aromatics containing in narrow shale oil fractions with a purity up to 99%.

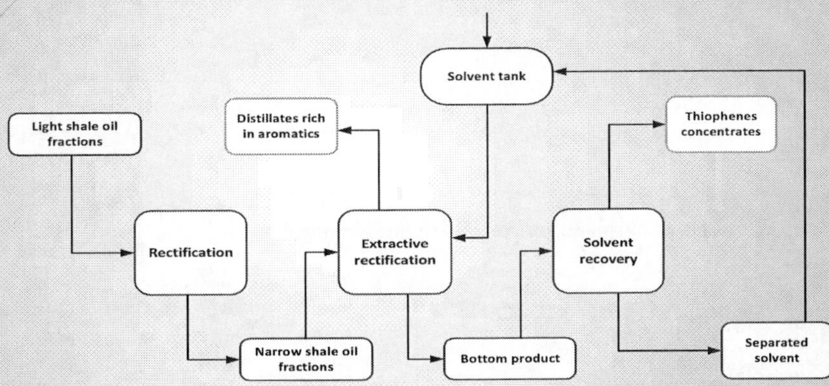

Scheme of thiophenes extraction from narrow shale oil fractions

Thank you!

Appraising the Jordan Oil Shale:

Example of Novel Subsurface Disciplines Integration

Wasfi Al Omari*, Mariela Araujo, Tom Fowler****

*Jordan Oil Shale Company (JOSCO)
** Shell International E&P (SIEP)

Definitions and Cautionary Note

The companies in which Royal Dutch Shell plc directly and indirectly owns investments are separate entities. In this presentation "Shell", "Shell group" and "Royal Dutch Shell" are sometimes used for convenience where references are made to Royal Dutch Shell plc and its subsidiaries in general. Likewise, the words "we", "us" and "our" are also used to refer to subsidiaries in general or to those who work for them. These expressions are also used where no useful purpose is served by identifying the particular company or companies. ''Subsidiaries'', "Shell subsidiaries" and "Shell companies" as used in this presentation refer to companies over which Royal Dutch Shell plc either directly or indirectly has control. Companies over which Shell has joint control are generally referred to "joint ventures" and companies over which Shell has significant influence but neither control nor joint control are referred to as "associates". In this presentation, joint ventures and associates may also be referred to as "equity-accounted investments". The term "Shell interest" is used for convenience to indicate the direct and/or indirect ownership interest held by Shell in a venture, partnership or company, after exclusion of all third-party interest.

This presentation contains forward-looking statements concerning the financial condition, results of operations and businesses of Royal Dutch Shell. All statements other than statements of historical fact are, or may be deemed to be, forward-looking statements. Forward-looking statements are statements of future expectations that are based on management's current expectations and assumptions and involve known and unknown risks and uncertainties that could cause actual results, performance or events to differ materially from those expressed or implied in these statements. Forward-looking statements include, among other things, statements concerning the potential exposure of Royal Dutch Shell to market risks and statements expressing management's expectations, beliefs, estimates, forecasts, projections and assumptions. These forward-looking statements are identified by their use of terms and phrases such as ''anticipate'', ''believe'', ''could'', ''estimate'', ''expect'', ''goals'', ''intend'', ''may'', ''objectives'', ''outlook'', ''plan'', ''probably'', ''project'', ''risks'', "schedule", ''seek'', ''should'', ''target'', ''will'' and similar terms and phrases. There are a number of factors that could affect the future operations of Royal Dutch Shell and could cause those results to differ materially from those expressed in the forward-looking statements included in this presentation, including (without limitation): (a) price fluctuations in crude oil and natural gas; (b) changes in demand for Shell's products; (c) currency fluctuations; (d) drilling and production results; (e) reserves estimates; (f) loss of market share and industry competition; (g) environmental and physical risks; (h) risks associated with the identification of suitable potential acquisition properties and targets, and successful negotiation and completion of such transactions; (i) the risk of doing business in developing countries and countries subject to international sanctions; (j) legislative, fiscal and regulatory developments including regulatory measures addressing climate change; (k) economic and financial market conditions in various countries and regions; (l) political risks, including the risks of expropriation and renegotiation of the terms of contracts with governmental entities, delays or advancements in the approval of projects and delays in the reimbursement for shared costs; and (m) changes in trading conditions. All forward-looking statements contained in this presentation are expressly qualified in their entirety by the cautionary statements contained or referred to in this section. Readers should not place undue reliance on forward-looking statements. Additional risk factors that may affect future results are contained in Royal Dutch Shell's 20-F for the year ended December 31, 2015 (available at **www.shell.com/investor** and **www.sec.gov**). These risk factors also expressly qualify all forward looking statements contained in this presentation and should be considered by the reader. Each forward-looking statement speaks only as of the date of this presentation, **XX Month 2016**. Neither Royal Dutch Shell plc nor any of its subsidiaries undertake any obligation to publicly update or revise any forward-looking statement as a result of new information, future events or other information. In light of these risks, results could differ materially from those stated, implied or inferred from the forward-looking statements contained in this presentation.

We may have used certain terms, such as resources, in this presentation that United States Securities and Exchange Commission (SEC) strictly prohibits us from including in our filings with the SEC. U.S. Investors are urged to consider closely the disclosure in our Form 20-F, File No 1-32575, available on the SEC website **www.sec.gov**. You can also obtain these forms from the SEC by calling 1-800-SEC-0330.

Outline

- Country entry May 2009, started operations May 2010
- Initial acreage 25% of Jordan (22,270 km[2])
- Jordan OSCA overview
- Extensive field operations: Coring, Wireline, Hydrology, 2D and 3D Seismic, Stress Test, Hot Wire Test
- Unrivaled geochemistry lab in Middle East
- Jordan Field Experiment (JFE)

Oil Shale Concession Agreement – Project Phases

Identify	Assess	Appraise	Pilot	Design	FID
Is Jordan of interest and can we secure acreage?	Does the play exist/work?	Is the materiality and productivity present?	Can we produce commercially?	How is the development defined?	Final Investment Decision
Regional resource evaluation: • Data integration: • Oil Shale wells • Petroleum wells • Water wells • Oil Shale Richness **Negotiate Concession Agreement:** • Investment Climate • Commercial Terms	**Explore for sweet spots:** • 9 months Venture Set Up • Drill/Learn/Drill campaign • Assess Richness, Thickness and Overburden • Regional G&G Evaluation • Define key risks	**Focused maturation:** • Define sweet spots - infill drilling • Define energy balance - kerogen conversion kinetics • Show heaters work & integrity can be maintained – rock mechanics • Show we have containment – hydrogeology • Demonstrate ICP works in Jordan – Jordan Field Experiment (JFE) • Characterize resultant products	**Field Trials to demonstrate:** • Key ICP parameters: • Well design & performance • Well spacing • Surface components • Product volumes • Waste products • Market requirements • Cost reduction	**FEED:** • Scope: • Plateau oil rate • Power generation • Oil & Gas treating • Refining strategy • Water • Cost • Efficiency • Learning Curve Application • Schedule • Drilling pace • Capital phasing	
✓ OSCA ratified	✓ Sweet spots identified	✓ Sweet spots appraised ✓ Successful JFE	○ ICP commerciality ○ Successful pilot	○ Commercial project defined	

JOSCO – Oil Shale Concession Agreement (OSCA)

- OSCA ratified by Jordan Parliament Aug 2009:
  - Initial acreage: 22,270 km^2
  - Final acreage: 1,000 km^2
- OSCA frames the path to a Commercial Project
- Drilling began in 2010; 365 wells drilled and evaluated
- **Appraise Phase Ended May 2016; reduced holding to 5,000 km^2. Focus: Define final 1,000 km^2 and executed JFE**
- Excellent Governmental support

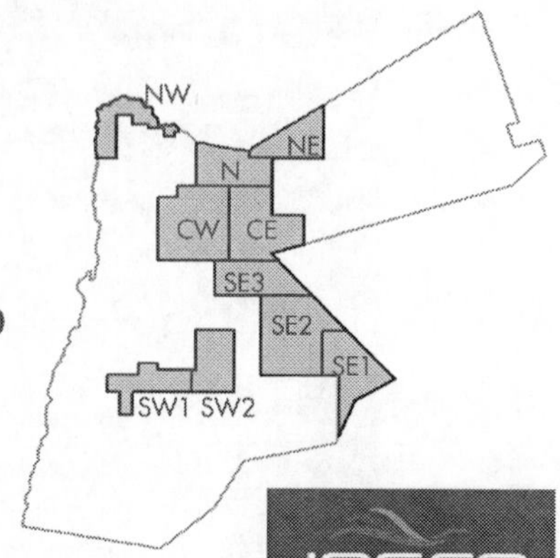

JOSCO
JORDAN OIL SHALE COMPANY

Drilling Campaigns

- Understand and de-risk subsurface geology
- Map the Oil Shale resource
  - The key parameters mapped were:
    - **Overburden Thickness**
    - **Oil Shale Richness**
    - **Oil Shale Thickness**

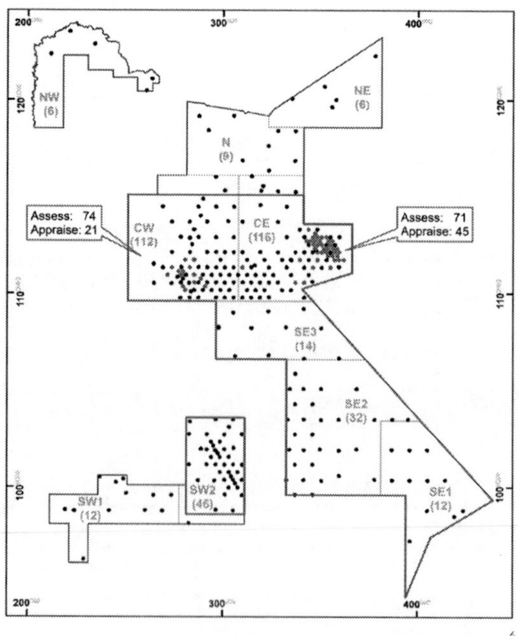

Footer 6

Stratigraphy: Lithofacies Characterization

- Improve geological **understanding** of the MCM
- Evaluate the correlatability and **lateral continuity** of the zones to be developed under ICP; and
- Represent one of the foundation elements which will be used to define the properties of the **HGZ** for development

URC	MCM	AHP	ASL
laminated *hemipelagic* limestone	kerogen-rich massive black marlstone	limestone (mudstone to wackestone)	SiO2-banded limestone
massive *hemipelagic* limestone	kerogen-rich laminated black marlstone	SiO2-banded black marlstone	well bedded Grainstone/Packstone
mottled *hemipelagic* limestone	low-energy limestone	phosphate-rich resedimented layers	ostreid layers and coquina beds
	moderate-energy limestone		massive to mottled limestone
	high-energy limestone		laminated limestone
	resedimented layers (*storm layers*)		resedimented layers
	calcarenite		
	muddy calcarenite		
	carbonate mudstone		
	greenish chalky marlstone		

Fracture Studies

- Collect and analyze fractures, **fault and damage** zone from cores, and logs.
- Analyze and interpret **fracture distribution** and mechanical stratigraphy.
- Correlate **mechanical/fracture** stratigraphy.
- Evaluate coring **induced fractures** and their potential impact on hydrology interval test results.

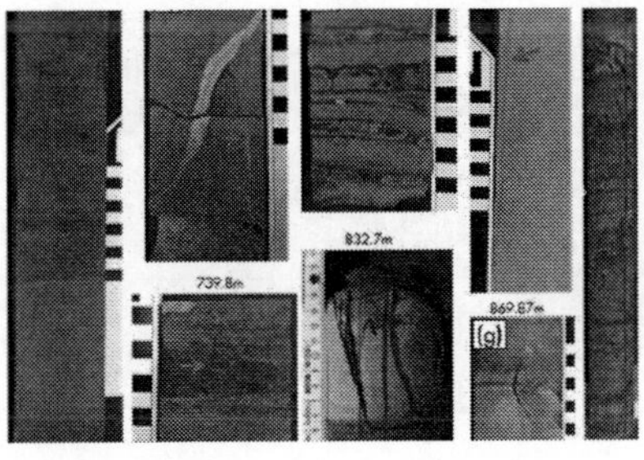

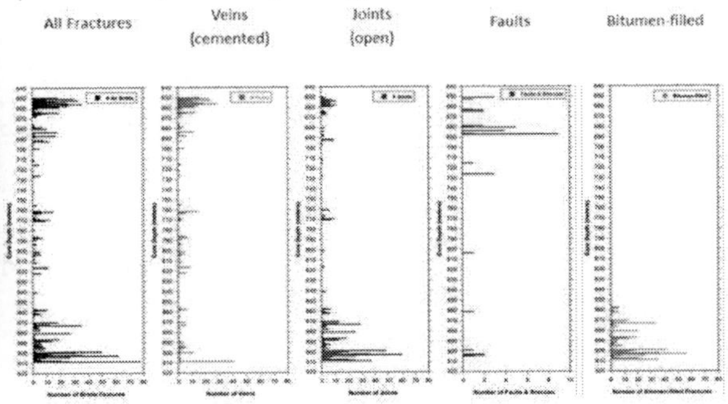

Seismic

- Interpret **depth conversion** to integrate the seismic structural framework and stratigraphic framework into a static model.
- Integrate seismic attributes with hydrological **water influx results**

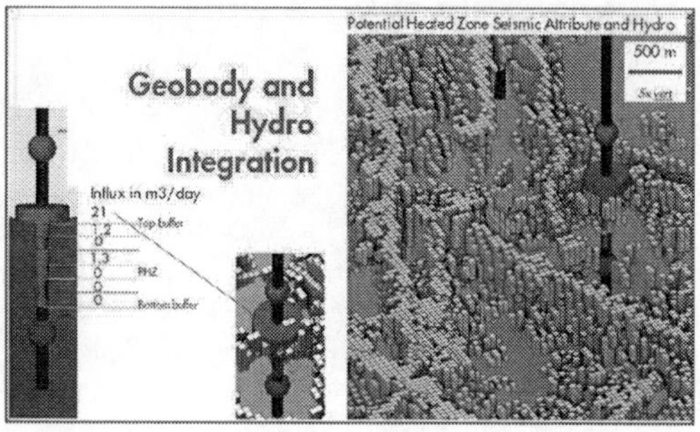

Integration between the hydrology influx measurements and the geo-body

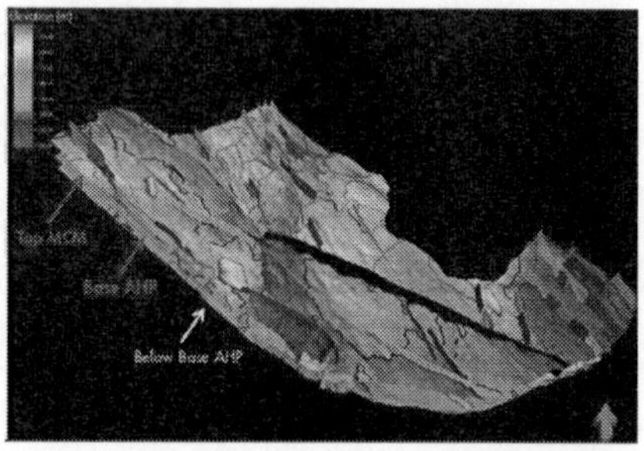

The depopulated fault set that is used for the base of the structural framework

Geomechanics

- Model impact of key rock failure mechanisms during ICP:
  - **Seal** rock integrity
  - **Tensile failure** of the rock
  - **Reservoir deformation** during the ICP
- Determine minimum **fault standoff** distance required to minimize the risk of fault activation/induced seismicity
- Determine dimensions of **pillars** to use under development
- Evaluate impact of surface heave/ **subsidence** on a commercial development

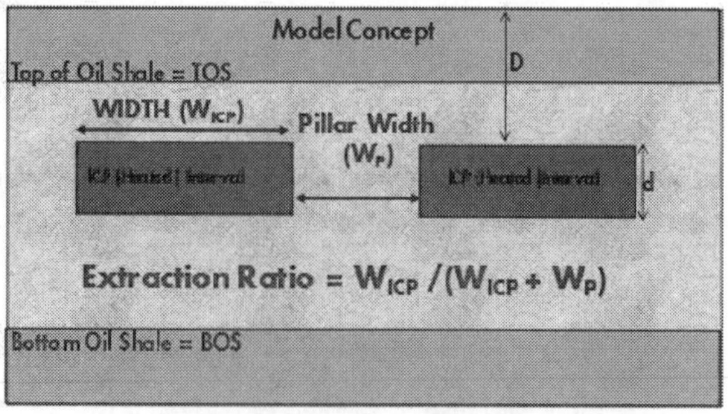

Concept of Geomechanical Model for Oil Shale Development

Petrophysics

- Geology was used to define guide lines separating areas of similar fluid properties: **4 areas were defined**

- Petrophysics was used to estimate the components of the reservoir rock including **kerogen concentration**, **bitumen in place**, **water saturation**, and **mineral content** based on core and logs data from the screened areas

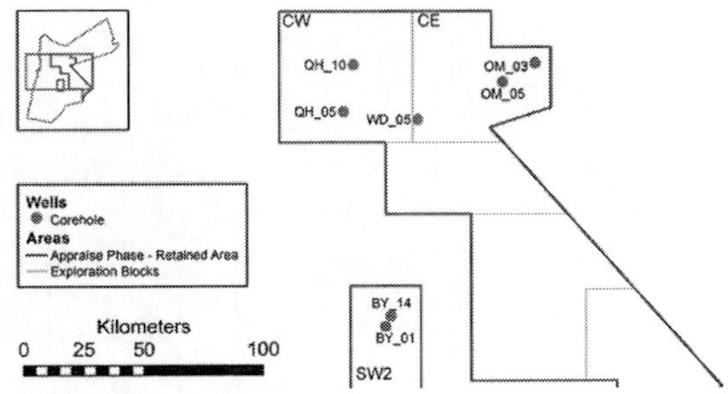

Hydrology

- Determine the **horizontal and vertical rock permeability** on drilled using multi-packer testing. Information used to locate seals for a potential development and assess potential water influx.

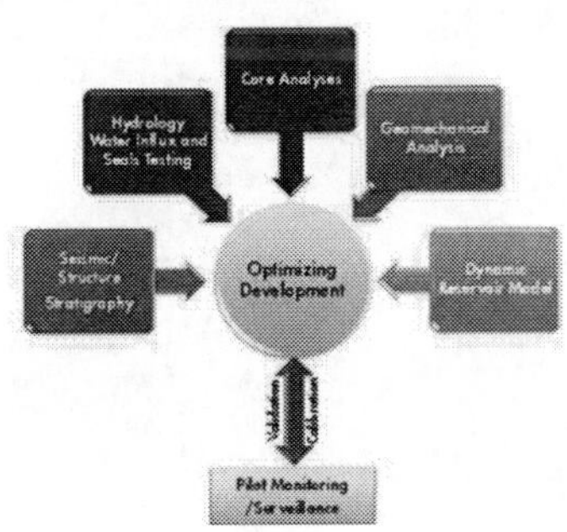

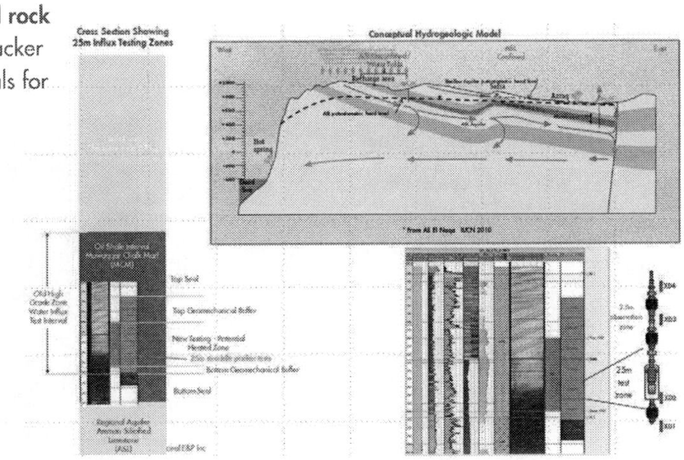

model showing typical testing intervals

 Integrated Seals and Containment Evaluation

Reservoir Engineering

- Evaluate **reservoir performance** and determine **production forecasting** either for field development planning (pilot or commercial) or history-matching of hydrocarbon field production

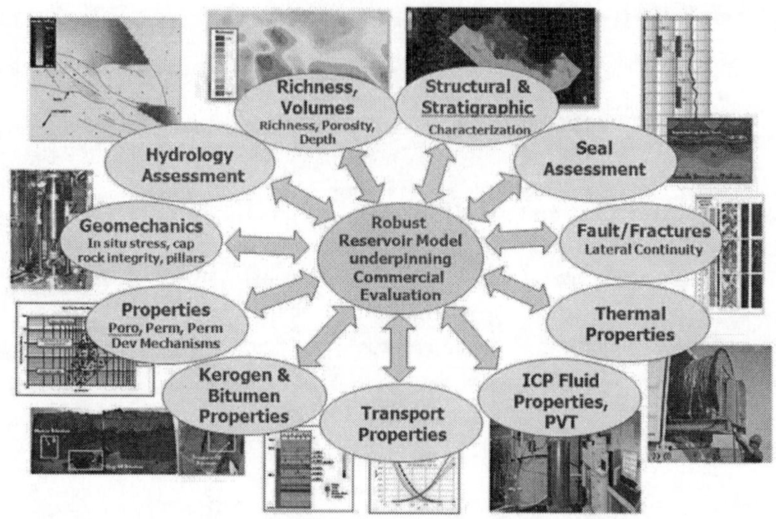

Key input to the dynamic model for integrated analysis

Results:

- Successfully able to quickly and efficiently integrate the results of all the geoscience disciplines (***geology, geophysics, hydrogeology, geochemistry, reservoir engineering and geomechanics***)
- The integrated subsurface workflow and geo-model development enabled:
  - The **project acceleration** by focusing the future development in limited areas bearing high resource intervals with minimum containment risk for maximum ICP return.
  - At the same time the subsurface geo-model served as basis for effective **well planning** and design, and drilling hazard prediction.

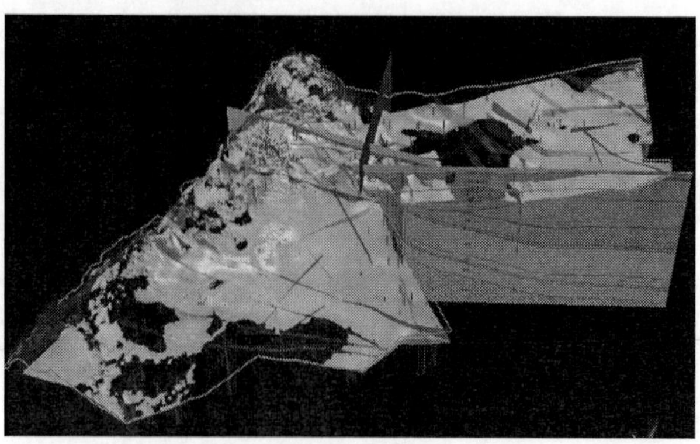

Key Learnings

1. **Time and Cost:** The early start of integration saved time and cost as it led to better understanding and execution

1. **Engagement:** Effective engagement between the leadership and team gave fruitful results. Focus on key de-risking steps, understanding of business objectives

2. **Coordination:** Despite having members of the technical team spread out around the globe, in different organizations, the team worked as one team with very high motivation, dedication and spirit

3. **Integration of global and local staff** led to rich and efficient knowledge transfer and capabilities developments

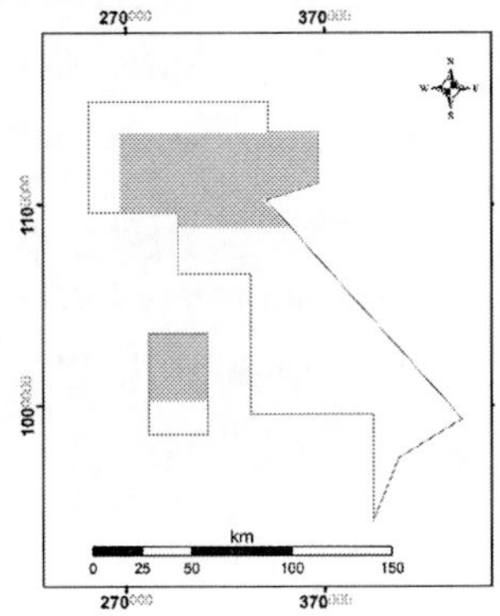

Copyright of Shell International   Footer   Date Month 2016   15

Thank You

清華大學

Tsinghua University

Critical Factors of Designs of Oil Shale Semi Coke Combustion in Circulating Fluidized Bed

Yiqun Huang

Key Laboratory for Thermal Science and Power Engineering of Education Ministry, Department of Thermal Engineering, Tsinghua university, Beijing, China 100084

CFB Group
Dept. of Thermal Engineering
Tsinghua University

- Introduction
- Oil shale and semi coke combustion in circulating fluidized bed
- Heat balance analysis
- Conclusion

Introduction: Background

- Oil shale is one of the most important unconventional oil and gas resources in China, playing a decisive role in helping to solve China energy shortage problem.

Energy sources	Exploitable amount	Amount equivalent to oil
Coal Bed Gas	10.9 trillion m3	10.2 billion tons
Shale Gas	25 trillion m3	23.4 billion tons
Tight Gas	8.8-12.1 trillion m3	8.2-11.3 billion tons
Oil Shale	250 billion tons	10 billion tons of shale oil
Oil Sand	3 billion tons	0.2 billion tons
Extra Check Heavy Oil	1 billion tons	5 billion tons
Tight Oil	1.3-1.4 billion tons	1.3-1.4 billion tons
Oil	24.58 billion tons	24.58 billion tons
Natural Gas	33.8 trillion m3	31.67 billion tons

Table 1 The exploitable amounts of main unconventional oil and gas in China

- The exploitable amount of oil shale is around 40% of the conventional petroleum resources in China.

Introduction: Background

- One of the main oil shale utilization technology is refining shale oil and shale gas by retorting oil shale, which produces a large amount of oil shale semi coke.
- One approach of comprehensive utilization technology of oil shale proposed in China is to use semi coke to generate electricity， instead of landfill.

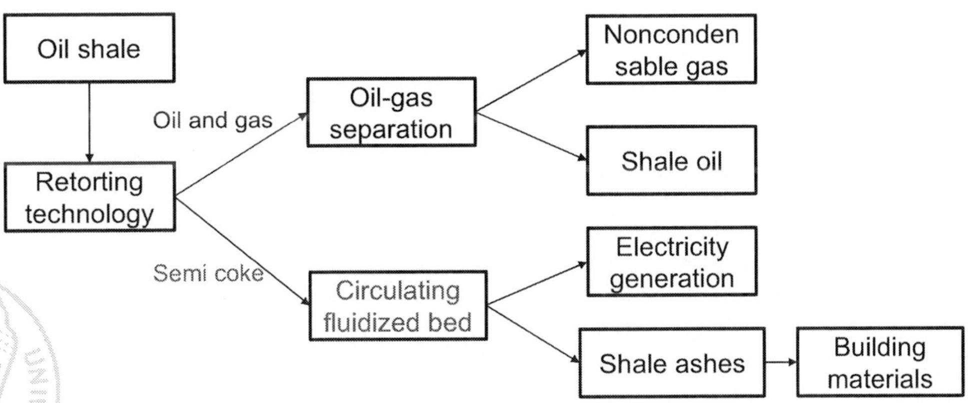

One approach of comprehensive utilization technology of oil shale

Introduction: Background

- One of the most important part of the above comprehensive utilization of oil shale is semi coke combustion in circulating fluidized bed boilers.
- Oil Shale semi coke: low calorific value, high content of ash, sulfide, phenols and polycyclic aromatic hydrocarbon.
- The circulating fluidized bed (CFB) combustion technology, up to now, is one of the most successful commercial clean solid fuel combustion technology.
  - Good fuel flexibility
  - High combustion efficiency
  - Widely load adjustment range
  - Low pollution emissions

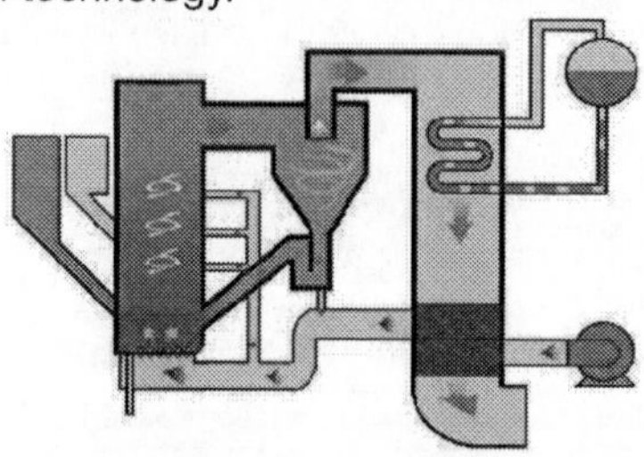

The schematic of circulating fluidized bed(CFB)

- It is necessary to verify the feasibility of semi coke combustion in CFB and to find out critical factors of designs and operations of oil shale semi coke combustion in CFB.

Introduction: Overview

- Semi coke characteristics
  - High content of ash
  - Low calorific value

	proximate analysis (%)				elemental analysis (%)				
	V_{ar}	A_{ar}	M_{ar}	FC_{ar}	C_{ar}	H_{ar}	N_{ar}	S_{ar}	O_{ar}
1	16.67	74.72	1.30	8.31	14.53	0.27	0.26	0.35	8.57
2	20.75	75.49	1.82	1.93	14.37	0.75	0.51	0.75	6.31
3	16.2	73.8	0.21	9.79	17.9	0.88	0.7	1.08	5.43

Proximate analysis and elemental analysis of oil shale semi coke

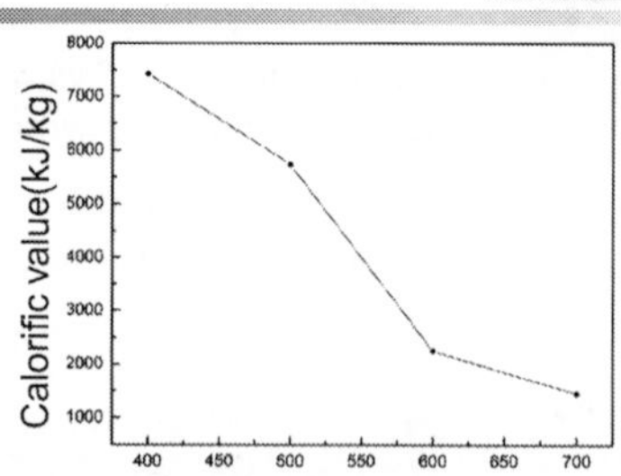

The relationship between retorting temperature and semi coke calorific value

- Blended combustion characteristics of oil shale and semi coke

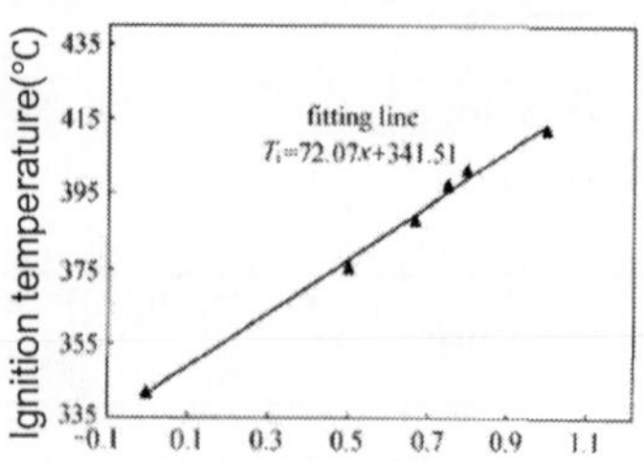

Ignition temperature of samples vs semi coke blended content

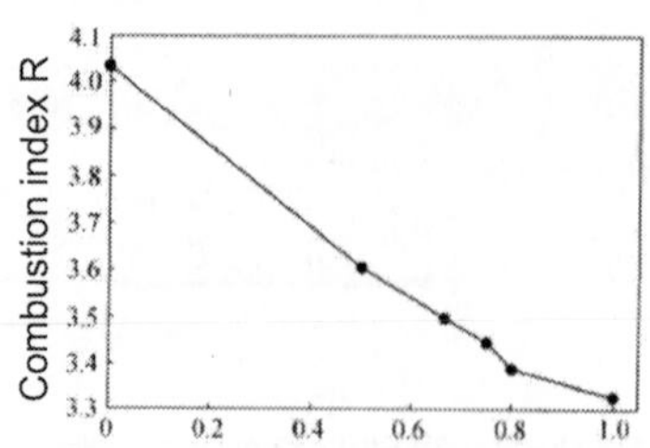

Combustion index vs semi coke blended content

Introduction: Overview

- Semi coke combustion in laboratory-scale circulating fluidized bed

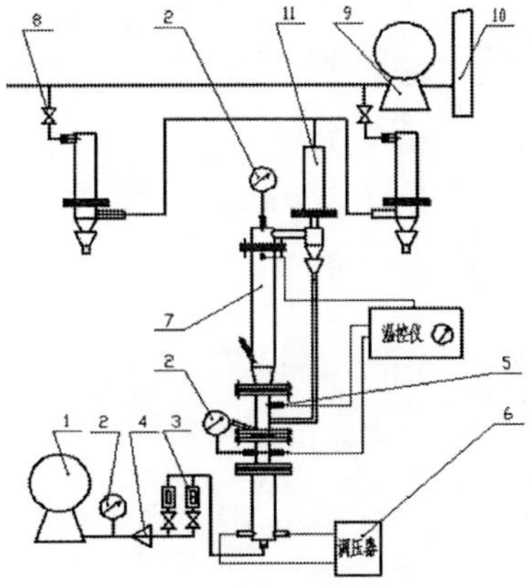

Schematic diagram of laboratory-scale circulating fluidized bed

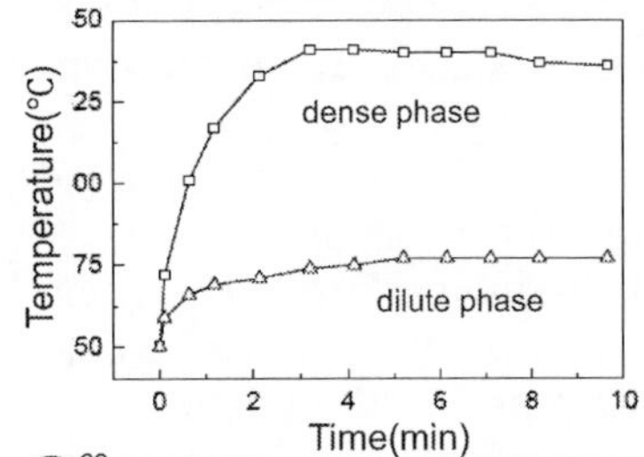

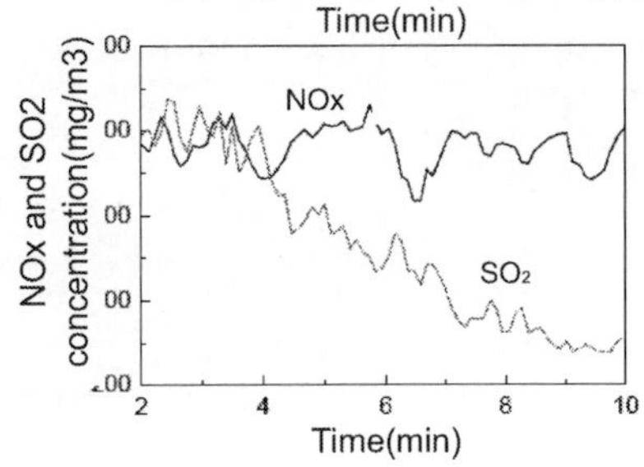

- ◆ The carbon contents of slag and fly ash were 2.156% and 4.378%.
- ◆ The combustion efficiency of CFB was 73.3%.

Introduction: Overview

- Semi coke combustion in pilot circulating fluidized bed

Kaljuvee(2004)

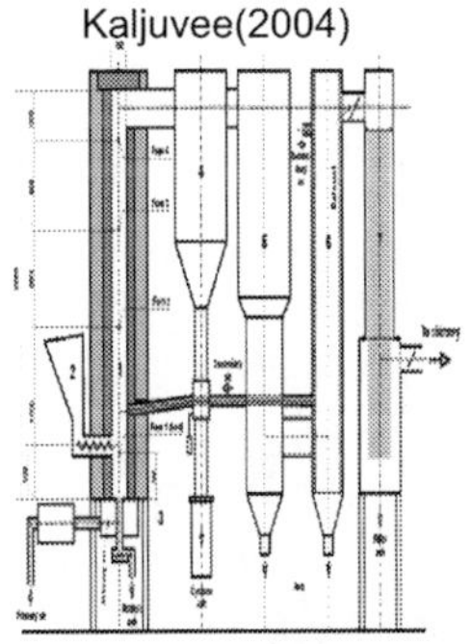

Stable combustion can be achieved when the moisture content of semi coke was less than 10%.

Wang(2011)

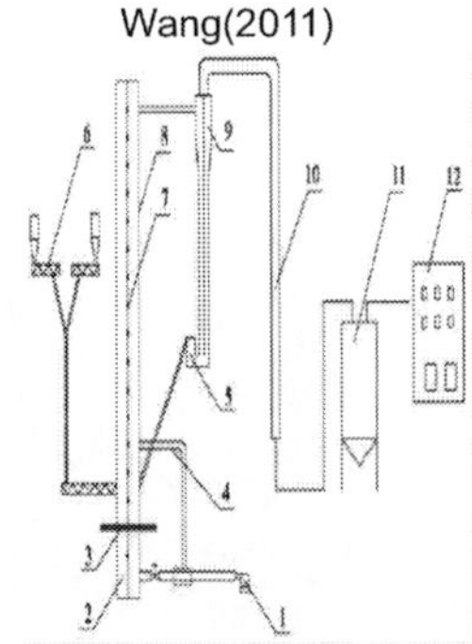

Bed temperature dropped rapidly along the height of furnace. Semi coke needed to be blended with oil shale.

Yan(2013)

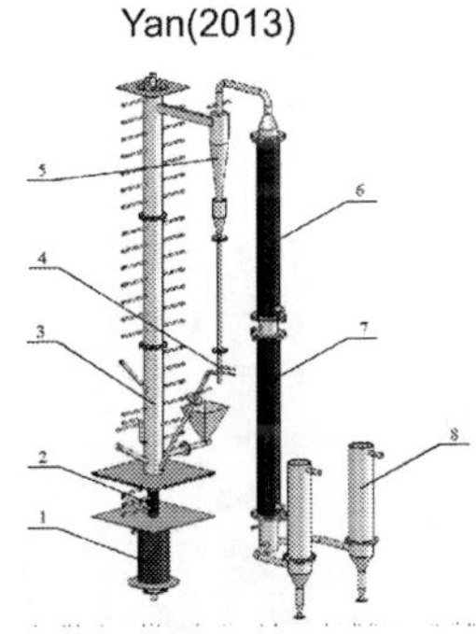

Semi coke was not easy to burn out in circulating fluidized bed boilers.

- Most existing researches showed that there lied great challenges when combusting semi coke in circulating fluidized bed boilers.
- Our work was to find out critical factors of designs of oil shale semi coke combustion in circulating fluidized bed through pilot CFB experiments and heat balance analysis.

Oil shale and Semi coke combustion in the CFB boiler

Experimental scheme

- Combusting bituminous coal (BC) to start up the CFB boiler
- Combusting oil shale (OS) and semi coke (SC) in the pilot CFB test rig successively
- Comparing the operation performances when combusting different fuels with the same boiler load, including bed pressure drop, bed temperature, oxygen concentration in flue gas and pollution emission.

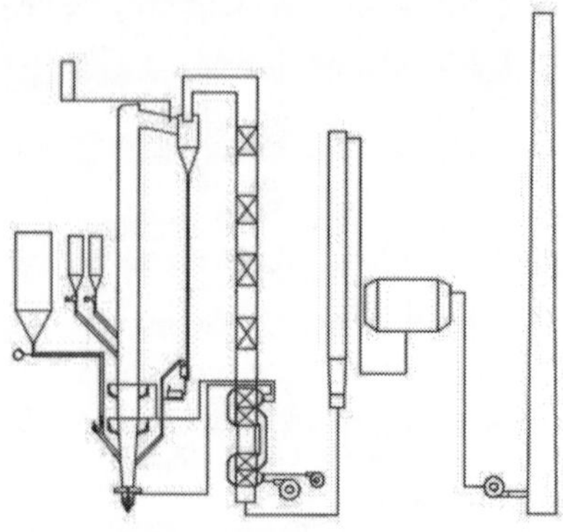

The schematic of 3MW pilot test rig

Proximate analysis and elemental analysis of fuels

	Proximate analysis /%				Calorific value /(kJ/kg)	Elemental analysis /%				
	M_{ar}	A_{ar}	V_{ar}	FC_{ar}	$Q_{ar,net}$	C_{ar}	H_{ar}	O_{ar}	N_{ar}	S_{ar}
Bituminous coal	3.07	38.34	17.41	41.18	17735.3	48.21	3.23	3.12	0.58	3.45
Oil shale	7.14	53.08	32.93	6.85	11139.7	33.58	3.14	1.99	0.33	0.75
Semi coke	14.06	66.3	13.57	6.07	3148.5	16.94	1.66	0.03	0.26	0.76

Oil shale and Semi coke combustion in the CFB boiler

- Start time: 18:00, May 22, 2016
- Time when BC combustion was stable:22:30 , May 22, 2016
- Time when OS combustion was stable:00:30, May 23, 2016
- Time when OS: SC= 3 : 1 combustion was stable:01:20, May 23, 2016

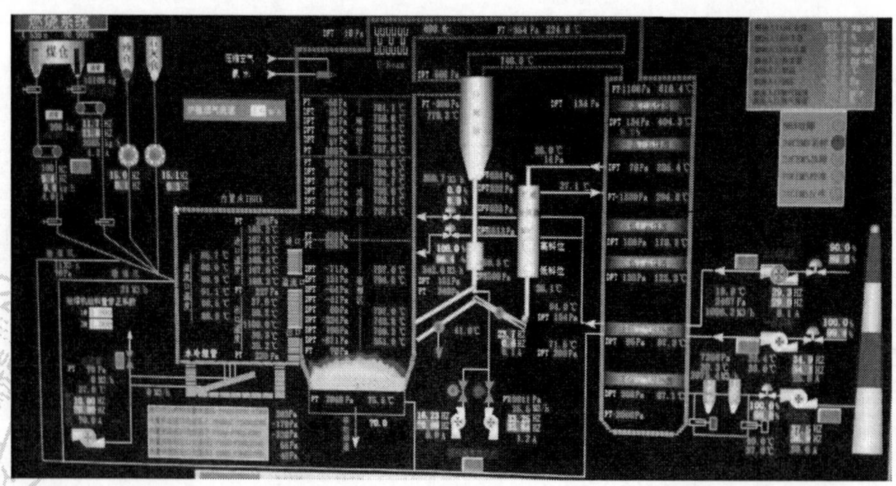

Oil shale and Semi coke combustion in the CFB boiler

The comparison of bed temperatures when combusting different fuel resources

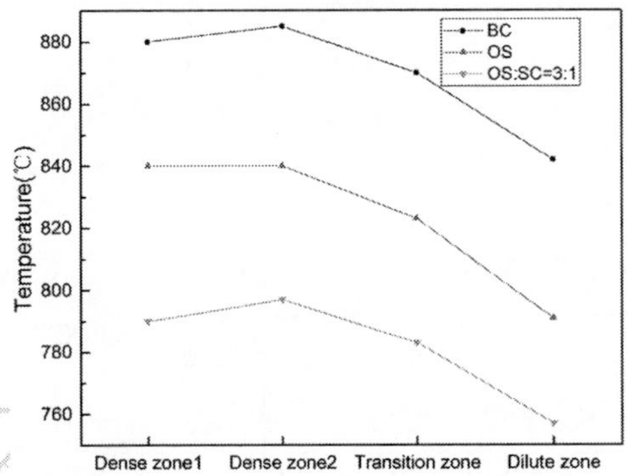

The comparison of oxygen concentrations in flue gas when combusting different fuel resources

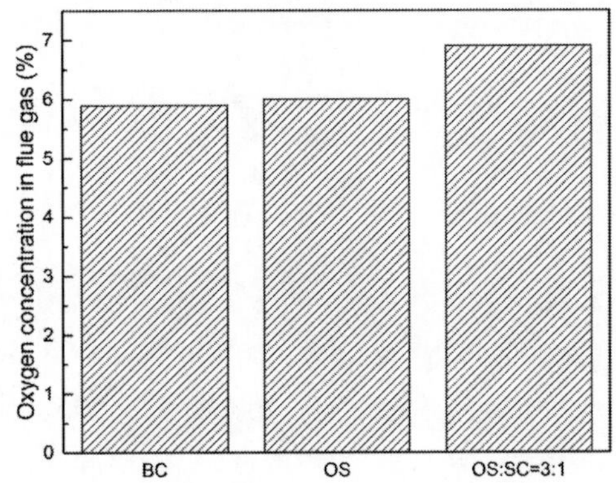

- The case of OS:SC = 3:1 had the lowest bed temperature and the highest oxygen concentration in flue gas.

Heat Balance Analysis

Calculation method

$$\eta = 1 - q_2 - q_3 - q_4 - q_5 - q_6$$

η :Boiler combustion efficiency
q_2:Heat loss due to exhaust gas
q_3:Heat loss due to unburned gas
q_4:Heat loss due to unburned carbon
q_5:Heat loss due to radiating
q_6:Heat loss of slag

$$q_2 = (I_{py} - a_{py} I_{lk})(1 - q_4) / Q_j$$

$$q_3 \approx 0$$

$$q_4 = 337.27 \frac{A_{ar}}{Q_j} \left[\frac{X_{pz} C_{pz}}{1 - C_{pz}} + \frac{(1 - X_{pz}) C_{fh}}{1 - C_{fh}} \right]$$

$$q_5 \approx 2.3\%$$

$$q_6 = X_{pz} A_{ar} c_{pz} (T_{ba} - 273) / Q_j$$

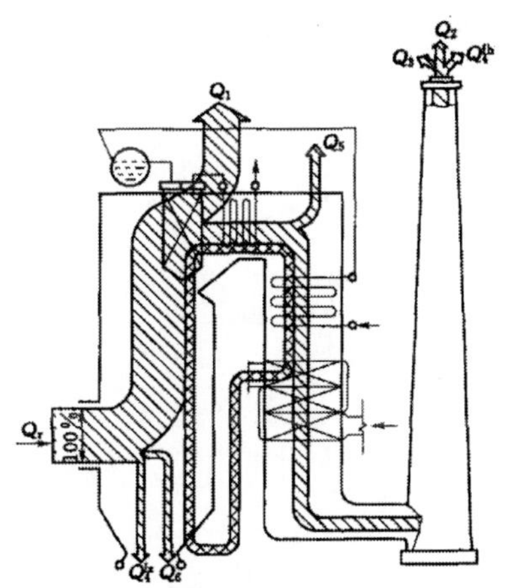

Boiler Heat Balance

Heat Balance Analysis

Boiler design parameters and elemental analysis of semi coke

	Unit	Value
D_{ss}	kg/s	19.44
P	MPa	1.098
t	°C	130.00
t_{gs}	°C	90.00
P_{gs}	MPa	1.198
C_{ar}	%	11.67
H_{ar}	%	0.80
O_{ar}	%	4.19
N_{ar}	%	0.73
S_{ar}	%	1.01
Cl_{ar}	%	0.00
A_{ar}	%	80.81
M_{ar}	%	0.79
V_{daf}	%	75.76
$Q_{ar,net,p}$	kJ/kg	3571

Calculation results

	Unit	Value
Q_j	kJ/kg	4247.56
q_2	%	7.27
q_3	%	0.00
q_4	%	4.14
q_5	%	2.30
q_6	%	9.22
η_{gl}	%	77.07
Q_1	kw	3287.19
BBJ	kg/s	1.15

Low calorific value → Great fuel consumption rate → Great slagging rate → Great heat loss due to slag

High slag temperature → Great heat loss due to slag

- Adjust heating surface distribution in the furnace to decrease the slag temperature and heat loss due to slag.
- Primary air and second air should be heated to provide a better combustion environment for semi coke.

Heat Balance Analysis

The effects of bed temperature on η and q_6

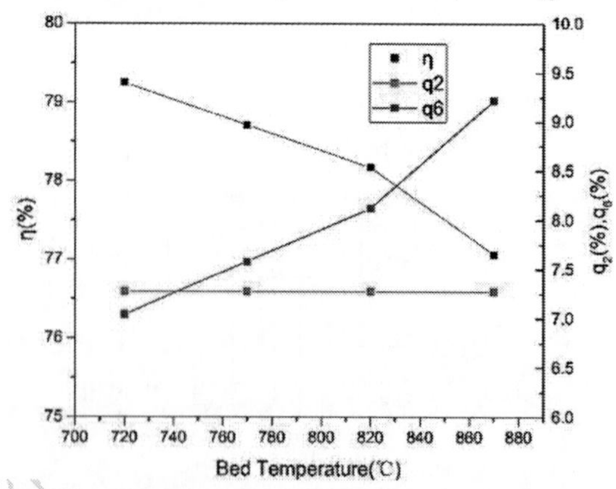

η :Boiler combustion efficiency
q_2:Heat loss due to exhaust gas
q_6:Heat loss of slag

- The decrease of heat loss due to slag can be attributed to the decrease of slag temperature caused by the decrease of bed temperature.
- The proportions of heat loss due to unburned gas q_3, heat loss due to unburned carbon q_4 and heat loss due to radiating q_5 were relatively small so that the total boiler efficiency increased with the decrease of heat loss due to slag.

Heat Balance Analysis

The effects of air temperature on η and q_6

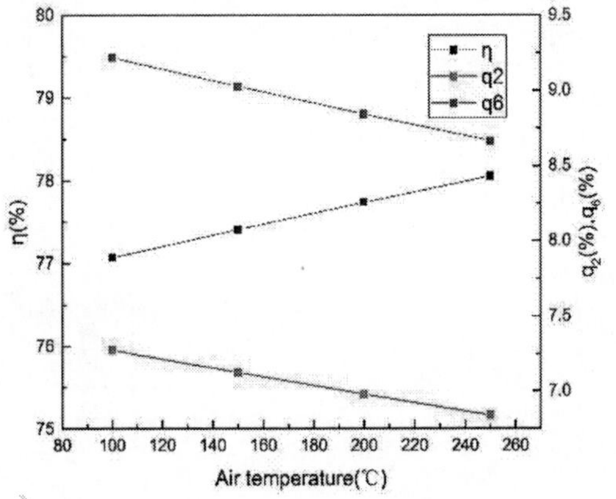

η :Boiler combustion efficiency
q_2:Heat loss due to exhaust gas
q_6:Heat loss of slag

- The increase of air temperature would increase the heat brought into furnace by unit mass fuel, leading to the decrease of fuel consumption rate and slagging rate.
- The decrease of fuel consumption rate can also decrease exhaust gas rate and heat loss due to exhaust gas.

Conclusion

- The ignition ability and combustion stabilization ability of semi coke are both worse than oil shale;
- It is feasible to combust mixed fuel of oil shale and semi coke in circulating fluidized bed, although causing the decrease of bed temperature and combustion fraction dilute zone.
- Using semi coke as independent fuel, heat loss due to slag of circulating fluidized bed is quite large. The boiler efficiency and combustion stabilization ability can be improved by reducing bed temperature and increasing air temperature.

Thank you!

PEPRIS

Sensitivity analysis of oil and gas production in the *in-situ* pyrolysis of Maoming oil shale

Yiwei Wang, **Peng Guo**, Youping Wang, Jianzheng Su

21/11/2016, Jordan

INTEGRITY COLLABORATION INNOVATION EXECUTION

Content

- Background
- Motivation
- Method & Specification
- Results & Discussion
- Conclusions

INTEGRITY COLLABORATION INNOVATION EXECUTION

PEPRIS

Background

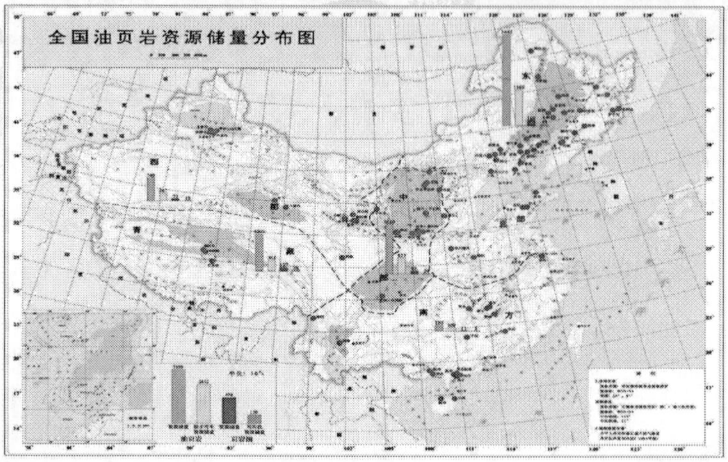

Oil shale resource distributes in 22 provinces, 47 basins and 80 mine areas.

Background

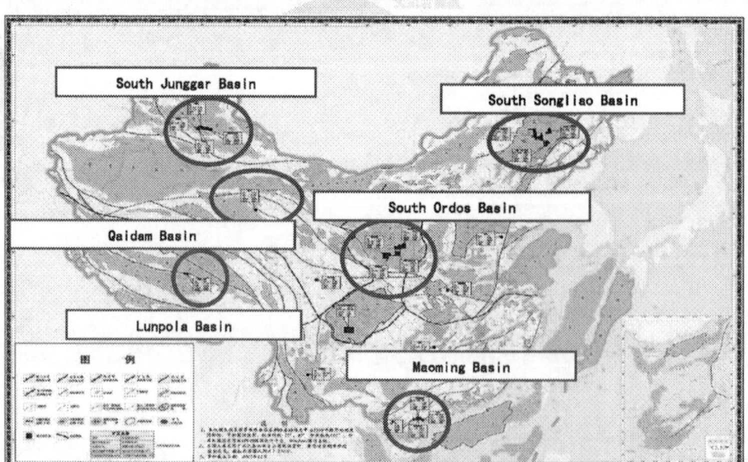

99.46% of SINOPEC's Oil shale resource distributes in the 6 basins.

Background

Outcrops of some oil shale mines in China

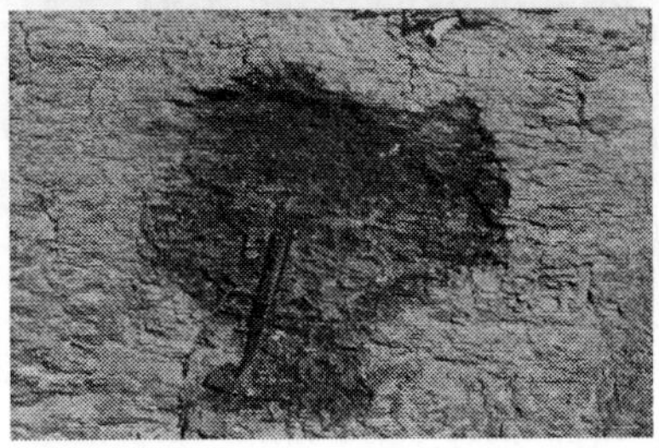

INTEGRITY COLLABORATION INNOVATION EXECUTION

PEPRIS

Background

Surface retorting by local factories is still the main exploitation method

Oil shale mines	Production in 2012 (10k tons)	Estimate in 2015 (10k tons)
Fushun	38.6	70
Rihui	2	2
Fengtai	2	12
Chengda	7	25
Beitaizi	2	3
Wangqing	6	12
Longkou	12	20
Tiefa	2	5
Yaojie	3.4	10
Beipiao	5	10
Aohanqi	1	15
Total	81	184

INTEGRITY COLLABORATION INNOVATION EXECUTION

PEPRIS

Background

Surface retorting	*In-situ* exploitation
√ Low technological requirement	× Highly complex mechanism
√ Quick economic returns	× Long-term running risk
× Environmental pollution	√ Environmentally friendly
× Inapplicability for deep resource	√ Feasible for deep deposits

The *in-situ* process has become an inevitable trend of the large-scale commercial exploitation of oil shale resources.

Background

Maoming Basin, with an area of 400 Km^2, locates in the southwest part of Guangdong province. Half of its area is oil shale buried.

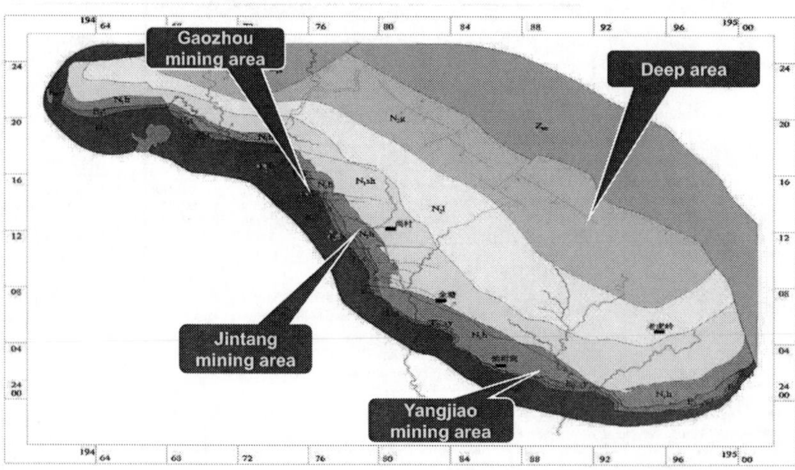

Background

Stratigraphical Sequence

代 (Era)	纪 （Period)	组 （Formation）	代号 （Symbol）	厚度 （Thickness）
新生代 （K_Z）	第四纪 Q （Quaternary)	/	Q	>25.5
	新近纪 N （Neogene）	Gaopengling	N_2g	83-1066.5
		Laohuling	N_1l	274-686
		Shangcun	N_1sh	2-23.55 oil shale
		Huangniuling	N_1h	71-404
	古近纪 E （Paleogene）	**Youganwo**	$E_{2\text{-}3}y$	19-46 oil shale
		Shangtong	E_2s	15-76
中生代 （M_Z）	白垩纪 K （Cretaceous）	Tongguling	K_2t	>926

Background

✓ **Shangcun Formation:**

- 3～4 layers of oil shale
- Oil content: 3.5-7%, 4.58% on average
- Moisture: 2.79%
- Thickness:2-23.55m, 15.58m on average

✓ **Youganwo Formation:**

- Single layer
- Thickness:19-46m, 22.75m on average
- Oil content:6.22-13.07%
- Moisture: 4.79－11.07%

Outcrops

Shangcun Formation

Youganwo Formation

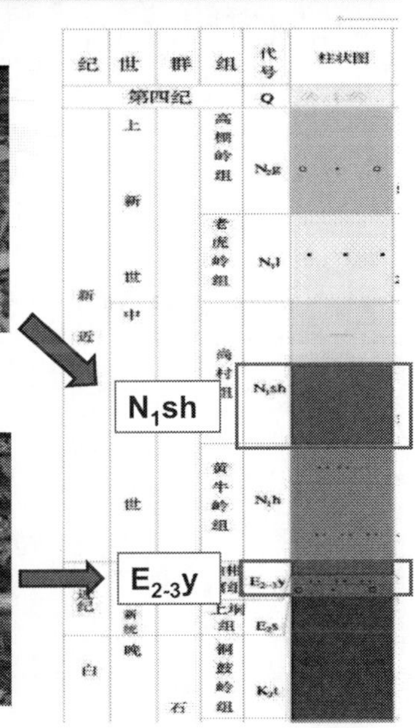

Background

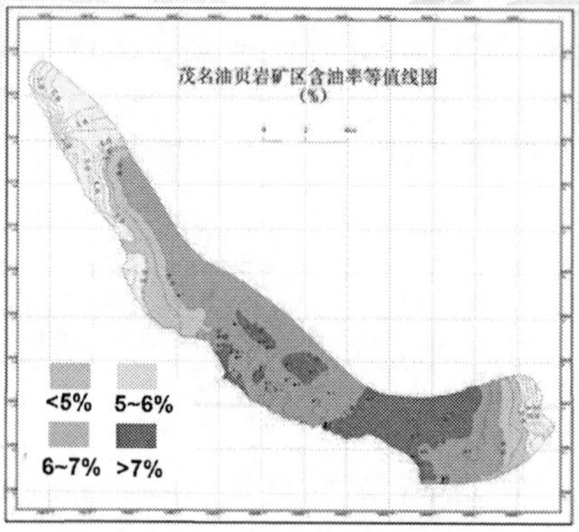

Contour of oil content in Maoming Basin

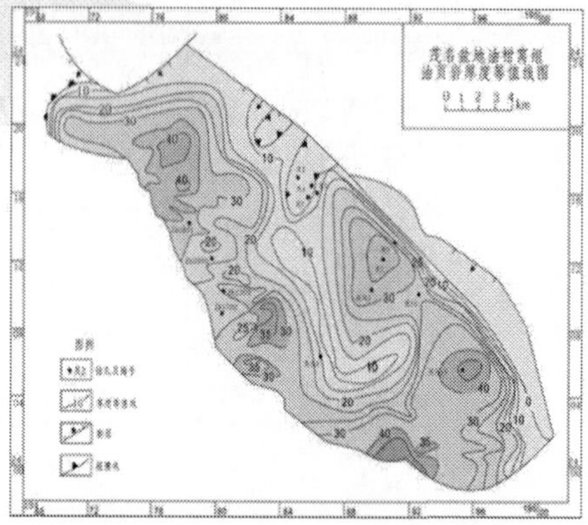

Contour of oil shale layer thickness in Youganwo Formation

INTEGRITY COLLABORATION INNOVATION EXECUTION

PEPRIS

Background

44.47 meters of oil shale cores were obtained as well as necessary geological data.

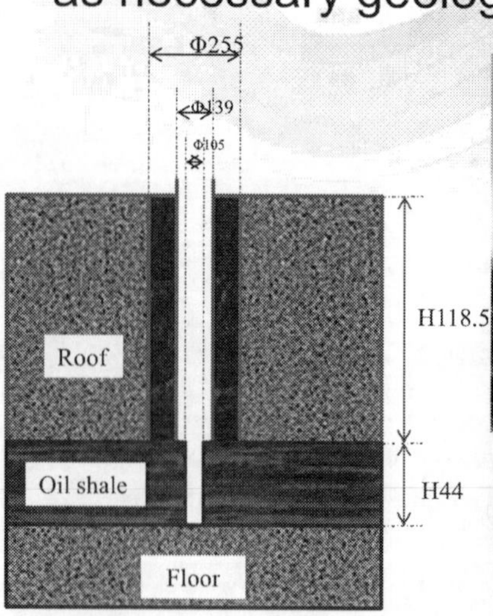

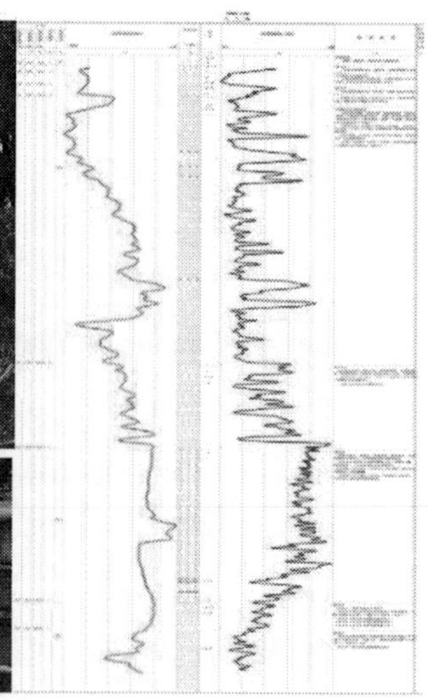

INTEGRITY COLLABORATION INNOVATION EXECUTION

PEPRIS

Motivation

- The widely used theories and analysis based on averaged temperature assumption cannot give credible prediction and instruction for the engineering practice.
- This work aims to reveal the most influencing factors and give suggestions for engineering practice, by studying the mechanism of *in-situ* processing of oil shale.

INTEGRITY COLLABORATION INNOVATION EXECUTION

PEPRIS

Method & Specification

The CMG software, version 2015, is applied to model the *in-situ* heating process of the target.

Parameters	Value	Unit
Density	2400	Kg/m3
Heat capacity of oil shale	0.94	KJ/Kg.K
Heat capacity of water	4.19	KJ/Kg.K
Heat conductivity of oil shale	3.17	W/m.K
Heat conductivity of water	0.62	W/m.K
Heat conductivity of oil	0.13	W/m.K
Heat conductivity of gas	0.03	W/m.K
Porosity	0.15	

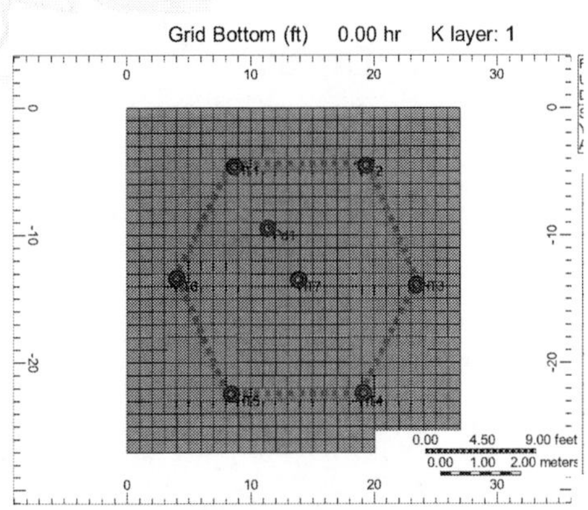

INTEGRITY COLLABORATION INNOVATION EXECUTION

PEPRIS

Method & Specification

Number	Species	Molecular weight	Critical pressure (psi)	Critical temperature (R)
1	Gas	52.0	660.3	522.7
2	LO (Light oil)	149.3	346.1	12891.2
3	HO (Heavy oil)	421.2	137.6	1728.7
4	CH_4	16.0	667.2	342.3
5	Char	12.4	/	/
6	CO_2	44.0	1070.1	547.5
7	Kerogen	2986.3	/	/

Reaction 1: Ker-->HO + LO + Gas + Char + CH_4 + H_2O + CO_2

Reaction 2: HO --> LO + Gas + Char + CH_4

Reaction 3: LO --> Gas + Char + CH_4

INTEGRITY COLLABORATION INNOVATION EXECUTION

PEPRIS

Method & Specification

Oil and gas productions are selected as objective functions, respectively.

$$\text{O\_OBJ001} = \sum_{i=1}^{27}\sum_{j=1}^{27}\sum_{k=1}^{10} \mathrm{V}Hoil_{ijk} + \sum_{i=1}^{27}\sum_{j=1}^{27}\sum_{k=1}^{10} \mathrm{V}Loil_{ijk}$$

$$\text{G\_OBJ001} = \sum_{i=1}^{27}\sum_{j=1}^{27}\sum_{k=1}^{10} \mathrm{V}(HCgas)_{ijk} + \sum_{i=1}^{27}\sum_{j=1}^{27}\sum_{k=1}^{10} \mathrm{V}(CH_4)_{ijk}$$

Considered controlling factors			
	initial kerogen concentration	formation permeability	activation energy of pyrolysis reaction
	heating power	heat loss	pre-factor
	heat conductivity	heat capacity	bottom pressure

INTEGRITY COLLABORATION INNOVATION EXECUTION

PEPRIS

Results & Discussion

Kerogen concentration
Activation energy
Heating power

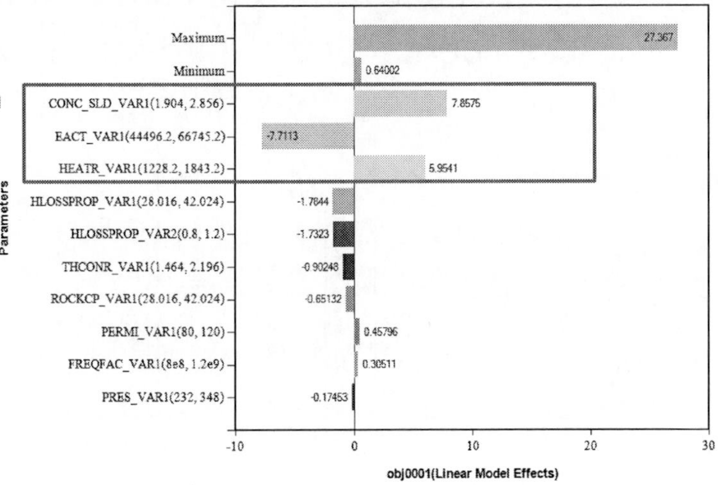

Influence of the single factor on the objective oil production.

Results & Discussion

Activation energy
Kerogen concentration
Heating power

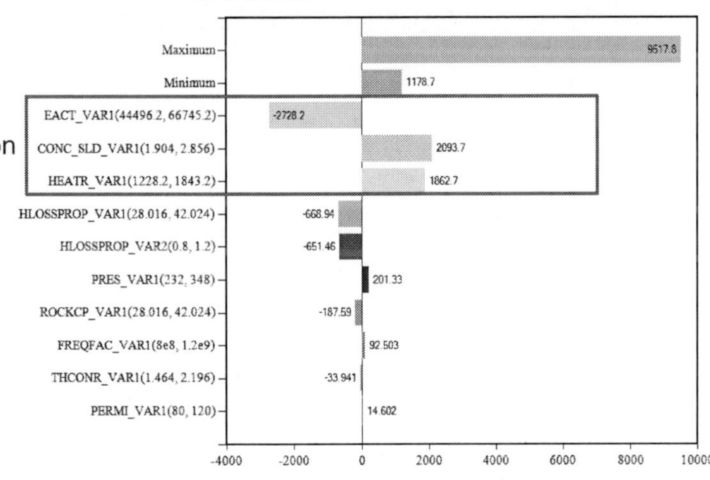

Influence of the single factor on the objective gas production.

Results & Discussion

- Heavy oil converting to light oil and gas;
- Fast influence of the third step reaction.

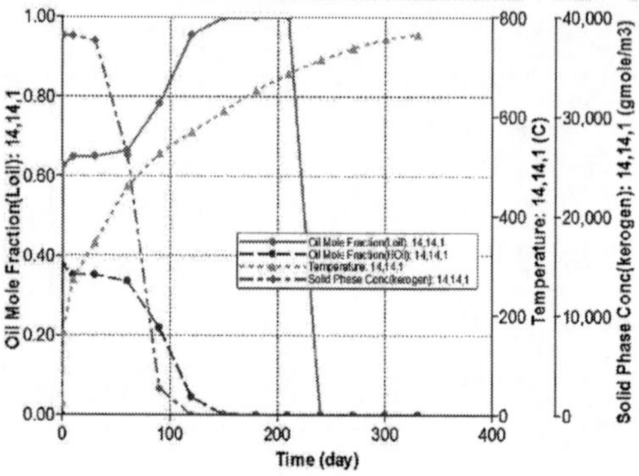

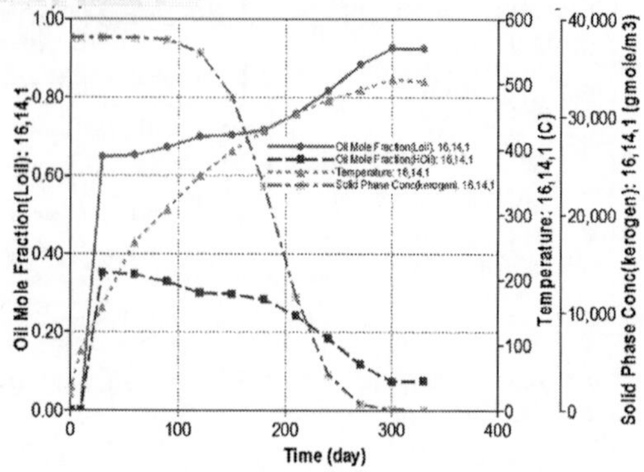

The evolution of mole fraction of LO and HO, temperature, kerogen concentration as a function of time, in one single grid block located at (14,14,1) and (16,14,1).

Results & Discussion

- Similar phenomenon is observed by increasing the formation permeability.

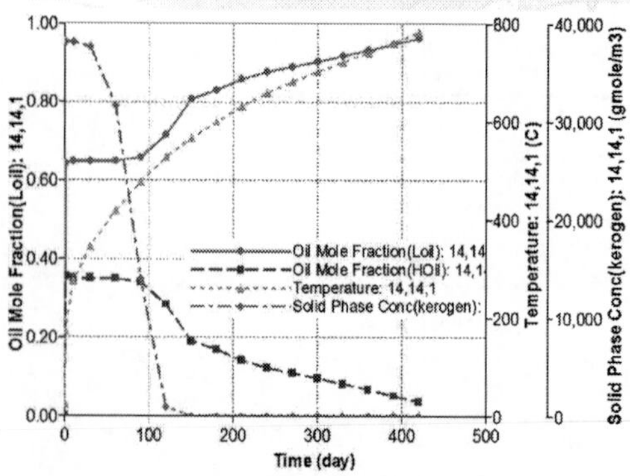

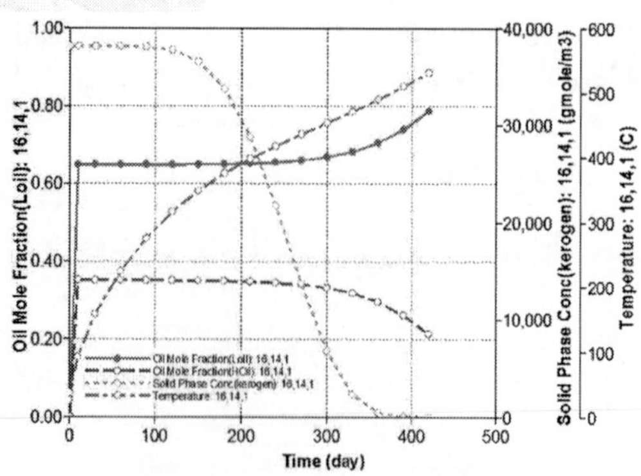

The evolution of mole fraction of LO and HO, temperature, kerogen concentration as a function of time, in one single grid block located at (14,14,1) and (16,14,1).

Conclusions

- The heating power, initial kerogen concentration and activation energy are the most influencing factors.
- High temperature zone near heating zone may cause the light oil production convert to gas and unwanted char.
- Properly controlling the temperature field can help to prevent the oil production loss due to secondary reaction.

INTEGRITY COLLABORATION INNOVATION EXECUTION

PEPRIS

PEPRIS

Thanks!

INTEGRITY COLLABORATION INNOVATION EXECUTION

Statoil

EIF Onshore Discharges Tool. A modelling tool to assess environmental risk to soil, groundwater and surface water from hypothetical *in situ* kerogen oil production

R. E. Hagemann and U.I. Musdalslien, Statoil ASA
S.K. Firth, Firth Consultants Ltd

Outline

- Introduction to Environmental Impact Factor (EIF) Onshore Discharges tool
- Case study using EIF Onshore Discharges tool to assess environmental risk associated with hypothetical *in situ* kerogen projects
- Other uses of the tool
- Conclusions

EIF Onshore Discharges tool

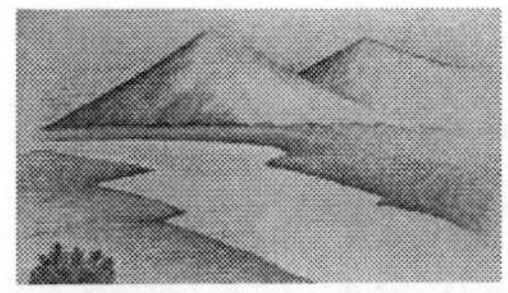

- One of a suite of EIF tools developed by Statoil to aid risk management decisions
- Calculates EIF value for accidental spills/leaks or planned discharges for 3 receptor compartments:
  - Soil
  - Groundwater
  - Surface water
- Methodology:
  - Estimate spatial distributions of contaminant concentrations in each receptor compartment and how these change with time
  - Compare these concentrations with environmental tolerance levels to determine volume of receptor compartment potentially impacted (EIF of 1 = 1000 m^3)
  - Plot EIF vs time for each receptor compartment
  - The tool does not estimate the probabilities of a spill to happen

3 Statoil

Fate and transport modelling

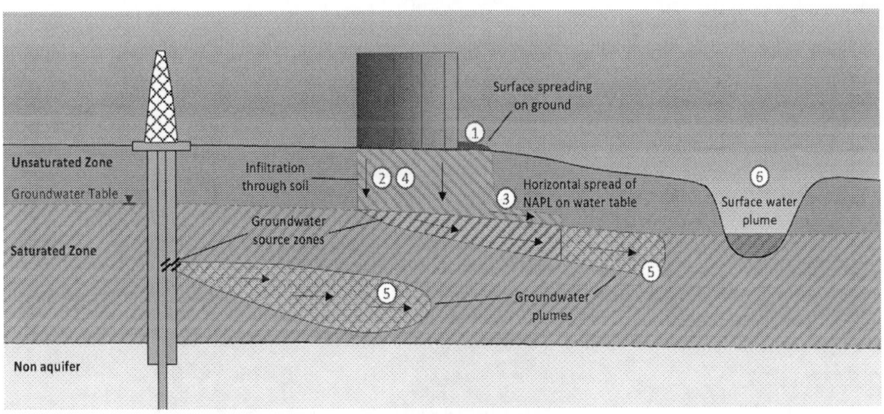

- Pathways 1 to 4: Used to predict concentrations in soil compartment
- Pathway 5: Used to predict concentrations in groundwater compartment
- Pathway 6: Used to predict concentrations in surface water compartment

4 Statoil

Parameter inputs

- Spill characteristics
  - Volume
  - Constituents of Potential Concern (=COPCs)
- Fate and transport data for COPCs
  - Solubility, sorption, biodegradation
- Environmental threshold levels for COPCs
  - Soil: Land-use guideline values or predicted no effect concs. (PNEC)
  - Groundwater: Drinking water standards
  - Surface water: Water quality standards or PNEC (acute, chronic)
- Environmental conditions
  - Soil, groundwater and surface water properties

Statoil

Results from EIF modelling

- Graphs and data tables for soil, groundwater and surface water:
  - Time from the spill happening until impact occurs
  - Duration of impact
  - EIF contributors = principal sources and COPC that contribute to EIF value
  - Severity of impact. Graphs of Risk Characterisation Ratio (RCR) = ratio of predicted concentration to environmental threshold level

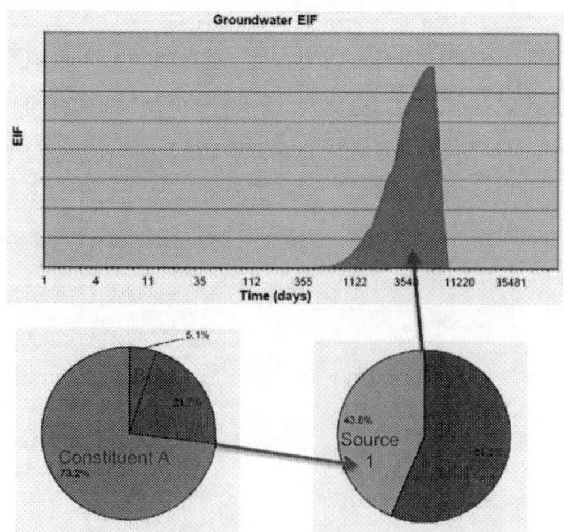

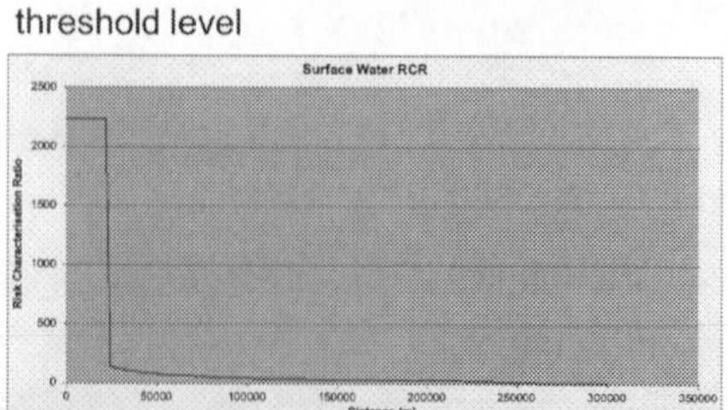

Statoil

Case study for hypothetical *in situ* kerogen oil extraction

- EIF tool used to assess risks from various potential releases associated with a hypothetical project in the Piceance Basin, Colorado, USA
- Scenarios considered:
  - Surface and sub-surface releases of molten "solar salt" considered as heating medium
  - Surface releases of kerogen oil compared with other oil types

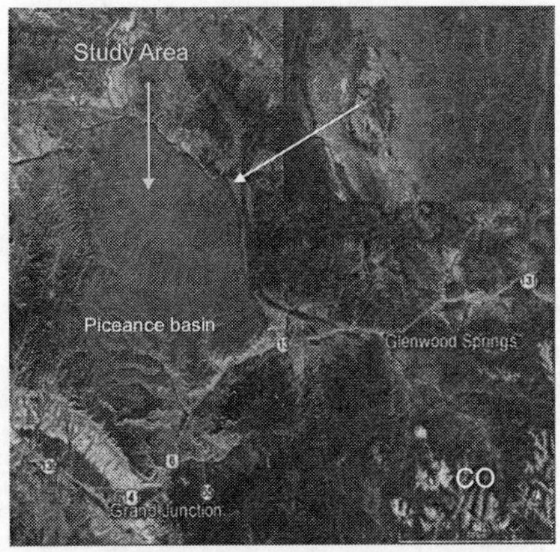

Statoil

7

Study area

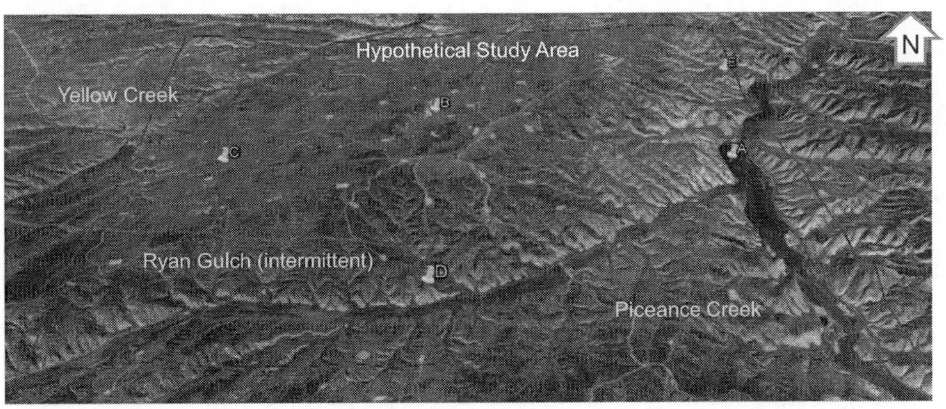

- Five hypothetical locations chosen to represent a range of environmental conditions in the study area

Statoil

8

Modelled hypothetical release scenarios

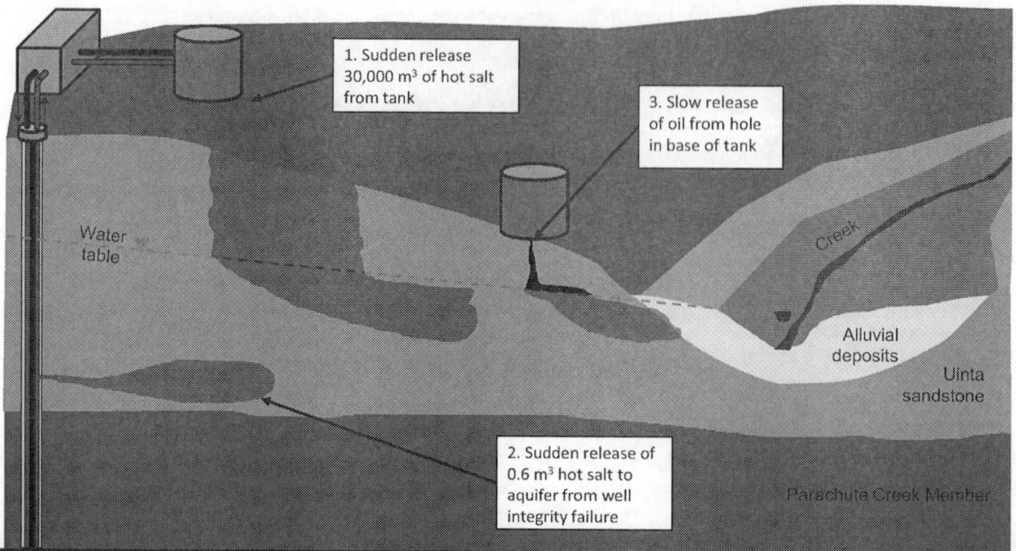

Statoil

Results for groundwater – effect of environmental conditions

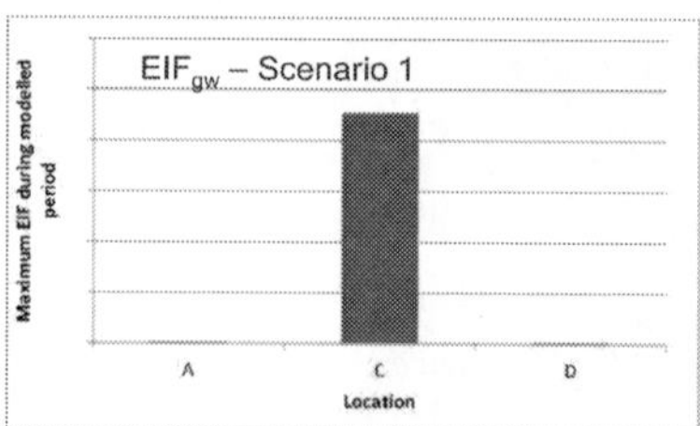

- Scenario 1 (large surface release of molten salt)
  - Large difference in max EIF$_{gw}$ between locations
  - Magnitude/duration of EIF$_{gw}$ principally related to areal extent of spill on ground surface

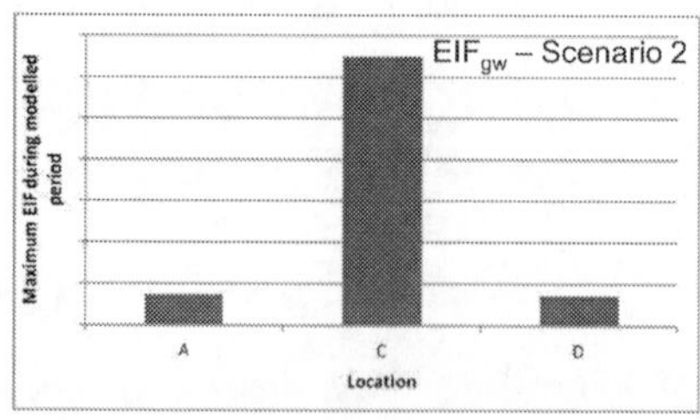

- Scenario 2 (leak of molten salt directly to aquifer)
  - Sites A and D, proximity to where groundwater feeds into surface water limits size of EIF$_{gw}$
  - Dilution in surface water results in EIF$_{sw}$ = 0

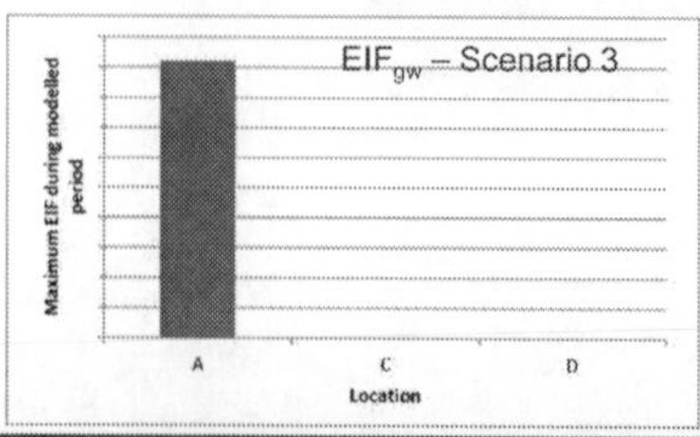

- Scenario 3 (slow leak of oil from base of above ground tank)
  - EIF$_{gw}$ only occurs for site A where groundwater is shallow
  - Large depth to groundwater at sites C and D mitigates risk

Statoil

Results for groundwater – comparison of oil types

- Scenario 3 (slow leak of oil from base of above ground tank)
  - EIF tool used to compare risks from releases of kerogen oil, shale oil (from a US shale play) and medium crude
  - Some uncertainty in oil composition but EIF results indicate risks are similar
  - Naphthalene is main contributor to EIF_{gw} for all three oil types

EIF_{gw} for Location A

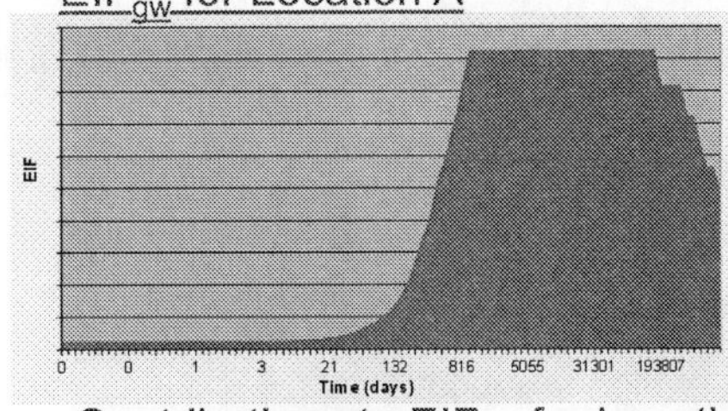

Contributions to EIF_{gw} for Location A

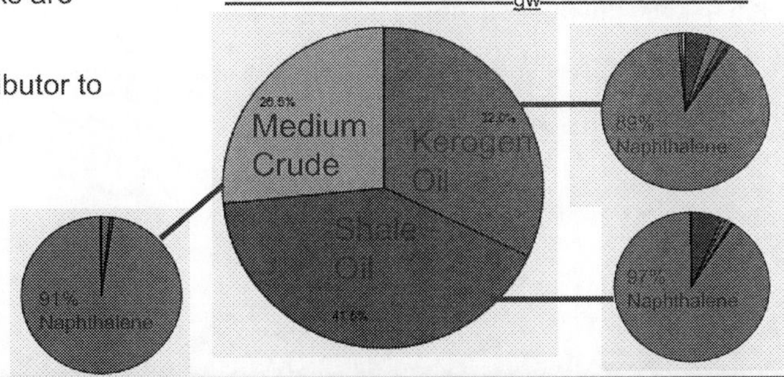

11 Statoil

Other uses of EIF Onshore Discharges tool

- Chemical selection
  - EIF tool was used to compare different types of hydraulic fracturing fluids for a US shale play
  - Largest EIF_{gw} predicted for wells using slickwater (due to bromate content)
- Risk management
  - EIF tool was used to assess risks from leakage from drilling waste pits
  - Results helped to identify optimal risk management strategy for site restoration
- Site selection
  - EIF tool was used to asses risks from landfilling of waste salt
  - Results helped to identify higher risk landfill locations and other risk mitigation considerations for landfill design

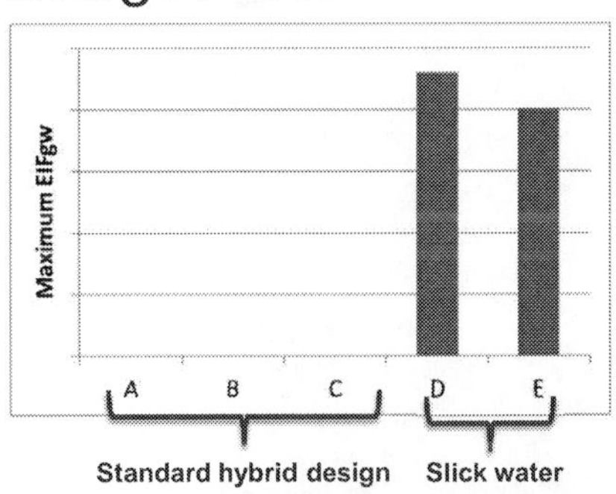

12 Statoil

Conclusions

- EIF Onshore Discharges tool helps understand the potential for environmental impacts should spills or leaks occur
- The tool can help to identify best measures for reducing environmental risk
  - Chemical substitution
  - Avoidance of high risk locations
  - Design risk mitigation measures
  - Optimizing spill response measures
  - Optimizing monitoring programme

Statoil

There's never been a better time for good ideas

Presentation title:EIF Onshore Discharges Tool. A modelling tool to assess environmental risk to soil, groundwater and surface water from in situ kerogen oil production

Presenters name: Randi Hagemann
Presenters title: Advisor SUS Env tech
E-mail address rhag@statoil.com
Tel: +4799163982

www.statoil.com

Statoil

Small Partnership Projects (Jordan Case)

Presented by
Osama Suliman
Director of PPP Unit / Ministry of Finance

November 21, 2016

Presentation overview:

- Introduction.
- Main Principles for Successful PPP Projects.
- Jordanian Legislative Framework.
- The Objectives of the Jordanian Legislative Framework.
- PPP Concepts.
- Entities and Sectors enclosed.
- Institutional Framework for PPP Projects.
- PPP Projects lifecycle.
- PPP Projects lifecycle Procedures.
- PPP Projects Committees .

2

Introduction:

- Jordan has embarked on an ambitious, national public-private partnership (PPP) program with the broad objective of creating a driving force for economic growth and employment through well-defined partnerships in the infrastructure, utilities, and service sectors.

- Increasing private sector participation—enhancing an environment conducive to private investments, supported by a strong political will, apt regulatory framework, and transparency—is at the core of the new era of private sector-led growth and development in Jordan.

3

Cont, Introduction:

- Jordan's economic growth has recently slowed despite remarkable achievements over the past decade. Regional crises, including the rising costs associated with supporting Syrian refugees and disruptions of Egyptian gas, have contributed to the decline in overall economic performance. Immediate measures targeting private sector participation in economic activities and planned development are required.

- Given chronic fiscal turbulence, Jordan needs to leverage private sector resources to meet infrastructure development requirements by exploiting public-private partnerships (PPPs). PPPs are recognized as a catalyst for growth and employment and a key ingredient for achieving the national development goals as outlined in the recently launched blueprint Jordan 2025

4

Main Principles for Successful PPP Projects:

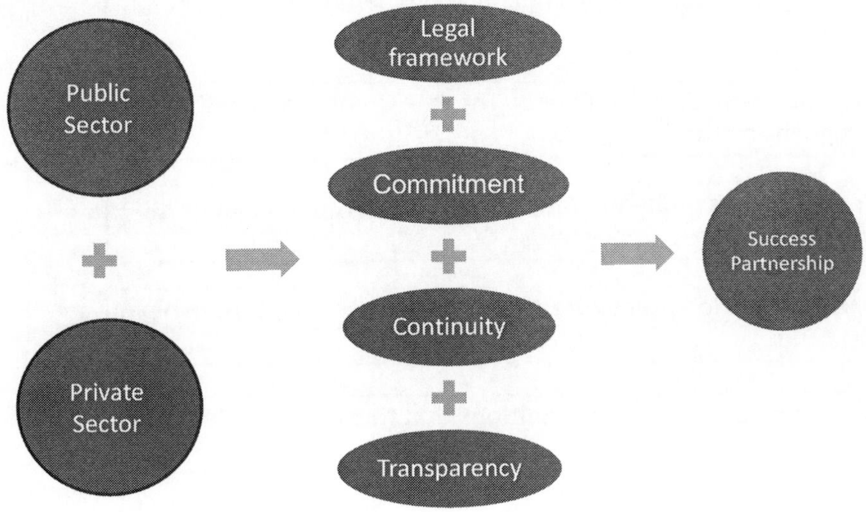

5

Jordanian Legislative Framework:

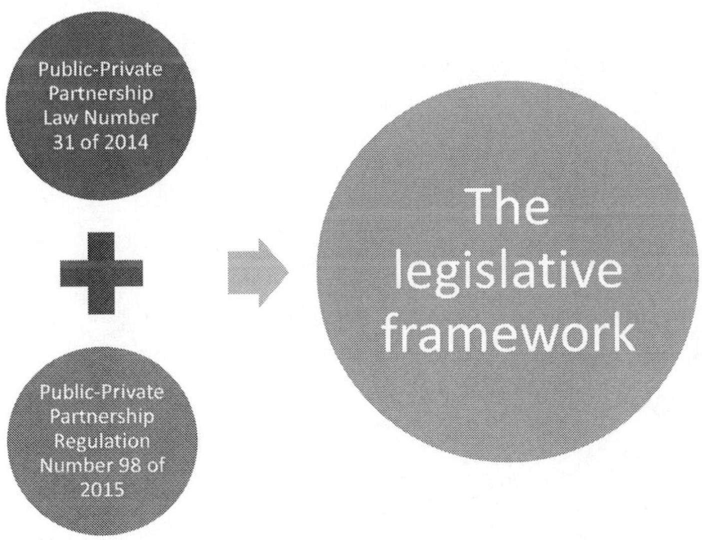

6

The objectives of the Jordanian Legislative Framework:

1. Determine the stages of the partnership between the public and private sectors.
2. Clarify the procedures carried out by the PPP Unit to carry out the tasks assigned to it under the law.
3. Identify the tenders implementation of the PPP projects that include the equality and transparency basis.
4. Identify the simplified procedures for the small-scaled partnership projects.
5. Determine the basic data and conditions that must be included in the partnership contract.

7

PPP Concepts:

Private Partner

- The company comprised of local or foreign private capital or both, which makes a bid for the partnership project tender and in which the Government, directly or indirectly, does not own more than (25%) of its capital; or a consortium consisting of more than one private company where the Government does not own more than 25% of the capital of the said consortium

Value for Money

- The private partner realizes an additional economic return for the contracting authority throughout the standardized project cycle, in terms of cost and price, quality and risk transfer, or a combination of any of them in accordance with the contract.

Direct Proposal

- The offer submitted by the private sector which is not based on a request from the contracting authority and in line with the terms stipulated in Paragraph (a) of Article (11) of the Law.

8

PPP Concepts:

Feasibility Study

- The analytical study prepared by the contracting authority addressing the institutional, legal, technical, environmental, social, economic, financial, and other aspects of the proposed partnership project during the project's standardized cycle. The study should also identify the need for government support to guarantee the sustainability and continuity of the project and clarify the realization of value for money, the ability to bear costs, risk analysis and diversification, and any other requirements based on the nature of the project.

Project advisor

- Natural or judicial person appointed by the contracting authority in accordance with the provisions of Paragraph (c) of Article (9) and Paragraph (b) of Article (18) of the Law, with a view to providing financial, technical and legal advisory services relating to the partnership project, and assisting the contracting authority in preparing the tender documents, the procedures for placing the tender, and any other matters required by the contracting authority.

Standardized Project Cycle

- All the phases of implementation of the partnership project shown in Article (5) of this Regulation

9

Entities and Sectors enclosed:

Government Bodies

- Any ministry, department, state-owned public institution, public institution, entity, council, authority, municipality, or firm that is wholly state-owned or of which the government holds at least 50 percent.

Contracting Party

- The Government Body that signs a Contract with any private-sector entity.

Sectors enclosed

- All economic sectors are fit for public-private partnership projects, except those identified by the Cabinet based on the Council's recommendations.

10

Institutional Framework for PPP Projects:

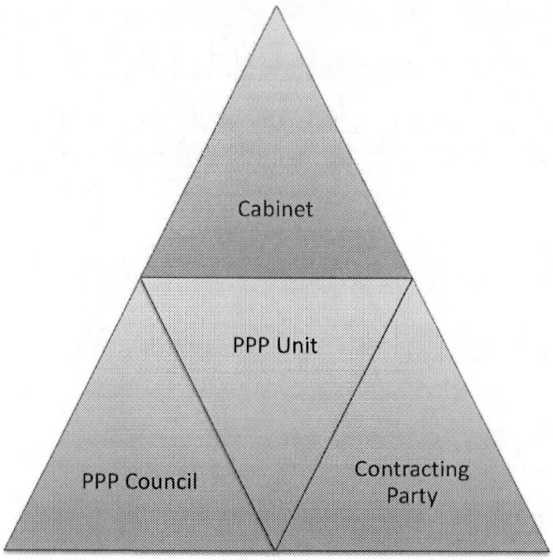

11

PPP Projects Lifecycle:

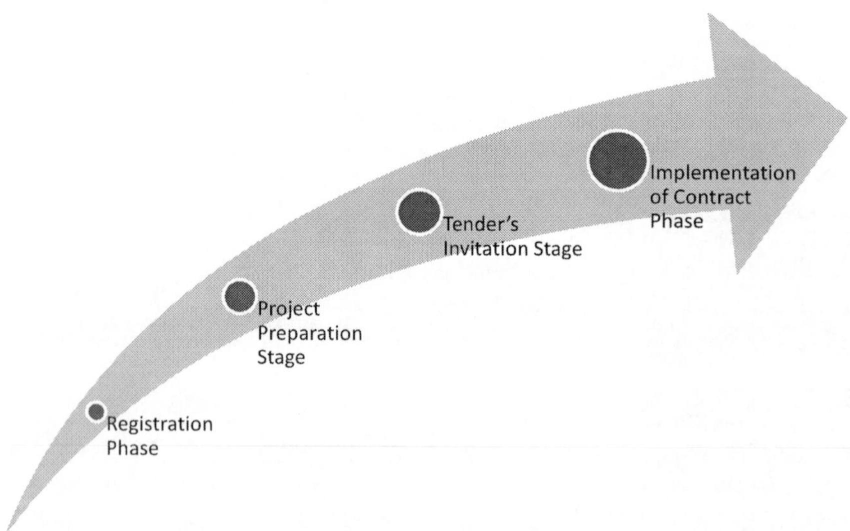

12

PPP Projects Lifecycle Procedures:

A) Registration Phase:

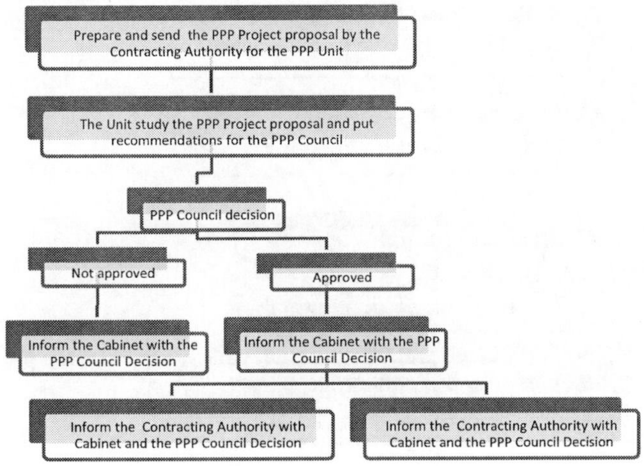

13

PPP Projects Lifecycle Procedures:

B) Project Preparation Stage:

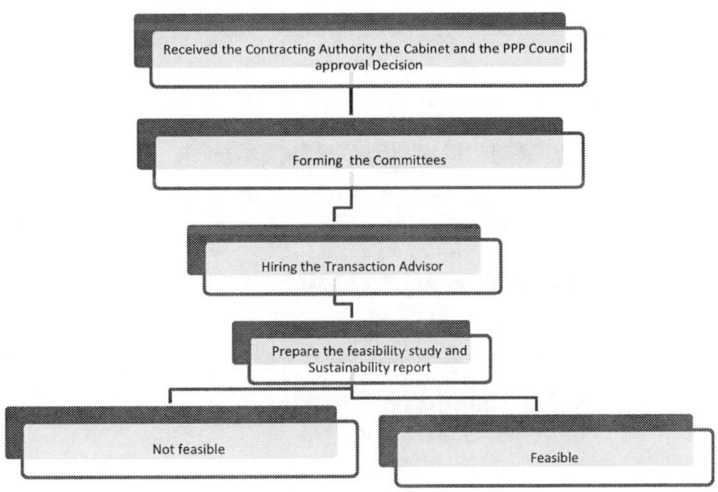

14

PPP Projects Lifecycle Procedures:

C) Tender's Invitation Stage:

1. The technical committee in coordination with the transaction advisor will prepare the tendering documents
2. Invitation to tender and interaction with bidders
3. Evaluation of tenders and select the winner bidder
4. Finalize the PPP contracts and financial closure
5. PPP Council adoption for the final agreement and authorize the contracting Authority for signing the agreement

15

PPP Projects Lifecycle Procedures:

D) Implementation of Contract Phase:

1. Start implementation of the project
2. Monitor and manage project delivery and service outputs
3. Manage changes orders on the PPP contract
4. Manage the PPP contract

16

PPP Projects Committees:

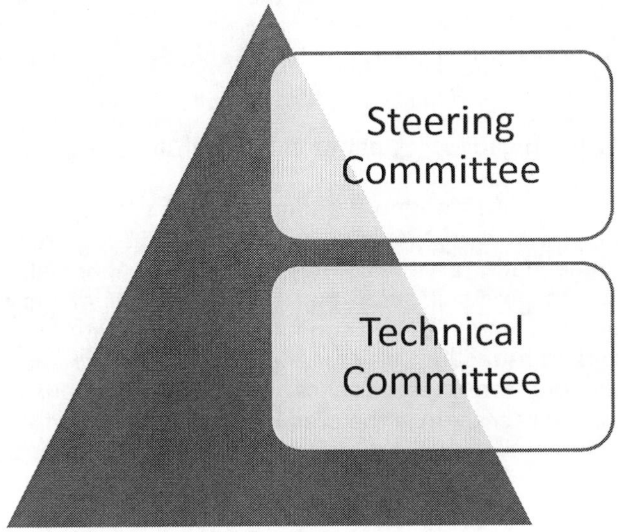

17

Steering Committee

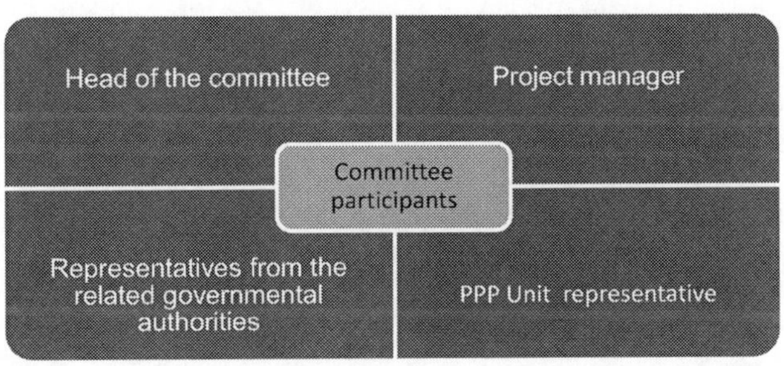

- **The PPP Council established the committee according to the contracting authority recommendation after the project listed as in the PPP Projects list.**

- **The main responsibility for the committee is to help the contracting authority during the project life cycle.**

Technical Committee

▪The PPP Council established the committee according to the contracting authority recommendation and with PPP Unit consolation.

▪The committee participants have the financial, technical and legal experience that the project required.

▪The Technical Committee has the following duties and responsibilities:

a. Determine the suggested qualification standards for the prequalification process. This will be reviewed by the steering committee and submitted along with necessary recommendations to the contracting party for approval.
b. Prepare the invitation to prequalification and the list of documents to be reviewed by the Steering Committee and submit them along with necessary recommendations to the contracting party for approval.
c. Respond to inquiries of interested parties concerning prequalification.
d. Receive responses and register them in the prequalification invitation.
e. Review responses stated in the prequalification invitation in accordance with the specified qualification standards and prepare a report in this regard.

19

For Further information about the Jordanian PPP Law and regulation, please visit the PPP Unit website:

www.pppu.gov.jo

•Or through the Jordanian Ministry of Finance website:

www.mof.gov.jo

20

Thank you...

21

th

trowers & hamlins

Presentation ——— November 2016

Jordanian Oil-Shale Industry

Making progress in difficult times

Middle East Strategy Group
مجموعة الشرق الأوسط للإستراتيجية

Introduction

th trowers & hamlins

Middle East Strategy Group
مجموعة الشرق الأوسط للإستراتيجية

Oil-Shale reserves in Jordan

- **Jordan's Oil shale reserves**
  - Jordanian oil shale is considered to be a kerogen-rich, **high-quality** marinite oil shale (comparable to Western US oil shale)
  - Oil shale deposits are estimated to underlie **more than 60%** of the Kingdom's territory
  - Jordan is estimated to have the **fourth largest** oil shale resources in the world
  - Resources are estimated at **40 to 70 billion tonnes** of oil shale (equivalent to 30 billion barrels of oil)
- **Domestic resource**
  - Potential for long-term supply of energy
  - Will support Jordanian **energy security**

trowers & hamlins

Middle East Strategy Group
مجموعة الشرق الأوسط للإستراتيجية

Locations

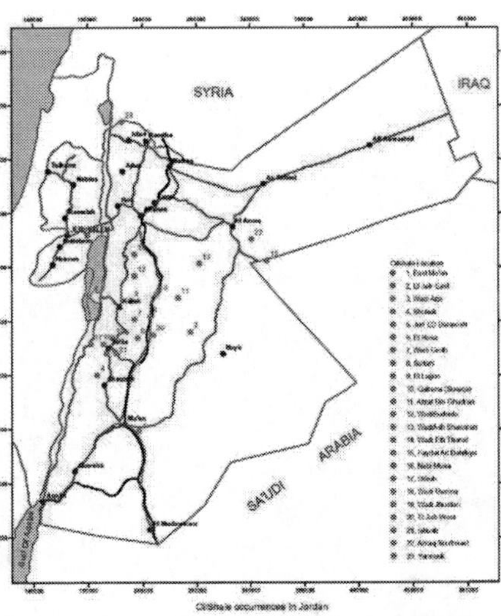

- **Location of oil shale deposits**
  - There are **26** known shale deposit locations in Jordan
  - The **8 most significant deposits** can be found in the **West-Central** region of the Kingdom:
    1. Juref ed Darawish
    2. Sultani
    3. Wadi Maghar
    4. El Lajjun
    5. Attarat Umm Ghudran
    6. Khan ez Zabib
    7. Siwaga
    8. Wadi Thamad
  - There are also 2 additional important deposits in the **Ma'an** district in the South, and **Yarmouk** in the North

trowers & hamlins

Middle East Strategy Group
مجموعة الشرق الأوسط للإستراتيجية

Extraction Technology

- Only three countries currently produce oil from shale – Estonia, China and Brazil
- Many technologies being developed – however most are at various stages of testing
- Process ranges from in-situ (below ground) to ex-situ (above ground)
- No one technology is preferred to another
- Each technology will require some modifications to suit the properties of the particular resource

trowers & hamlins

Middle East Strategy Group

The current situation – an update

trowers & hamlins

Middle East Strategy Group

Concession Companies

There are currently four companies with Concession Agreements in place:

1. **Jordan Oil Shale Company (JOSCO) (Shell)**
   - Concession Agreement signed in 2009 / 1000 km^2 after final relinquishment
   - In-Situ Conversion Process
2. **Jordan Oil Shale Energy Company (JOSE) (Esti)**
   - Concession Agreement signed in 2010 / 72 km^2 - Attarat
   - Estonian Solid Heat Carrier Technology 470MW Direct Burning / 38,000 bpd
3. **Karak International Oil (KIO)**
   - Concession Agreement signed in 2011 / 33 km^2 - Lajjun
   - ATP / 30,000 bpd
4. **Saudi Arabian Corporation for Oil Shale (SACOS)**
   - Concession Agreement signed in 2014 / 11 km^2 - Attarat
   - Own Technology / 30,000 bpd

trowers & hamlins

Middle East Strategy Group

مجموعة الشرق الأوسط للإستراتيجية

MoU Companies

1. **Aqaba Petroleum Co. Ltd**
   - MOU signed 2009 / Wadi Naddeya
   - UTT3000
2. **Fushun Mining Group Limited**
   - MOU signed 2014 / Wadi Naddeya
   - Fushun Technology
3. **Global Oil Shale Holdings Inc**
   - MOU signed 2012 / Attarat Umm Al Ghudran
   - PRIX
4. **Jordan Oil Shale Energy Company (JOSECO)**
   - MOU signed 2010 / Sultani
   - UTT3000

trowers & hamlins

Middle East Strategy Group

مجموعة الشرق الأوسط للإستراتيجية

MoU Companies

5. **Karak International Oil**
   - MOU signed in 2012 / Wadi Naddeya
   - ATP
6. **Al-Lajjun Corporation for oil shale**
   - MOU signed 2010 / Al – Lajjun
   - UTT3000 / Power Generation
7. **Al-Qamar for Energy & Infrastructure Ltd**
   - MOU signed 2014 / Attarat
   - UTT3000
8. **Whitehorn Resources Inc**
   - MOU signed 2012 / Wadi Abu Al Hamam
   - Ecoshale "In Capsule" Technology

trowers & hamlins

Middle East Strategy Group
مجموعة الشرق الأوسط للاستراتيجية

Oil-Shale Fired Power Plant

- **Attarat Power Plant**
  - The agreement for **Attarat Power Company** to develop a 470 MV **oil shale fired power plant** was signed in **October 2014**
  - The project is being developed in partnership with **YTL Power International**
  - NEPCO and the developer are continuing to work towards achieving financial close on this project
  - Construction is expected to commence shortly and the plant is expected to be operational in **2020**
  - Once completed, **Attarat Power Plant** will be one of the **largest** power plants in Jordan (second only to Aqaba Thermal Power Plant)

trowers & hamlins

Middle East Strategy Group
مجموعة الشرق الأوسط للاستراتيجية

Current difficulties

trowers & hamlins

MESG
Middle East Strategy Group
مجموعة الشرق الأوسط للاستراتيجية

Oil Prices

- Sustained level of record low oil price – less than $50 a barrel
- OPEC has agreed to cut oil production to stabilize prices
  - Some non-OPEC nations following suit – such as Oman
  - Others increasing production – such as Libya
  - OPEC countries of Iran, Iraq and UAE seeking exemptions to increase capacity
  - The lack of concrete progress since the Algiers declaration has led to further weakening of prices
- Lower oil price has made oil shale a less attractive resource and sources of funding are harder to find

trowers & hamlins

MESG
Middle East Strategy Group
مجموعة الشرق الأوسط للاستراتيجية

Investment

- Investment in oil & gas is expected to decline by more than $300bn over 2015 and 2016
- Iran plans to award oil contracts to international companies for the first time in 2017
- Large volumes of Australian LNG are coming onto the market
- GCC contract awards were down 38% in the first three quarters of 2016 compared to 2015
  - Governments cut budgets to deal with oil price deficits
  - Private sector drop due to low oil prices impacting finance arrangements

Regional Instability

- Libya has been able to increase production after reaching deals with military commanders
- Easing on sanctions on Iran has added to the availability of Iranian oil on the market
- Continuing instability in Yemen, Syria and Iraq

What is the Government doing to help?

trowers & hamlins

Middle East Strategy Group

Continued development

- The government recognises that these are unusual circumstances
- The government's priority is to ensure the sustainable development of Jordan's oil shale
- Jordan will need a mix of energy sources in the future, oil shale should be part of this as it is a domestic resource
- This will not be achieved unless a flexible approach is taken to the issues which are currently arising
- A draft "Oil Shale Development" is published alongside this conference

trowers & hamlins

Middle East Strategy Group

Flexible and Accommodating Approach

- Has been flexible and accommodating in relation to existing obligations on oil shale companies
  - Has signed, and is willing to sign extensions to MoUs with companies as the need calls for
  - Reviewing spending amounts in light of size of project
  - Flexible approach to which technology is to be used
- Companies must still demonstrate an ability to proceed and must also continue to progress their projects
- Open doors policy and close involvement through joint TAC committees

trowers & hamlins

Middle East Strategy Group
مجموعة الشرق الأوسط للاستراتيجية

Contact

Tom Wigley – Trowers & Hamlins
Partner

d +44 (0)20 7423 8034
e twigley@trowers.com

Eyas Shuaibi – Middle East Strategy Group
Partner

d +962799500099
e eshuaibi@strategygroup.me

trowers & hamlins

Middle East Strategy Group
مجموعة الشرق الأوسط للاستراتيجية

Challenging Oil Shale Geology of Central Jordan 2006-2016

Prof. emeritus Väino Puura, University of Tartu
Prof. Alvar Soesoo, University of Tartu
Hardi Aosaar, Eesti Energia AS

Preface

- No economically eminent (= unquestionably profitable) OS deposits are known in the world
- Without proving of economic and technological feasibility, the reported vast oil shale volumes in the world remain just on the level of hypothetical and speculative ressources, positive assessment of which is still lacking. It is a pity, that failured in past feasibility studies here and there did not grow optimism
- The Estonian experience of OS utilization helps to understand forthcoming advantages and problems on Attarat Um Ghudran (AUG) Deposit, Central Jordan

About a look of a geologist

- A geologist's view may differ from views of any other professionals... from miners to investors.
- After the happiness of discovery of a deposit a geologist feels like a midwife following the fate of his or her newborn...
- An economic geologist Voelker 1979 said : „The approach of a geologist to resource evaluation should be aggressive and basically optimistic - even imaginative and daring".
- ...also a dare geologist has to prove his estimates and their representativeness and reliability.

OS subbasins in Jordan and the JOSE concession area

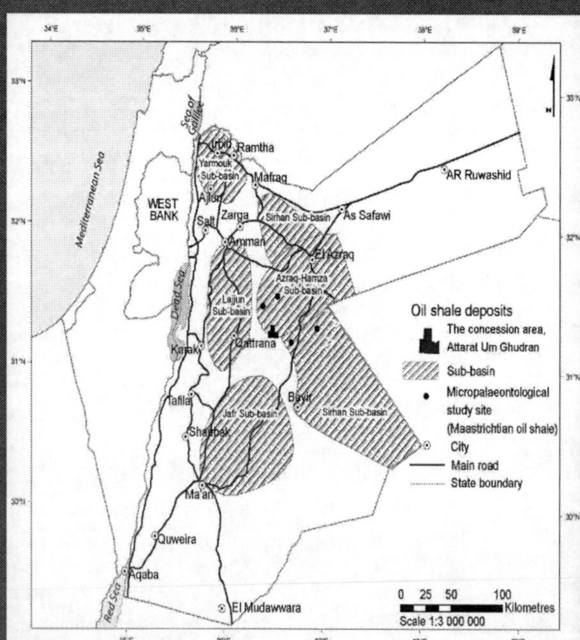

Modified from Alqudah et al. 2015

Attarat Um Ghudran Deposit,

JOSE explor field

At early stage of exploration project

the projected simple model of the deposit broke

(i) instead of a uniform, homogenious OS unit distinct layering with variable composition and grade was established. Good NEWS: the regular internal layered structure well serves for the studies and assessment of the fuel variability!

(ii) Bad NEWS: the lateral discontinuity of the oil shale seam was established. What happened with it?

Answer: deeply penetrating oxidation of OS!

In order to understand the problems, special research was applied.

A fragment of principal profile of layered and partially weathered OS unit in the JOSE field (AUG)

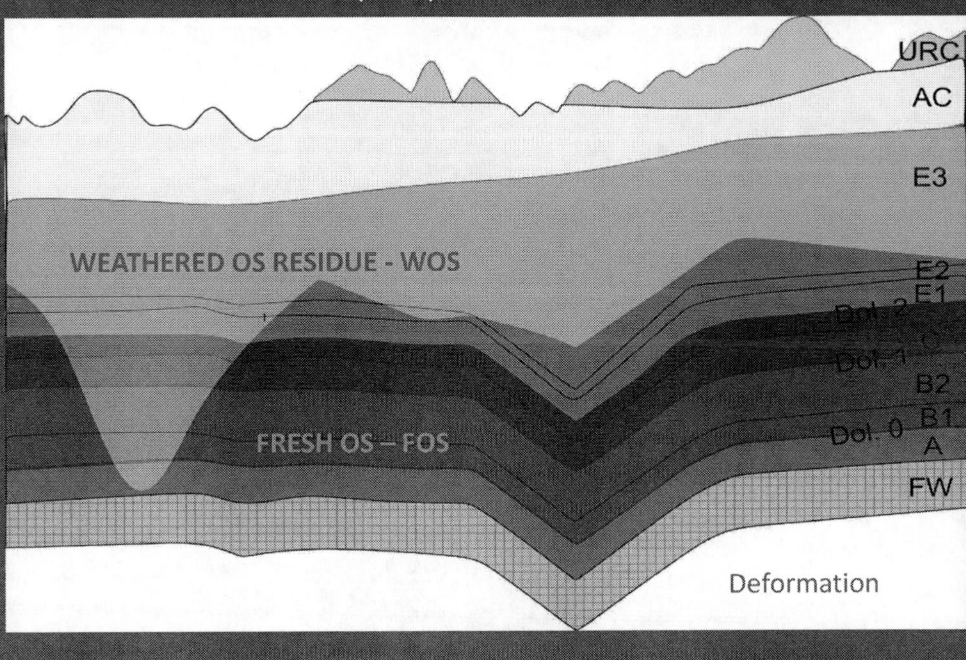

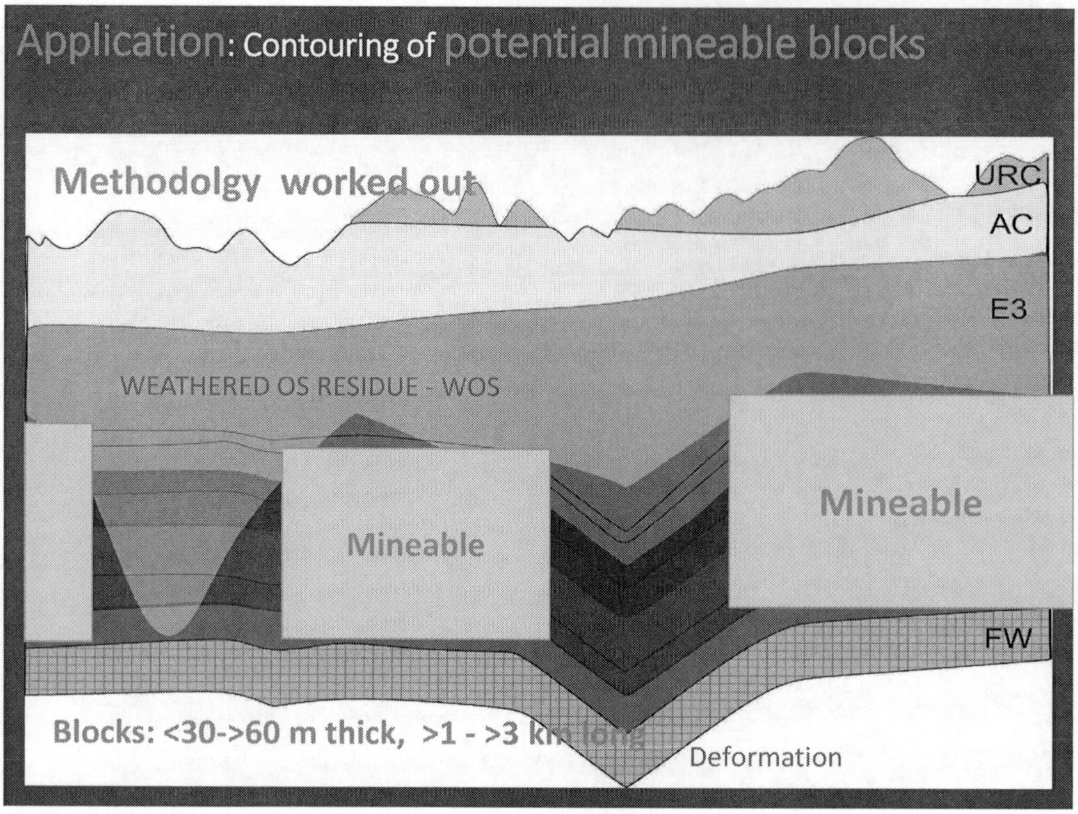
Application: Contouring of potential mineable blocks
Methodolgy worked out
URC
AC
E3
WEATHERED OS RESIDUE - WOS
Mineable
Mineable
FW
Blocks: <30->60 m thick, >1 - >3 km long
Deformation

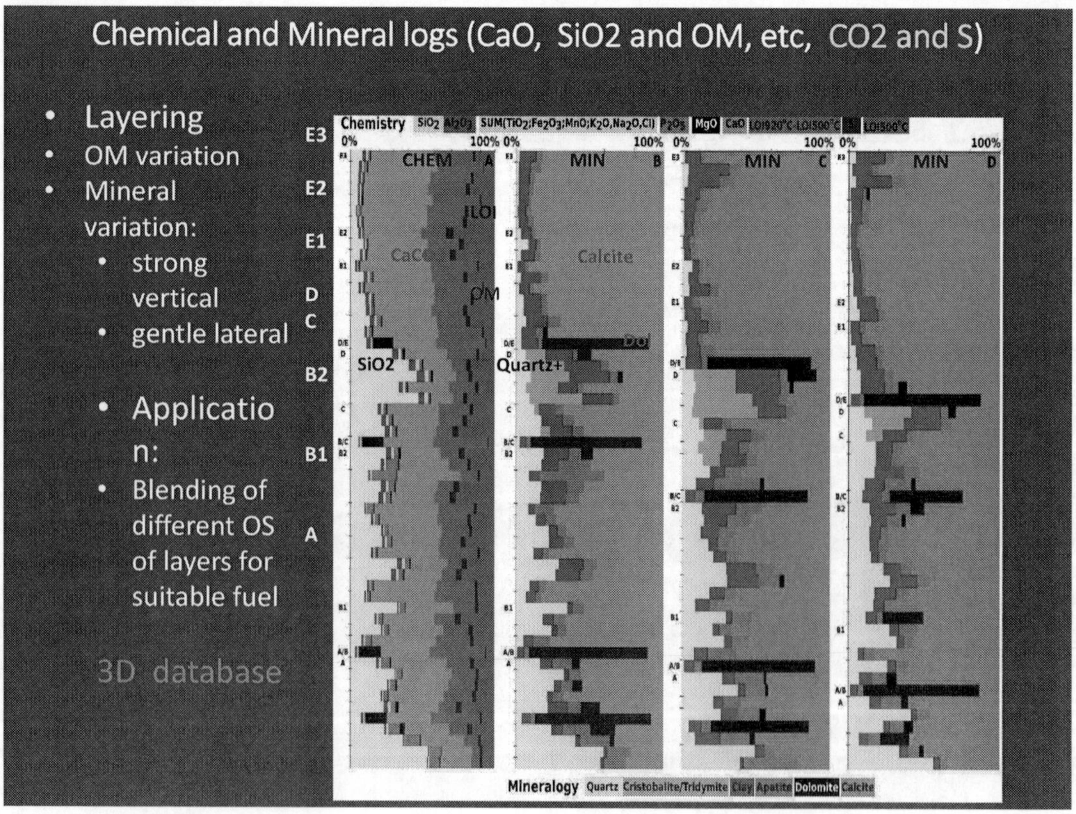
Chemical and Mineral logs (CaO, SiO2 and OM, etc, CO2 and S)
• Layering
• OM variation
• Mineral variation:
• strong vertical
• gentle lateral
• Application:
• Blending of different OS of layers for suitable fuel
3D database
E3
E2
E1
D
C
B2
B1
A
Chemistry
CHEM A
MIN B
MIN C
MIN D
LOI
CaCO3
OM
SiO2
Calcite
Dol
Quartz+
Mineralogy Quartz Cristobalite/Tridymite Clay Apatite Dolomite Calcite

Integrated results of special research and exploration practice:

- Specific original research-based methodologies have been worked out and applied:
  - For identification and indexation of layers;
  - For contouring of FOS blocks and linear zones and funnels of WOS.
  - Representative visual, mineralogical, chemical, natural gamma-logging **field and laboratory techniques** served for **layering and weathering identification/mapping** and served for **reliable** exploration and evaluation of the **resources/reserves within the discontinuous OS area**

Look into the OS lithological composition

- Oil shales are different in visual appearance
- Types and variations of composition of both organic (OM) and mineral matter (MM) can be determined in laboratory etc tests.

FOS lithological varieties

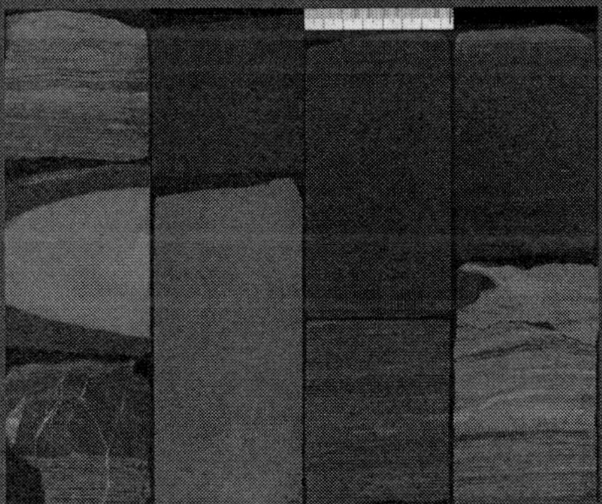

Finely laminated mudstone OS (richest)

Grain-bearing laminated OS

Barren carbonate and chert inclusions

Barren carbonate and chert interlayers

Chemistry: **Depletion** of components in WOS compared to FOS

- OM, S, Mo ... are depleted in WOS
-
- Main mineral/geochemical components of FOS layers survive also in WOS:
  - Main minerals (calcite, quartz, clays) = macrochemical components (CaO, SiO2, Al2O3 etc)
  - Elevated concentrations of trace elements and their specific ratios (Zn, Ni, Cu, Cr, V, Cd etc)

Geochemistry enables:
(I) to identify the oil shale layers in both FOS and WOS,
(ii) to distinguish barren rocks that are really weathered OS

Appearance of the WOS from the transition zone FOS to WOS

WOS block from the 1.4 m drill core

Layer A
(a) FOS (74 m)
(b) (b) WOS (62,5 m)

Input for regional geology and OS survey

- Combination of the present layering studies of FOS and WOS with micropaleontological studies (2015) by Prof Mutterlose and his assistants opens a new window into OS geology and ressource assessment in central Jordan (and possibly elsewhere).

OS subunits hosted by MCM, main data from Alqudah, Hussein, Mutterlose et al 2015

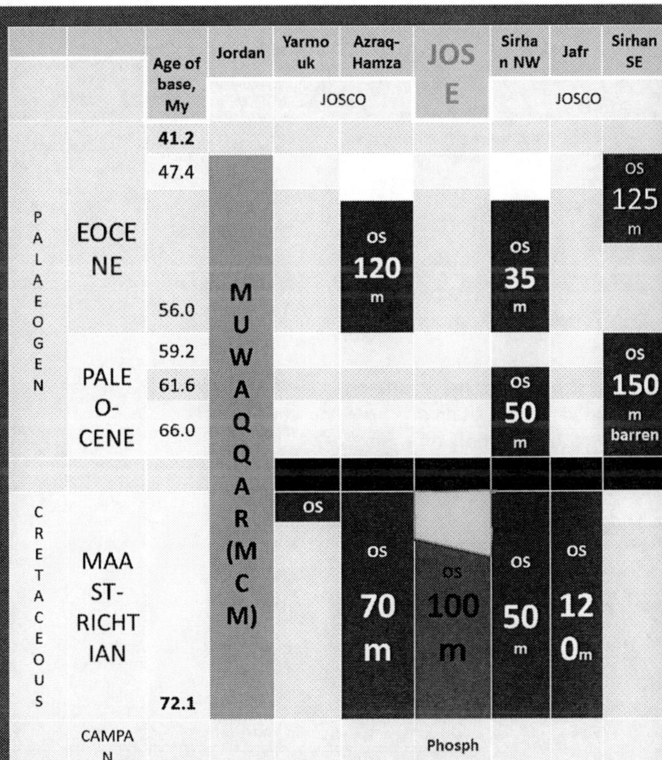

Application of research results:

Dominance of CaCO3 is favorable for solving environmental problems :

- Minimization of sulfur and trace elements in gaseous emissions
- Minimization of trace element contents leaching from ashes

CaCO3-rich ash will safely lock trace elements into ash dumps – forever

NOTE: an amount of CO2 emanated due to thermal processing will be bound back into ash during tens of years .... Ca(OH)2 (slaked lime) +CO2 -> CaCO3 Not studied, not assessed

Dealing with high S content in OS and shale oil requires additional technological and financial efforts

Lessons from history : Estonia (1916 -2016) & Jordan (2006-2016): upgrading technologies .... subjectively

Development ativities	Estonia	Jordan
Exploration	1920-30, 1950/1970	2008/2014*
OS mining techologies: opencast –underground -	1918, 1945, 1974 1920, 1945, 1972, 2010	2016 project*
Processing: for el. power, for shale oil complex shale oil + electr power ENEFIT	1946, 1956, 1980, (1991!), 2016 1924, 1946, 1980, (1991!), 2010, 2015...	2016 project*
Minimization of envir impact from emissions and wastes. Nature protection, land remediation	(1991)... ongoing modernization	
	Technology transfer Estonia -> Jordan	

Fate of OS industry: to be or not to be - ....subjectively

Long life prerequisites :

- Rich resource base with stable quality;
- Social motivation, ...shortage of domestic energy carriers;

- Stubborn consistency of on-site developers, and investors:
  - Industry and science together holding the line of economic efficiency and developing the best technologies;
  - Development of parallel and combined technologies;
  - International cooperation: transfer of best technologies;
    - Last decades - Sustaining the environment
      - Nowadays: Matching with use of renewables;

- (Strong) governmental support, at least in certain critical times;

Development of a new OS resource has to be based on adequate deep research, transfer of best experiences, and ... Stubborn attitude

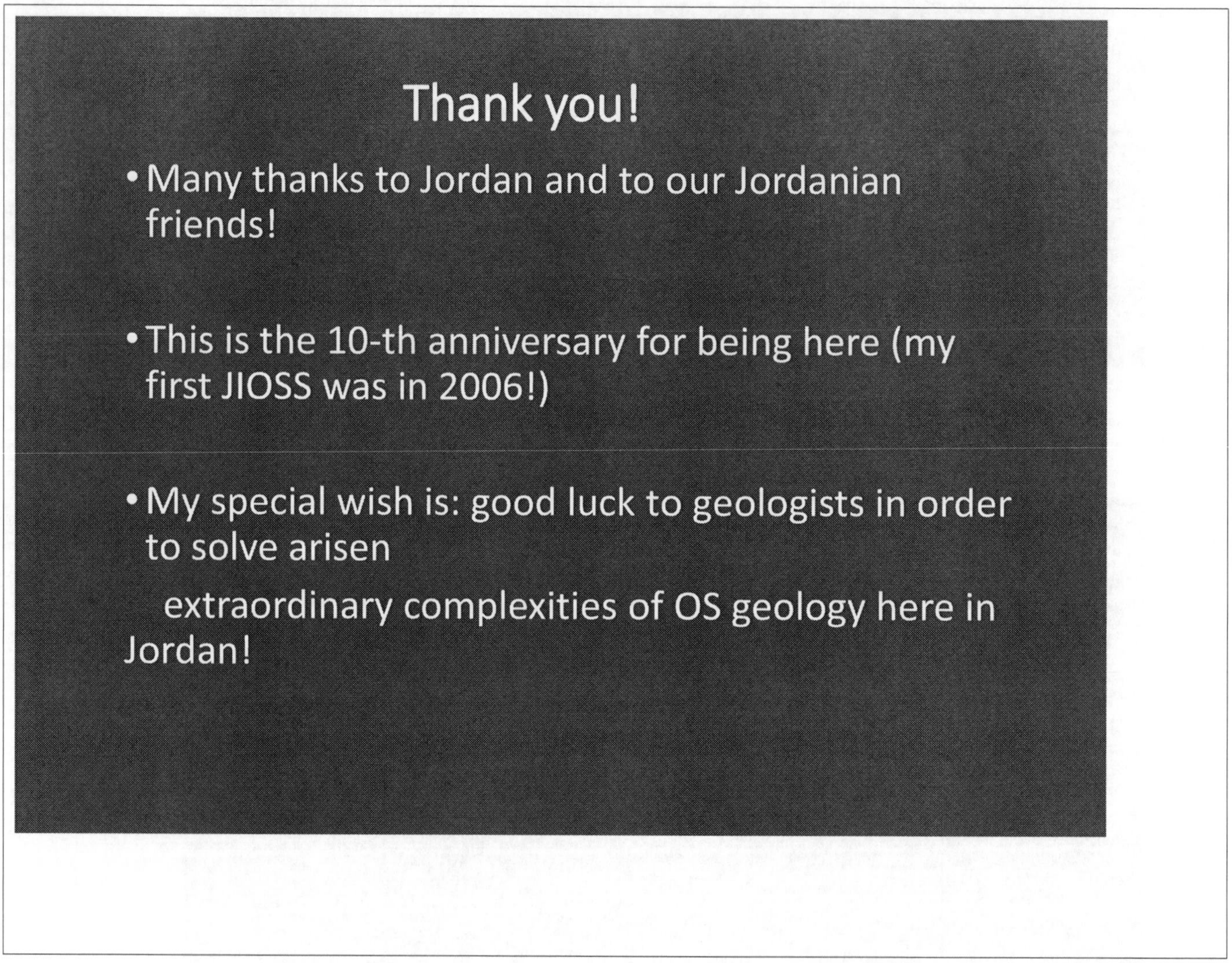
Thank you!
• Many thanks to Jordan and to our Jordanian friends!
• This is the 10-th anniversary for being here (my first JIOSS was in 2006!)
• My special wish is: good luck to geologists in order to solve arisen
extraordinary complexities of OS geology here in Jordan!

Script to accompany "Oil Shale Economics - Updating a Legacy"
Presented by Thomas Sladek
To JIOSS 2016 - The 3rd International Oil Shale Symposium
Dead Sea Hashemite Kingdom of Jordan 22 November 2016

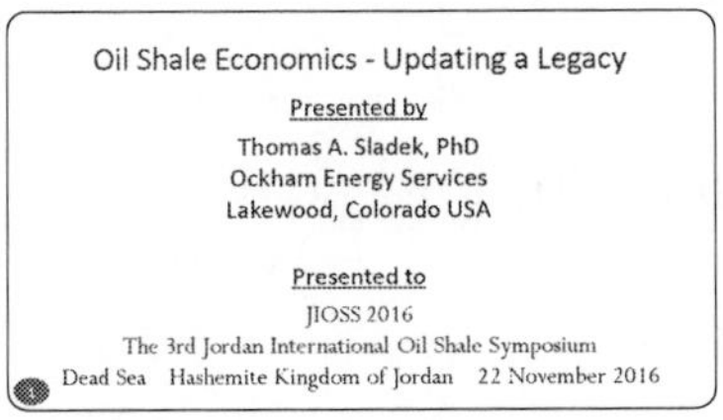

Good afternoon. I bring greetings and best wishes from the Colorado School of Mines, which, like Estonia, has been involved with oil shale for 100 years, more or less. That may seem like a long time, but it is but a blink of an eye compared with Jordan's oil shale activities, which extend back more than 2000 years. Although I have been chasing oil shale for only 50 years, I have seen four cycles begin and end. I am hopeful that the current interest in Jordan's oil shale resources will lead to an efficient and economical industry. However I am convinced that the industry will emerge only if we learn from the teachings of the past, acknowledge the realities of the present, and be creative in how the projects are developed and supported by both industry and the Government.

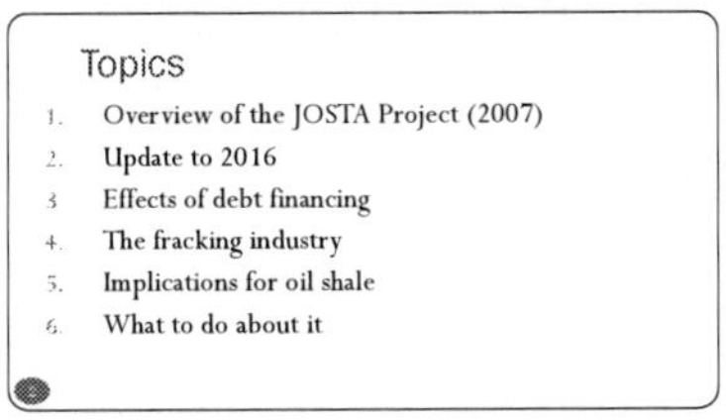

I will begin with a brief discussion of the Jordan Oil Shale Technical Assistance project, which I worked on in 2007. I will present the financial projections from that study updated to current conditions, with emphasis on the effects of alternate financing strategies on the required oil price. Then I will describe the fracking industry -- an entirely new approach to the recovery of oil and gas -- which has revolutionized the petroleum industry in the United States and which will define the maximum price that an oil shale project can hope to obtain. Finally, I'll present suggestions on how Government participation could help an oil shale project reduce its financing costs.

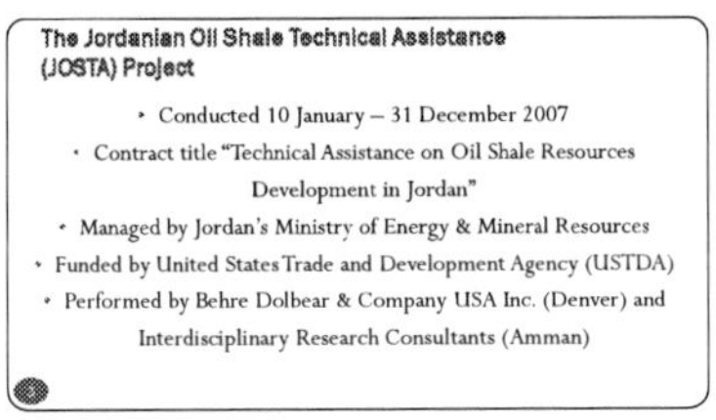

The objective of the JOSTA project was to help the Government of Jordan attract foreign investments in Jordan's oil shale industry. It was conducted by Behre Dolbear and Company of Denver and Interdisciplinary Research Consultants of Amman. Funds were provided by the US Trade and Development Agency, a branch of the US Department of Commerce, and the contract was managed by the Ministry of Energy and Mineral Resources.

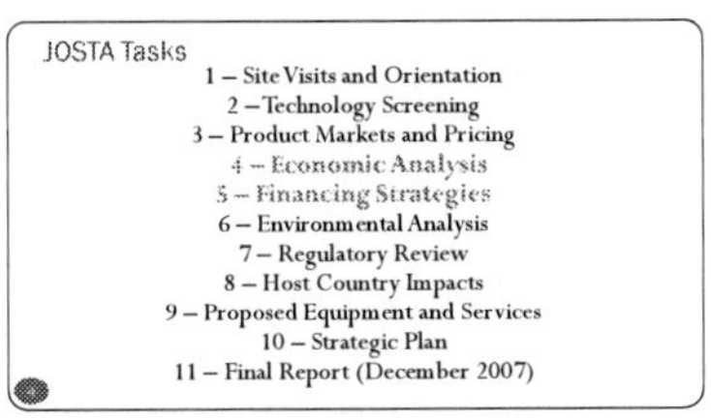

There were eleven major tasks. The two tasks relevant to today's discussion were Number 4 -- Economic Analysis and Number 5 -- Financing Strategies.

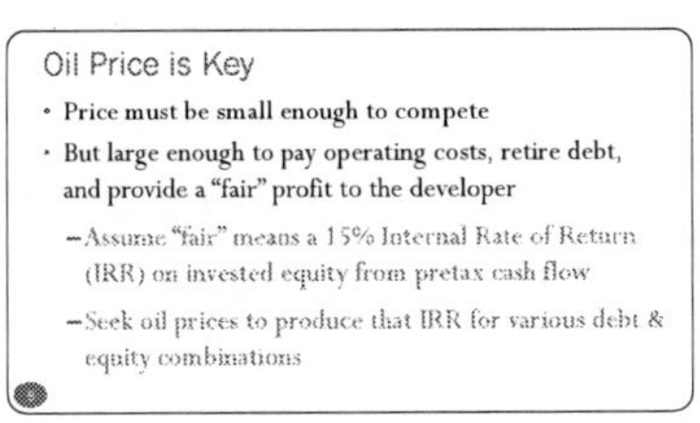

We assumed that an oil shale project in Jordan could be financed with a combination of debt and equity and that the price obtained for the oil product would be large enough to pay expenses, retire the debt portion of financing, and provide a 15% Internal Rate of Return on the invested equity from the cash flow before taxes. We then sought oil prices that would provide that IRR for various combinations of debt and equity.

The plant we modeled had a design capacity of 50,000 barrels per day of high-

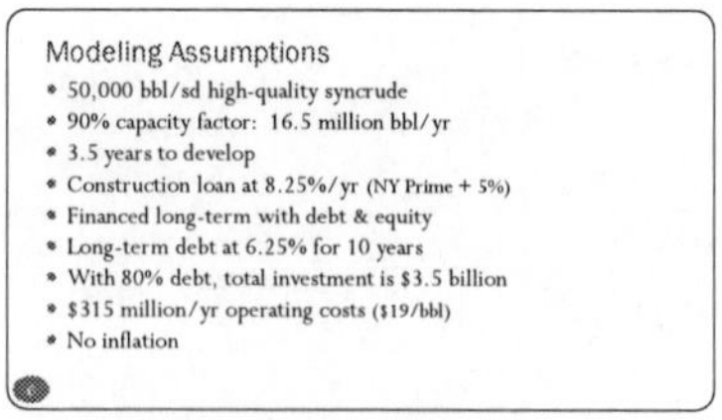

quality synthetic crude oil, plus byproducts such as elemental sulfur. Construction costs were financed with a short-term loan with interest accruing at the New York prime rate plus 5 percentage points. Long-term debt was at New York prime plus 3 percentage points and was retired with level payments over ten years. There was no inflation.

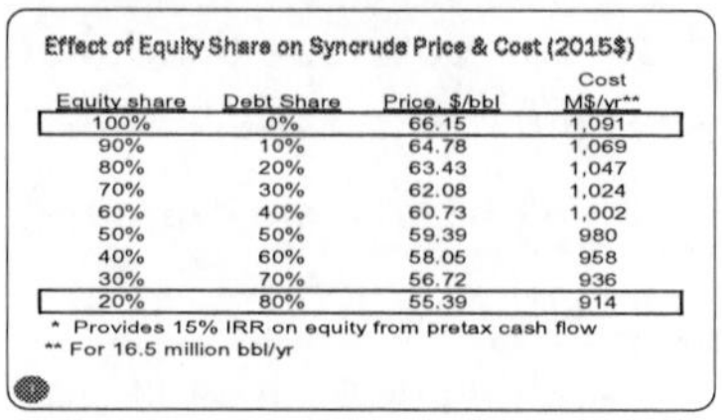
Effect of Equity Share on Syncrude Price & Cost (2015$)

Equity share	Debt Share	Price, $/bbl	Cost M$/yr**
100%	0%	66.15	1,091
90%	10%	64.78	1,069
80%	20%	63.43	1,047
70%	30%	62.08	1,024
60%	40%	60.73	1,002
50%	50%	59.39	980
40%	60%	58.05	958
30%	70%	56.72	936
20%	80%	55.39	914

* Provides 15% IRR on equity from pretax cash flow
** For 16.5 million bbl/yr

These are the results. With all-equity financing, a price of $66 per barrel in 2015 US dollars is required to provide the target IRR. With 20% equity, only $55 per barrel is required.

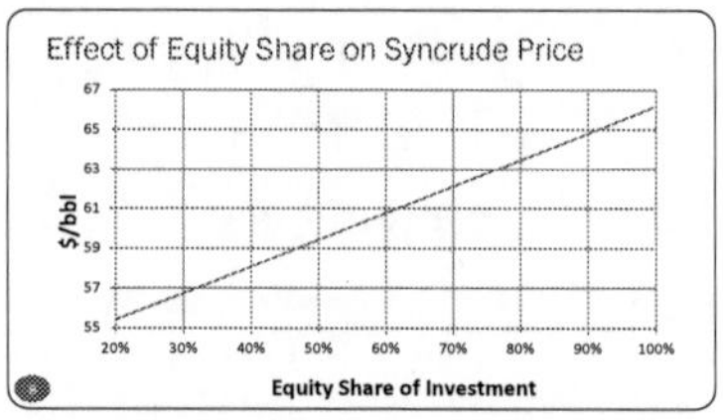

These are the findings in graphical form. 100% equity $66 oil. 20% equity $55.

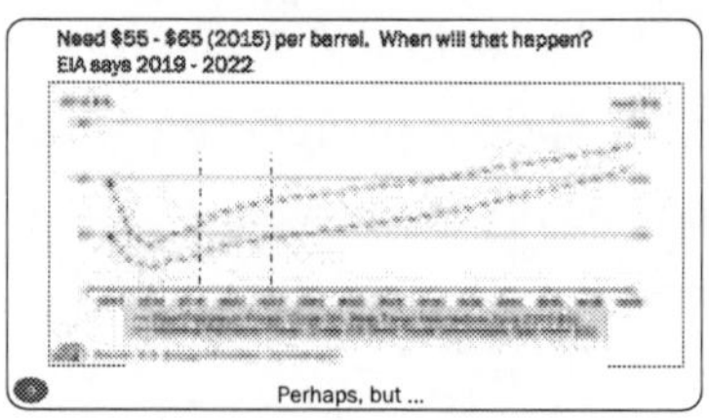

So when will we see $55 to $65 oil again? The US Energy Administration forecasts $55 oil by 2018 and a half and $65 oil by 2022. That may be correct, but I do not think EIA paid enough attention to the role of the fracking industry.

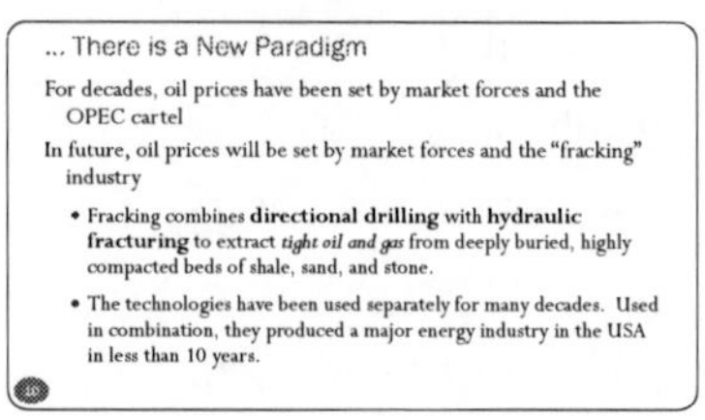

I believe fracking sets a new paradigm and will continue to determine oil prices for many years. Fracking uses a combination of directional drilling and hydraulic fracturing to extract "tight" oil and gas from deeply buried, highly compacted beds of shale, sand, and stone. Fracking produced a major new energy industry in the USA in less than 10 years.

One of the early applications of the hybrid technology was for recovering natural gas from lenses of tight sands in western Colorado. The lenses are remnants of ancient sand bars and beaches, and they contain natural gas trapped in pores between the sand grains. In 1969, the U.S. government liberated some of the gas with a nuclear explosion. Forty years later, the energy industry liberated much more gas with steel pipe and water.

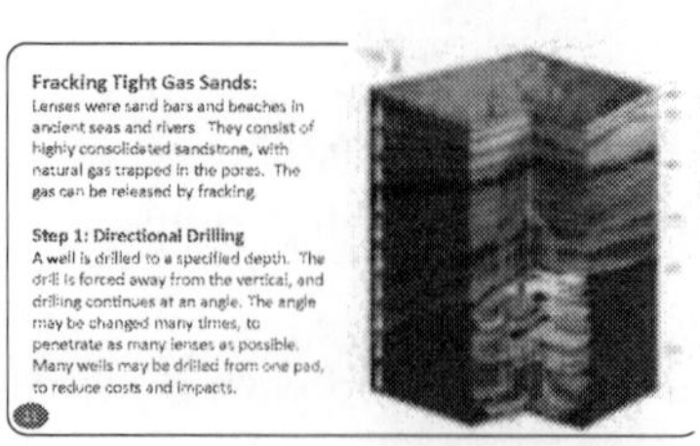

In Step 1 of the fracking process, directional drilling is used to push pipe through the sand lenses. It is not clear from the figure, but each of the branches of pipe is a separate well, although all of them are drilled from the same pad, which reduces time, costs, and impacts.

In Step 2, the walls of the pipes are perforated with an explosive "gun" where they pass through the lenses. Then a sand slurry is forced through the perforations, to expand both natural fissures and cracks created by the gun. Most of the slurry is withdrawn, but sand remains in the cracks, propping them open. Gas migrates to the cracks and is sucked to the surface. This gas gathering process is carried out 2 to 4 kilometers underground.

This slide shows the stages in the fracturing step. The center pipe is has penetrated a tight gas sand lens. The pipe on the left has been perforated, and the pipe on the right has been treated with the sand slurry.

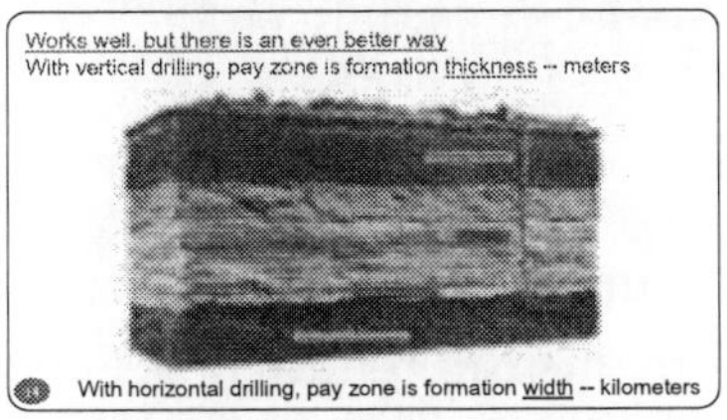

Opportunities to frack sand lenses with vertical drilling are geographically limited, but there are many other fracking opportunities in other locations. This slide shows an even more productive approach -- horizontal drilling parallel to the beds of the target formation. With vertical drilling (through a sand lens, for example) the size of the pay zone is measured in meters. With horizontal drilling, it is measured in kilometers.

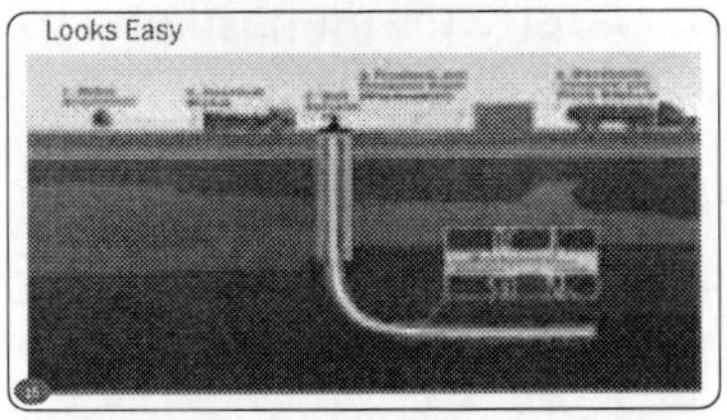

The process looks simple. A well is drilled and capped and the derrick is removed. Water is mixed with sand and chemicals and injected into the borehole, under very high pressure. When the formation is fractured, water is withdrawn and removed for disposal, usually by injection into a deep well.

But it's really quite complicated and expensive. The yellow circle identifies the unfortunate borehole. All the vehicles are there to serve its needs.

The derricks that drill the wells are also sophisticated and expensive. These are "Ironman" derricks drilling into a formation in the state of Oklahoma. Each rig can cost many millions of dollars. They have feet, and they can walk. After a wellbore is drilled, the rig will lift itself, move to one side, and begin to drill a second well. And then a third, and so on.

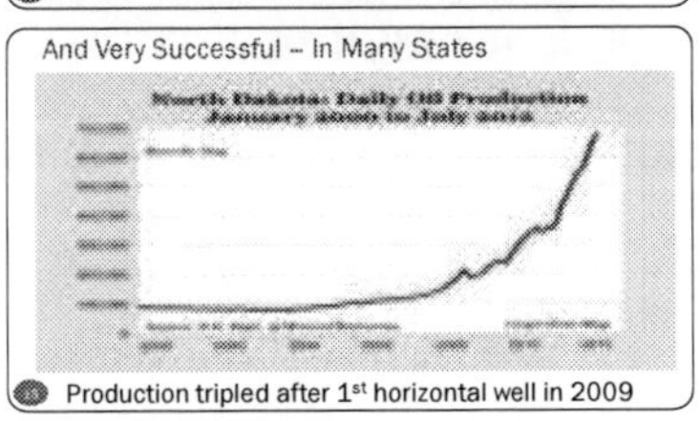

The fracking industry really began to take off in 2009. The effect on oil production in the state of North Dakota is shown here. Oil production tripled in three years. Other states -- Colorado, Utah, Oklahoma, Texas, Pennsylvania -- experienced similar booms.

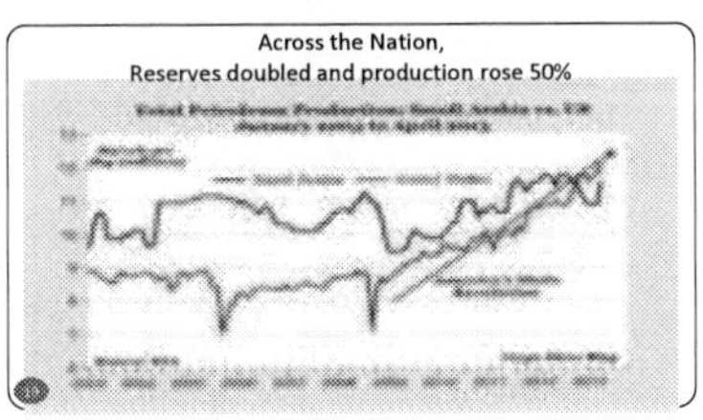

Nationwide, petroleum reserves doubled and oil production increased by 50%. for a few months in 2013, the USA produced more oil than Saudi Arabia. Very similar results were obtained for natural gas.

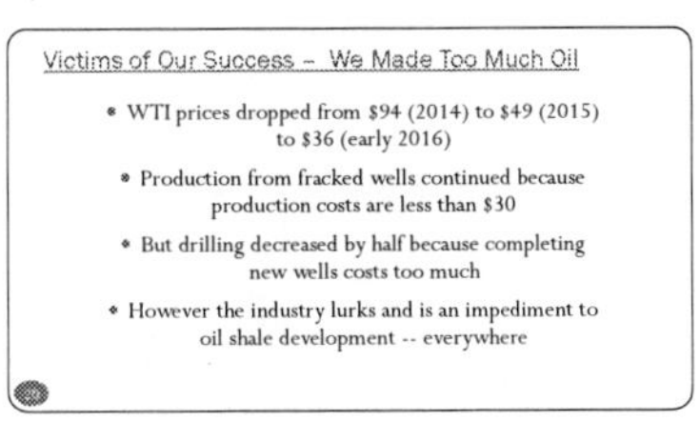

But we got carried away. We produced too much oil, and prices dropped -- from $94 in 2014 to a low of $36 in 2016. Production from existing wells continued, because production is fairly cheap. But drilling decreased by half. The industry is resting right now, but it has not gone away. It will continue to affect the global price of oil, and it will continue to be an impediment to the development of unconventional fuels worldwide.

Fracking has taken hold in the United States, and although there are concerns

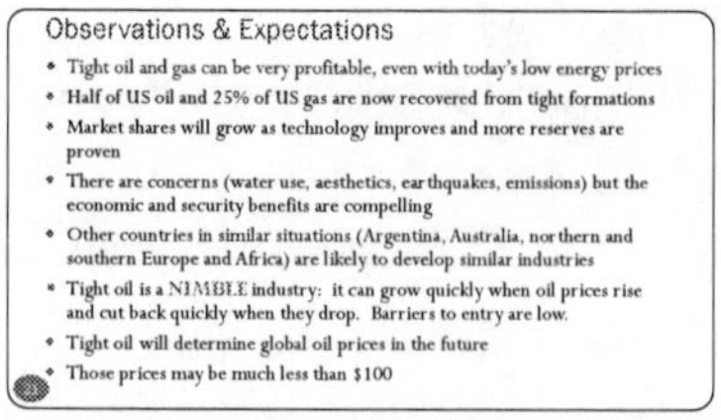

about water requirements and other factors, the economic and national security benefits are compelling. It is very important to recognize that fracking is a **NIMBLE** industry. It can grow quickly when prices are high and retreat quickly when they decrease. Barriers to entry are low. Any dentist or sports hero with $5 million can become an oil tycoon. Tight oil will determine global oil prices in the future. I doubt those prices will exceed $100 in today's currency for a long time.

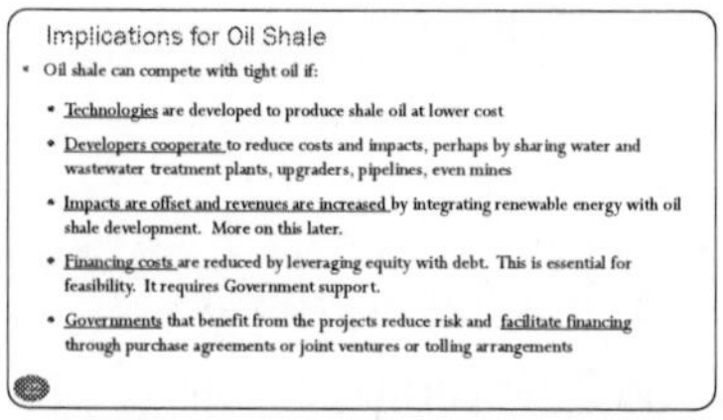

Oil shale can still play a major role, especially in countries such as Jordan where the need is so urgent. Technological developments can reduce costs. Developers could cooperate and share facilities from water treatment plants to mines. Impacts could be reduced by integrating renewable energy with development of oil shale. Most importantly, financing costs could be substantially reduced by leveraging equity with debt. To obtain such debt, however, a government that will benefit from a project must facilitate financing by reducing risk. This could be done through purchase agreements (also known as offtake agreements) or tolling arrangements, or joint ventures.

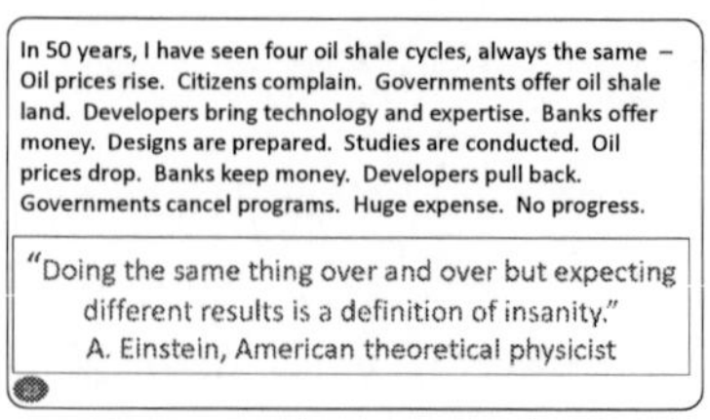

I have been involved with oil shale for only 50 years, but in that time I have seen four oil shale cycles, and they were all played out in much the same way. Prices rise. People get excited. Commitments are made and money is spent. Prices drop, and everybody leaves. A very smart man once said that doing the same thing over and over but expecting different results is a definition of insanity. We know better now. Let's not do it again.

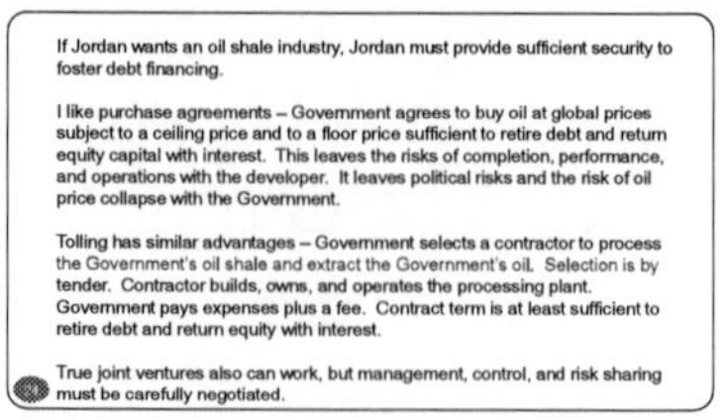

If Jordan wants oil shale, Jordan must provide security. I like purchase agreements because they distribute risks to the entities most capable of dealing with them. Tolling can also work well, given open accounting records and a diligent operating contractor. Joint ventures can work, but caution is advised. Thank you for your attention.

MINISTRY OF ENERGY AND MINERAL RESOURCES

COLORADO SCHOOL OF MINES
EARTH • ENERGY • ENVIRONMENT

3rd Jordan International Oil Shale Symposium | 36th CSM Oil Shale Symposium

• 21–22 November 2016 • King Hussein Bin Talal Convention Center • Dead Sea, Jordan

Converting Challenges into Opportunities in Oil Shale Industry

The Ways to Achieve Independence from the Oil Price Volatility for Oil Shale Projects in Jordan

Ruslan Salikhov
Chief engineer for design
TTU Galoter L.P.
www.oil-shale.ru www.galoter.com

TTU

Company profile

TTU – company possessing patents for methods and installations for processing oil shale by solid heat carrier (UTT)

- Established by specialists in the field of oil shale processing with long experience in design, building, start and operation of UTT
- Goal of the company – development and distribution of technology for the processing of low-calorific solid fuels and wastes by high speed pyrolysis method Galoter
- Number of patents – 24
- Number of inventors involved – 37

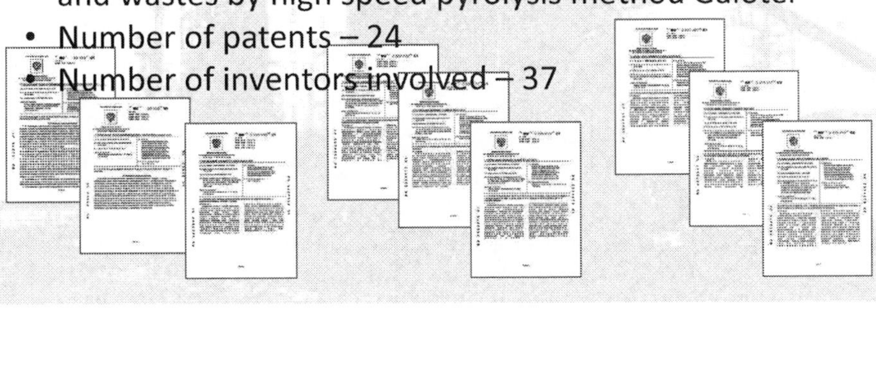

Company profile

TTU Ltd. (Russia) a company act as a general designer and technology provider for Solid Heat Carrier oil shale processing technology used in UTT-3000 installations. Patent owner for original technology.

TTU Galoter L.P. (UK) a company appointed by TTU Ltd. (Russia) as an international representative and service provider for clients outside of Russia.

TTU Galoter Ltd. (Israel) established for development oil shale project at the Mishor Rotem deposit in the State of Israel

Stake in Zonatec S.A., Moroco, Tarfaya project

Company profile

Appointed Representatives in countries:

Jordan
Uzbekistan
Kazakhstan
China
USA
South America
Italy
Morocco
Egypt

Projects

Country	Projects
Russia	3
Uzbekistan	1
Serbia	1
Morocco	1
Jordan	3
Israel	3
Egypt	1
Brazil	1
Myanmar	1
USA	1
China	4
Kazakhstan	1
Brazil	1
Estonia	1

References

2xUTT-3000 in Narva (Estonia)
Enefit-140

3xUTT-3000 Kohtla Yarve (Estonia)
Petroter 1,2,3

What is Petroter

What is Petroter

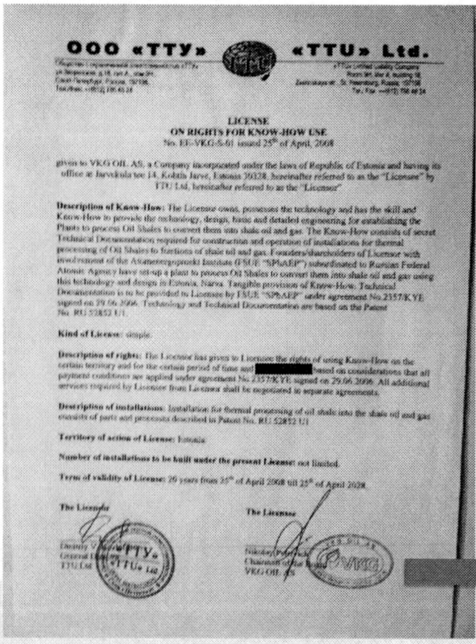

ООО «ТТУ» **«TTU» Ltd.**

LICENSE
ON RIGHTS FOR KNOW-HOW USE
No. EE-VKG-S-01 issued 25th of April, 2008

given to VKG OIL AS, a Company incorporated under the laws of Republic of Estonia and having its office at Jarveküla tee 14, Kohtla Jarve, Estonia 30328, hereinafter referred to as the "Licensee" by TTU Ltd, hereinafter referred to as the "Licensor"

Description of Know-How: The Licensor owns, possesses the technology and has the skill and Know-How to provide the technology, design, basic and detailed engineering for establishing the Plants to process Oil Shales to convert them into shale oil and gas. The Know-How consists of secret Technical Documentation required for construction and operation of installations for thermal processing of Oil Shales to fractions of shale oil and gas. Founders/shareholders of Licensor with involvement of the Atomenergoproekt Institute (FSUE "SPbAEP") subordinated to Russian Federal Atomic Agency have set-up a plant to process Oil Shales to convert them into shale oil and gas using this technology and design in Estonia, Narva. Tangible provision of Know-How, Technical Documentation is to be provided to Licensee by FSUE "SPbAEP" under agreement No.2357/KYE signed on 29.06.2006. Technology and Technical Documentation are based on the Patent No. RU 52852 U1.

Kind of License: simple.

Description of rights: The Licensor has given to Licensee the rights of using Know-How on the certain territory and for the certain period of time and [redacted] based on considerations that all payment conditions are applied under agreement No. 2357/KYE signed on 29.06.2006. All additional services required by Licensee from Licensor shall be negotiated in separate agreements.

Description of installations: Installation for thermal processing of oil shale into the shale oil and gas consists of parts and processes described in Patent No. RU 52852 U1

Territory of action of License: Estonia

Number of installations to be built under the present License: not limited.

Term of validity of License: 20 years from 25th of April 2008 till 25th of April 2028.

The Licensor

General Director
TTU Ltd

The Licensee

Chairman of the Board
VKG OIL AS

VKG is operating under **the License for Know-how Use** issued by technology provider TTU Ltd allowing to build unlimited number of installations on the territory of Estonia only.
VKG has no patent for the process.

What is Petroter

What is Petroter

Recent developments 2014-2016

- Modified process Galoter for processing Municipal Solid Wastes (UTT-500 and UTT-3000)
- Modified technology for processing coals and shungit with gasification of coke
- Modified technology UTT with using Multi Stage Flash distillation facilities for desalination of sea water
- Elaborated, tested and designed technology for thiophenes extractive rectification
- Elaborated small scale mobile installation UGT-100 with gaseous heat carrier for wastes processing
- Elaborated hot asphalt mixes plant on UTT-500

Recent developments

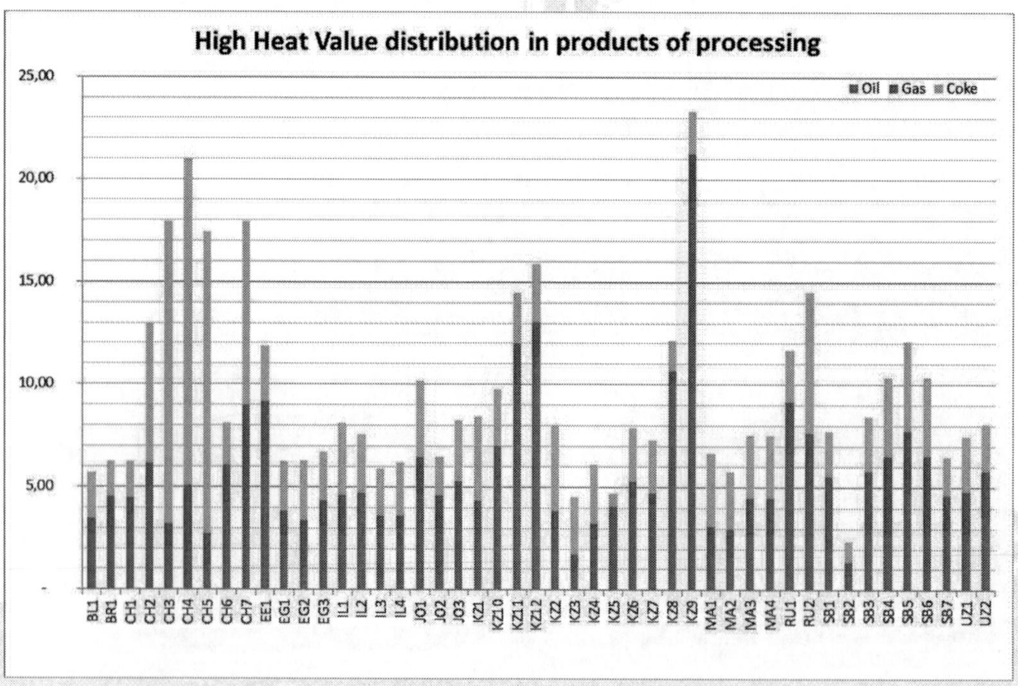

Recent developments

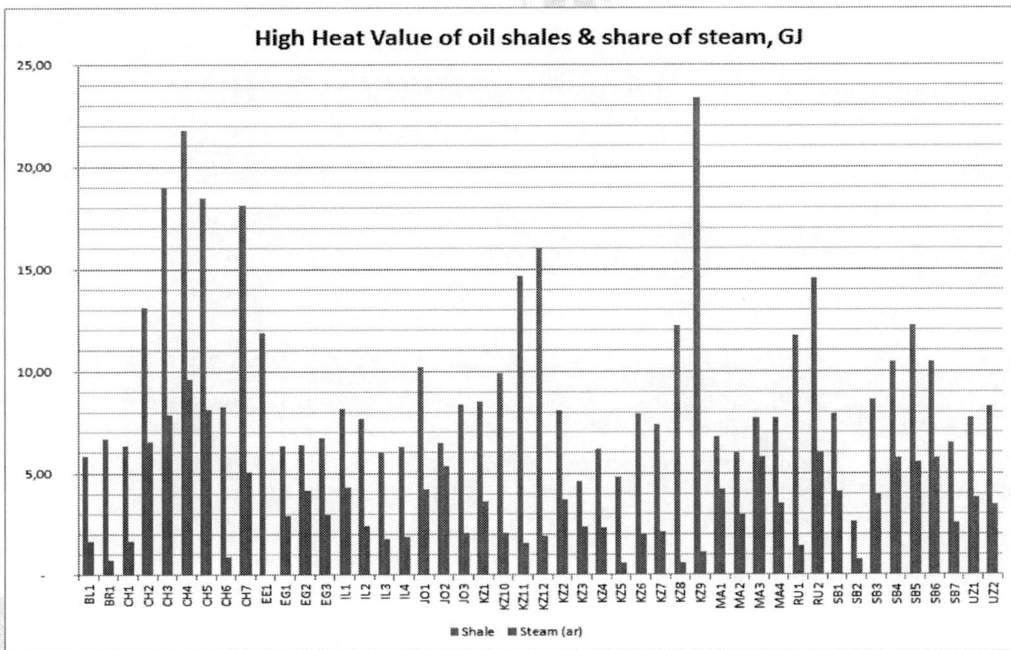

Thiophenes extraction

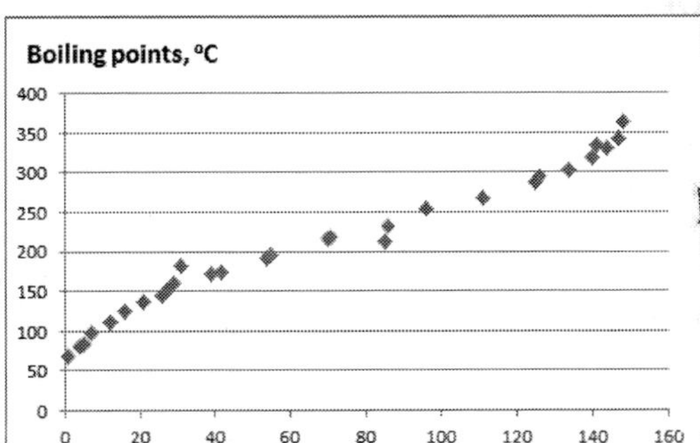

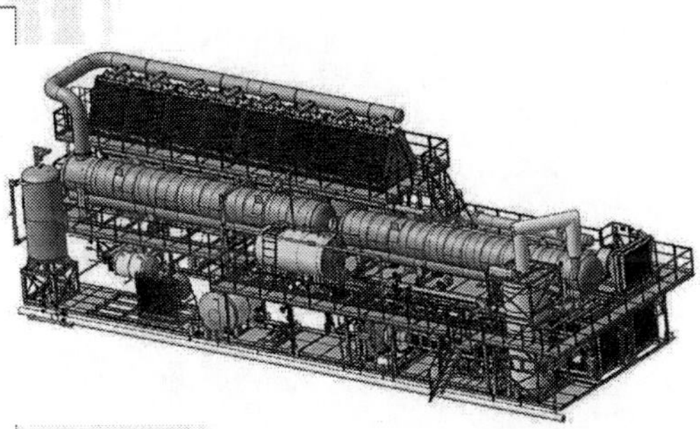

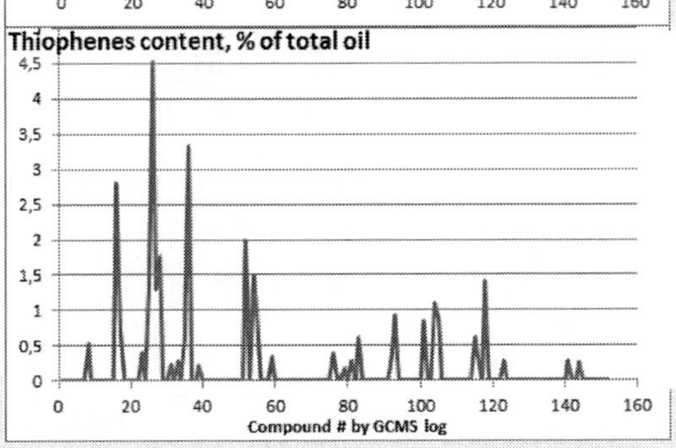

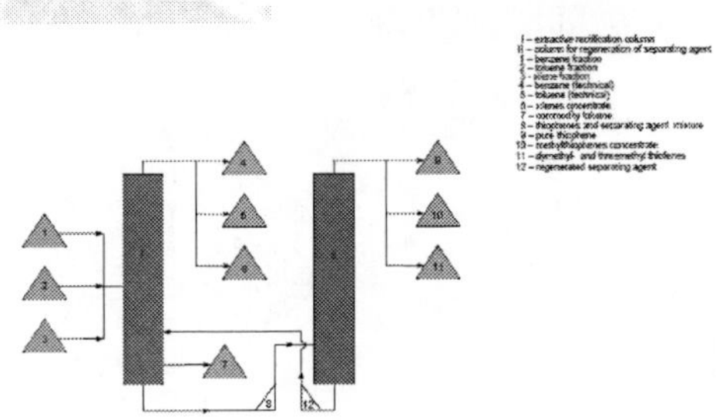

Thiophenes extraction

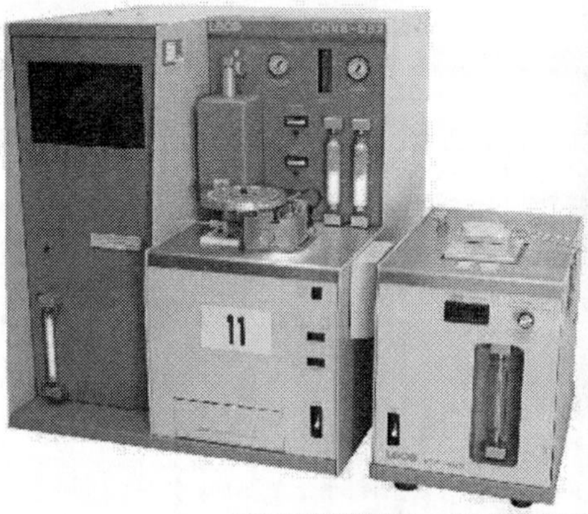

General view Leco CHNS(O) 932

General view of Fisher retort

Thiophenes extraction

General view of bench scale installation UTT.

Heating Control processor.

Thiophenes extraction

Reactor of bench scale installation UTT. *Condensation facility*

Thiophenes extraction

Condensation section of bench scale installation, drive with control board.

Thiophenes markets

1. production high polymers,
2. lubricants including special used in supercritical low and high temperatures as well as in aeronautics and astronautics,
3. in synthesis of biologically active compounds,
4. in pharmaceutical,
5. in perfumery,
6. in nanotechnologies, in production of fullerens and nanotubes and nanofibers,
7. production of composite bipolar membranes used for separation H and OH ions that is very important for cheap hydrogen production,

Thiophenes markets

8. production polymeric semiconductors,
9. displays of computers,
10. ROM (read-only-memory) elements,
11. synthetic dyes,
12. fluorescent inks,
13. solar cells,
14. in oil refineries
15. electrical conductors,
16. nonlinear optical devices,
17. polymer LEDs,
18. Light emitting films,

Thiophenes markets

19. electrochromic or smart windows,
20. photoresistors,
21. antistatic coatings,
22. sensors,
23. batteries,
24. electromagnetic shielding materials,
25. artificial noses and muscles,
26. electrodes,
27. microwave absorbing materials,
28. new types of memory devices,
29. nanoswitches,

Thiophenes markets

30. optical modulators and valves,
31. imaging materials,
32. polymer electronic interconnects,
33. nanoelectronic and optical devices, and transistors

Thiophenes costs

Table Asia Thiophene Capacity Production Price Cost Production Value

Year	2010	2011	2012	2013	2014	2015
Capacity (T)	4626.80	4645.95	4729.09	4749.92	4782.85	4820.34
Production (T)	3044.60	3125.10	3218.93	3273.83	3347.48	3421.65
CapacityUtilization	65.80%	67.27%	68.07%	68.92%	69.99%	70.98%
Price (USD/T)	5495.04	5391.20	5321.33	5172.68	5064.85	4953.13
Production value (K USD)	16730.22	16848.03	17128.97	16934.49	16954.49	16947.84
Cost (USD/T)	4754.86	4700.05	4698.20	4595.41	4519.36	4431.56
Gross (USD/T)	740.18	691.15	623.13	577.27	545.48	521.56
Gross Margin	13.47%	12.82%	11.71%	11.16%	10.77%	10.53%

Source: QYR Thiophene Research Center, Sep 2015

Simple scheme of production

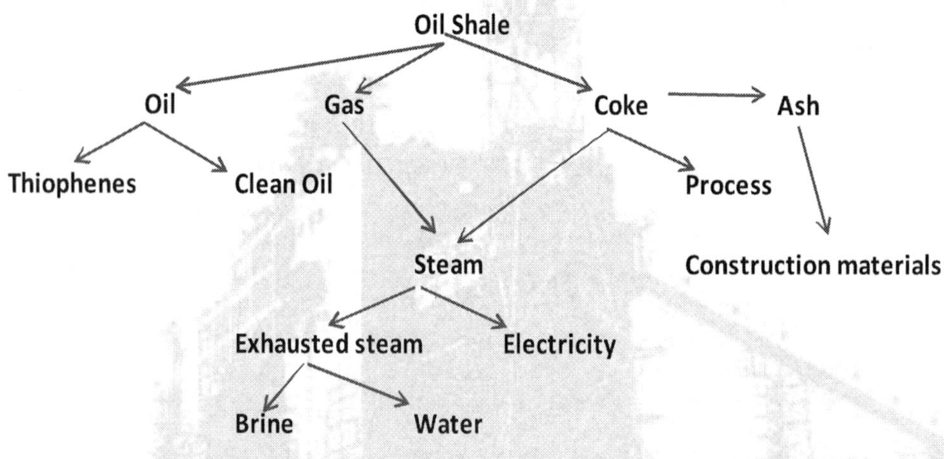

UTT-3000 with sea water distillation

Variant 1 - Sales of oil produced

Product	Unit of measure-ment	Price, \$/unit	1st phase, 1 x UTT-3000		
			Quantity per day	quantity per year	Sales, ths.\$/year
Oil	barrels	50	932	279 715	13 986
Electricity	MW-h	75	538	161 330	12 100
Thiophenes	ton	2000	9	2 700	5 494
Water	cu.m.	1,02	49 416	14 824 862	15 121
Ash	ton	10	1 663	498 925	4 989
Total					46 701

For oil shale with oil yield 50 kg/t

UTT-3000 with sea water distillation

Variant 1, Sales of oil

№	Position	2020	2021	2022	2023	2024	2025	2026	2027	2028
	Production specific costs									
	Operational specific costs (without depreciation)									
	oil, \$/t	85	75	77	76	75	74	73	73	73
	oil, \$/bbl	14	12	12	12	12	12	12	12	12
	electricity \$/kW-h	0,02	0,04	0,04	0,04	0,04	0,04	0,03	0,03	0,03
	Full production costs									
	oil, \$/t	266	179	181	180	179	178	156	73	73
	oil, \$/bbl	42	29	29	29	28	28	25	12	12
	electricity \$/kW-h	0,06	0,09	0,09	0,09	0,09	0,08	0,07	0,03	0,03

Red Sea – Dead Sea Water Conveyance Project

In a jointly signed letter to the World Bank dated May 9, 2005, the State of Israel, the Hashemite Kingdom of Jordan and the Palestinian Authority requested the World Bank to undertake a Study Program to investigate the feasibility of reversing the environmental degradation of the Dead Sea by transferring water from the Red Sea.

The shared vision of the Study Program centered on:

- Saving the Dead Sea from environmental degradation;
- Desalinating water and generating energy at affordable prices;
- Building a symbol of peace and cooperation in the Middle East

Red Sea – Dead Sea Water Conveyance Project

Table 9.2: Potable Water Demands Approved by the Technical Steering Committee

Beneficiary Party	Potable Water Supplied from the Project - Million m^3/year				
	2020	2030	2040	2050	2060
Jordan	230	310	370	460	560
Israel	60	60	60	60	60
Palestine Authority	60	60	60	60	60
Available for Further Allocation	0	0	50	90	170
Total	**350**	**430**	**540**	**670**	**850**

Red Sea – Dead Sea Water Conveyance Project

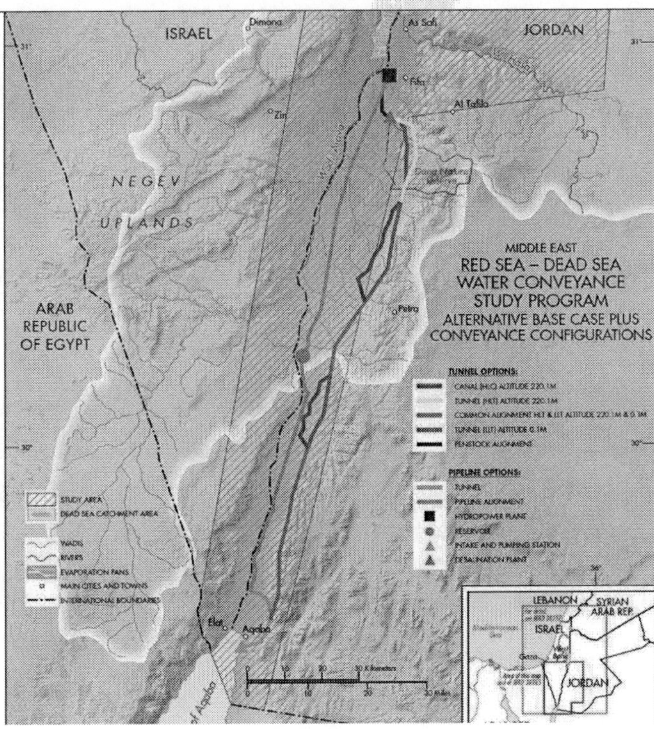

Red Sea – Dead Sea Water Conveyance Project

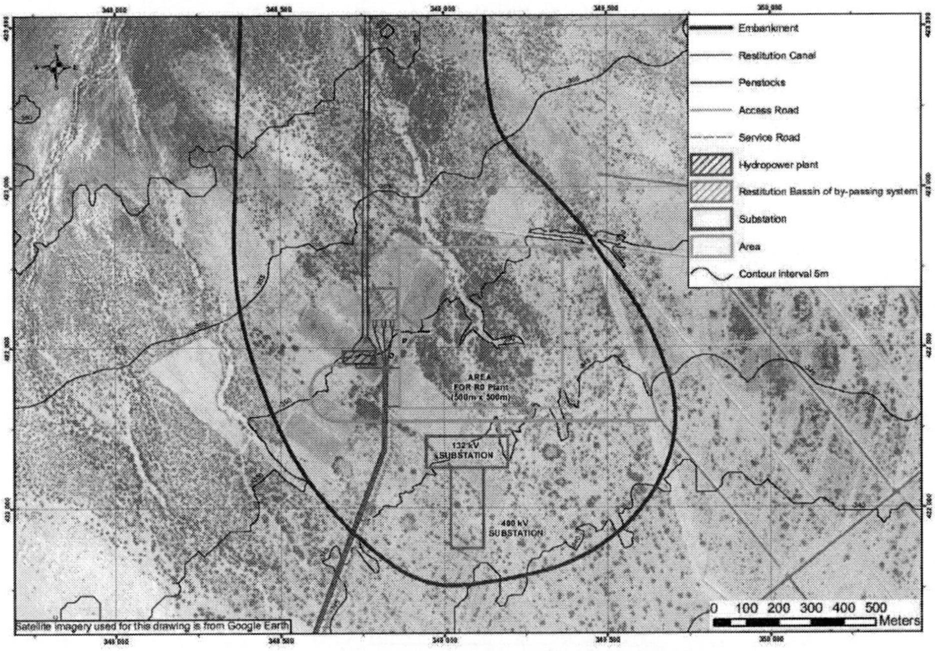

Figure 17.3: General Layout of the Hydropower Plant Area

Red Sea – Dead Sea Water Conveyance Project

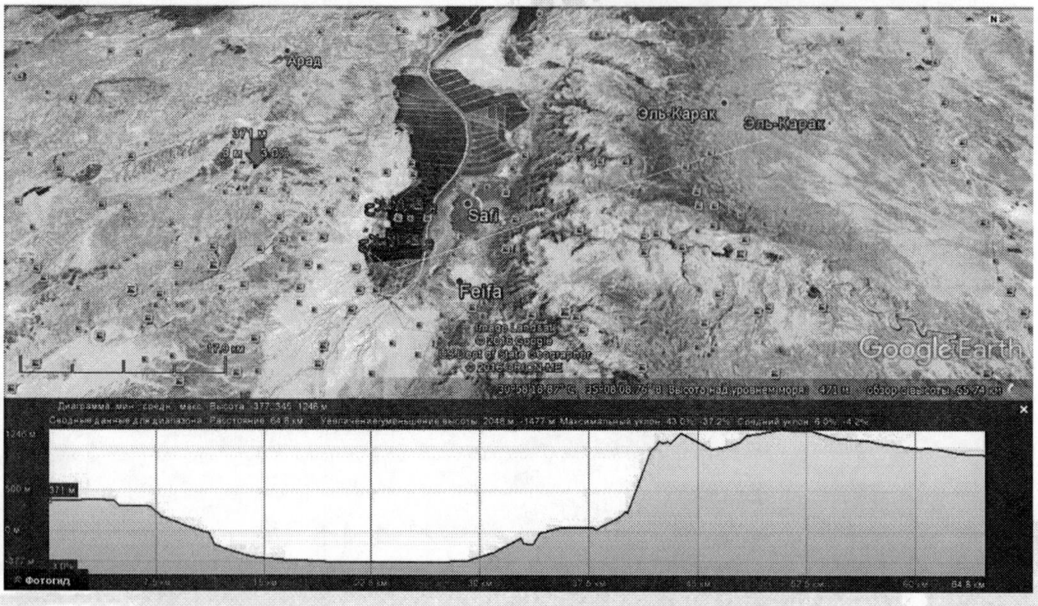

Red Sea – Dead Sea Water Conveyance Project

Table 26.3: Key Schedule Milestones

Milestone		Date	
		Initial Schedule	**Realistic Schedule**
1	Final Feasibility Study Report	Jun-11	Jul-13
2	Beneficiary Parties Go / No Go decision	Sep-11	Dec-14
3	Enabling legislation enacted by all Beneficiary Parties	Mar-12	Sep-15
4	Geotechnical site investigation contract awarded	May-12	Dec-15
5	Implementation organisational structure in place	Sep-12	Apr-16
6	Project Management Consultant appointed	Sep-12	Apr-16
7	Contract for desalination pilot trials awarded	Sep-13	Apr-17
8	Financing is secured and available for drawdown	Sep-13	Apr-17
9	EPC Contract for conveyance tunnel awarded	Jan-14	Aug-17
10	EPC Contract for conveyance pipelines awarded	Mar-14	Oct-17
11	Results of desalination pilot trials available	Jan-15	Aug-18
12	EPC contract for sub-station and power transmission awarded	Jan-15	Aug-18
13	EPC contract for desalination plant awarded	Mar-15	Oct-18
14	EPC contract for pump station awarded	Jun-15	Jan-19
15	EPC contract for Hydropower plant awarded	Jan-16	Aug-19
16	EPC contract for intake works awarded	Jan-17	Aug-20
17	Start up and commencement of commissioning	Oct 20	May-24

Red Sea – Dead Sea Water Conveyance Project

Table 30.2: Breakdown of the Initial Capital Costs of Project (US$ Million)

Item	Saving the Dead Sea	Provision of Potable Water	Generation of Hydropower	Total
Conveyance	3,831.5	983.9	128.5	4,943.9
Desalination Plant	0	1,104.7	0	1,104.7
Hydro Power Plant	0	0	241.4	241.4
Restitution canal	206.9	53.1	6.9	266.9
HV Connections to national grid	205.9	52.8	6.9	265.6
Project management and development of institutional structure	195.6	50.2	6.6	252.4
Land	107.7	38.7	9.1	155.5
Environmental & Social Management Plan	0	75.1	0	75.1
Sub-Total (excluding transmission of potable water)	4,547.6	2,358.5	399.4	7,305.5
Potable water transmission system to **Jordan**	0	1,663.1	0	1,663.1
HV Connections to national grid	0	131.4	0	131.4
Project management	0	64.4	0	64.4
Land	0	64.5	0	64.5
Sub-Total	0	1,923.4	0	1,923.4

Red Sea – Dead Sea Water Conveyance Project

Table 30.3: Potential Sources of Initial Finance

Project Objective	Financing Package	Finance Required US $ billion	Potential Sources of Finance
Saving the Dead Sea	Conveyance	3.831	– Beneficiary government equity. – Grants. – Donations
	Dead Sea Restitution Canal	0.207	
	HV Power Connections	0.206	
	Institutional and Organisational Costs	0.196	
	Land	0.108	
	Sub-Total	4.548	
Producing Potable water	Conveyance	0.984	– Government equity / Public funding – Private equity. – Soft loans and export credit. – Multilateral loans. – Contractor finance.
	Desalination	1.105	
	Dead Sea Restitution Canal	0.053	
	HV Power Connections	0.053	
	Institutional and Organisational Costs	0.050	
	Land	0.039	
	ESMP	0.075	
	Sub-Total	2.359	
Generating Hydropower	Conveyance	0.129	– Government equity / Public funding – Private equity. – Soft loans and export credit. – Multilateral loans. – Contractor finance.
	Hydropower plants	0.241	
	Restitution canal	0.007	
	HV Power Connections	0.007	
	Institutional and Organisational Costs	0.006	
	Land	0.009	
	Sub-Total	0.399	
Sub-Total – Multinational Project		**7.306**	

Red Sea – Dead Sea Water Conveyance Project

Parameters of Recommended Optimum Configurations by WB

Desalination Plant	
Location	In wadi Araba / Arava Valley 50km north of Grarandal saddle and 40km south of the southern tip of the chemical industry evaporation ponds
Pre-Treatment	Multi-media filtration and dissolved air flotation
Desalination process	Sea water reverse osmosis (SWRO)
Post-treatment of potable water	Re-mineralisation with disinfection
Capacity	320 million m^3/year at start up increasing to 850 million m^3/year in 2060
Installed power capacity	247 MW (2020) – 556 MW (2060)
Average specific energy consumption	3.33 kWh/m^3 of potable water

System Energy Balance	
Total energy generated	2020: 1,817 GWh / 2060: 1,642 GWh
Total energy consumed	2020: 4,347 GWh / 2060: 7,782 GWh
Net energy balance	2020: -2,530 GWh / 2060: -6,140 GWh

To cover demand on the level 2020 it may be required 4-5 UTT-3000

Reserves evaluation example

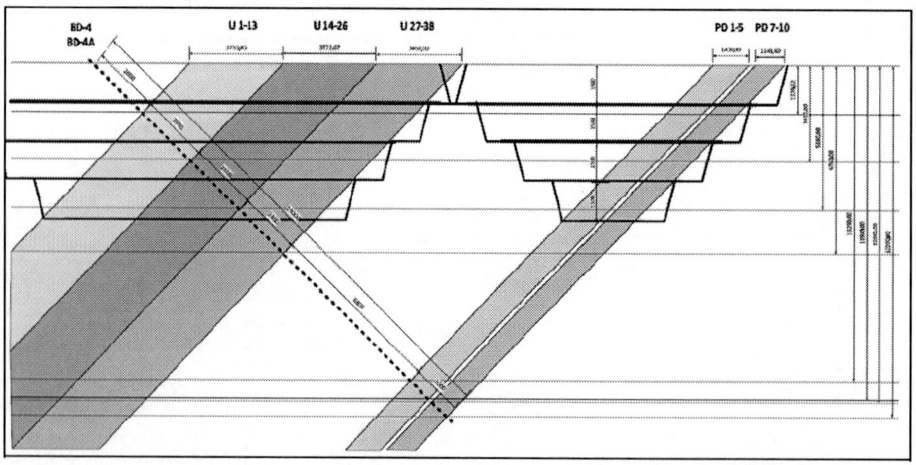

Reserves evaluation example

hole rval h, m	Area, sq.km	Strip-ping ratio	Thick m	Reserves, mln t (mln kW-h)				Energy reserves, mln. BOE			Priduc-tivity BOE/m2	Production cost		Profitability of reserves		Net Value mln $
				OS	Oil	Ash	Electri-city	OS	Oil	Electri-city		oil $/bbl	Electri-city $/kWh	$/t	$/sq.m	
R-1	140		-	-	-		-	-	-	-	-	-	-	-	-	-
229	140	76,33	3,0	840	68	610	199 147	824	440	109	3,92	19,81	0,016	21,80	130,82	18 315
129.5	140	31,48	4,1	1 152	41	840	246 043	783	263	134	2,84	35,49	0,028	5,94	48,89	6 844
R-2	2,5		-	-	-	-	-	-	-	-	-	-	-	-	-	-
51	2,5	12,75	4,0	20	1	13	4 217	18	10	2	4,78	21,73	0,017	18,64	149,10	373
9.5	2,5	1,90	52,5	263	8	198	60 742	180	54	33	34,74	28,61	0,023	8,47	889,28	2 223
9.5	2,5	66,33	1,5	8	0	6	1 834	6	2	1	1,28	33,11	0,026	8,43	25,28	63
R-3	3		-	-	-	-	-	-	-	-	-	-	-	-	-	-
47	3	11,75	4,0	24	1	16	4 767	16	6	3	2,77	35,67	0,028	5,84	46,71	140
1.65	3	49,48	1,7	10	0	7	2 286	8	3	1	1,42	33,02	0,026	8,49	28,03	84
R-4	12		-	-	-	-	-	-	-	-	-	-	-	-	-	-
R-5	5,5		-	-	-	-	-	-	-	-	-	-	-	-	-	-
61	5,5	7,63	8,0	88	3	60	19 458	62	19	11	5,37	35,18	0,028	5,95	95,27	524
60	5,5	30,00	2,0	22	1	15	5 764	19	8	3	1,96	27,28	0,022	12,82	51,28	282
61	5,5	20,33	3,0	33	2	23	9 664	30	10	5	2,85	25,79	0,020	13,48	80,85	445
116	5,5	58,00	2,0	22	1	15	5 409	17	7	3	1,72	30,79	0,024	9,63	38,54	212
R-6	90		-	-	-	-	-	-	-	-	-	-	-	-	-	-
12.5	90	32,14	3,5	630	27	471	208 557	583	172	114	3,18	28,18	0,022	11,93	83,48	7 513
190	90	7,31	26,0	4 680	119	3 582	909 940	2 741	766	496	14,03	49,08	0,039	0,43	22,38	2 014
317	90	4,23	75,0	13 500	370	10 298	2 941 427	8 623	2 386	1 604	44,34	40,35	0,032	3,52	528,44	47 560
355	90	161,36	2,2	396	17	264	86 042	294	112	47	1,77	34,24	0,027	7,38	32,48	2 923
R-7	18		-	-	-	-	-	-	-	-	-	-	-	-	-	-
44	18	22,00	2,0	72	5	53	16 629	65	31	9	2,20	24,87	0,020	15,52	62,09	1 118
130	18	3,61	36,0	1 296	45	881	406 757	1 121	290	222	28,44	29,84	0,024	9,73	700,85	12 615
				22 972	**706**	**17 296**	**5 107 203**	**15 317**	**4 549**	**2 786**	-	-	-	-	-	**102 085**

Reserves evaluation example

Content of metal compounds in ashes:

ell	Interval	Ash%	Reserves of Ash, mln t	Al_2O_3	TiO_2	Fe_2O_3	MgO	CaO	Na_2O	K_2O	P_2O_5	S	V	Zn	Ag ppb	U	Th	REE+Y
R-1	-																	
·	227-229	72,67	610	5,5	0,2	2,8	17,7	0,9	0,5	0,6	2,3	3,0	747	998	1	28	6	164
·	97.5-129.5	72,96	840	4,0	0,2	1,9	29,0	1,5	0,3	0,6	0,5	1,6	156	193	0	17	5	139
R-2	-	-	-															
·	47-51	66,40	13	7,2	0,2	2,6	1,5	49,6	0,2	0,5	3,7	2,5	-	709	1 618	32	3	118
·	47-99.5	75,29	198	11,4	0,4	4,0	2,7	33,5	0,2	0,6	1,8	2,6	-	163	325	15	5	131
·	98-99.5	78,78	6	12,8	0,5	4,1	8,2	18,8	0,3	0,9	3,7	2,3	-	361	360	25	7	195
R-3	-	-	-															
·	43-47	67,50	16	6,2	0,2	2,1	7,6	48,8	0,2	0,3	2,4	1,9	88	166	259	24	4	124
·	80-81.65	68,70	7	7,6	0,4	4,1	3,8	37,5	0,4	1,0	8,9	3,5	1 332	1 910	1 307	98	5	216
R-4	-	-	-															
R-5	-	-	-															
·	53-61	68,35	60	16,9	0,6	5,8	1,3	29,8	0,5	1,0	1,7	3,0	388	531	884	28	10	256
·	59-60	69,20	15	16,3	0,5	6,6	1,2	30,1	0,5	1,0	2,0	3,3	421	572	1 029	34	10	258
·	59-61	68,20	23	17,0	0,6	6,2	1,2	28,4	0,5	1,0	1,7	3,2	396	698	1 226	33	10	255
·	114-116	68,10	15	12,7	0,5	5,8	1,1	41,3	0,3	0,6	2,8	3,5	264	872	1 171	34	6	177
R-6	-	-	-															
·	109-112.5	74,81	471	12,48	0,37	4,87	1,83	26,28	0,89	1,04	2,35	2,38	1 581	638	3 677	35	8	195
·	164-190	76,54	3 582	8,58	0,29	2,63	1,12	45,83	0,70	0,75	0,78	1,31	111	165	203	12	5	169
·	242-317	76,28	10 298	12,42	0,47	4,73	0,88	36,20	0,69	0,46	1,32	2,56	137	141	106	14	6	158
·	352-355	66,60	264	6,62	0,30	4,55	1,10	43,18	0,81	0,39	23,02	3,47	912	1 044	1 363	86	4	270
R-7	-	-	-															
·	42-44	73,31	53	11,2	0,4	3,7	1,3	35,5	0,3	1,0	4,7	1,9	1 632	1 188	6 873	41	7	197
·	94-130	68,00	881	3,8	0,1	1,7	3,2	63,1	0,1	0,3	7,0	1,5	75	354	346	39	3	175
			17 296															

Examples

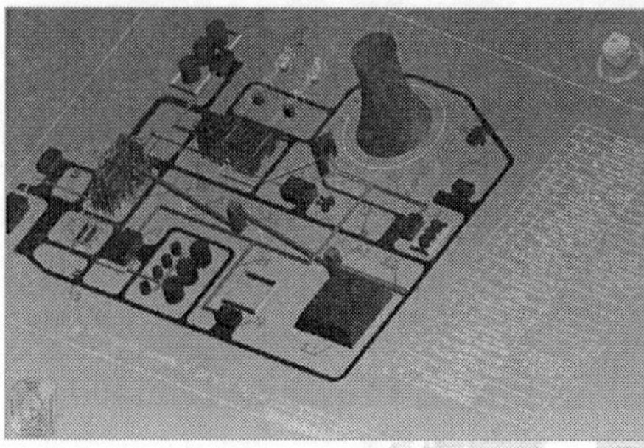

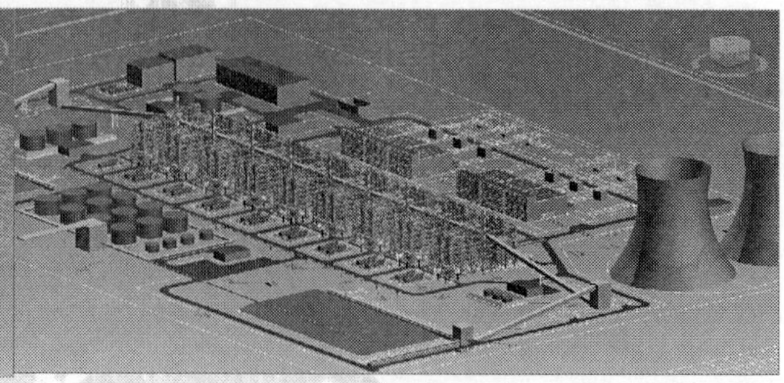

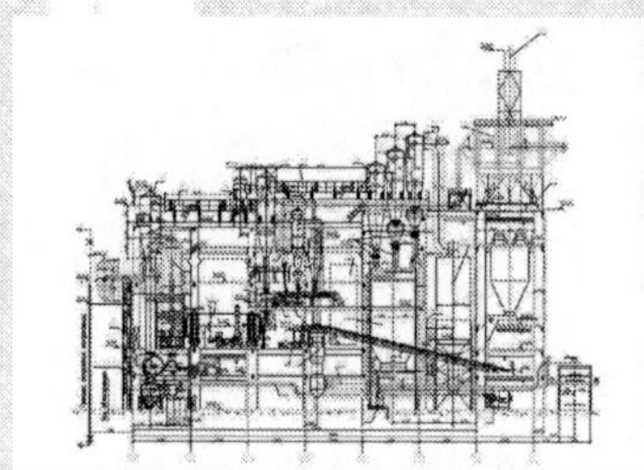

Schedule

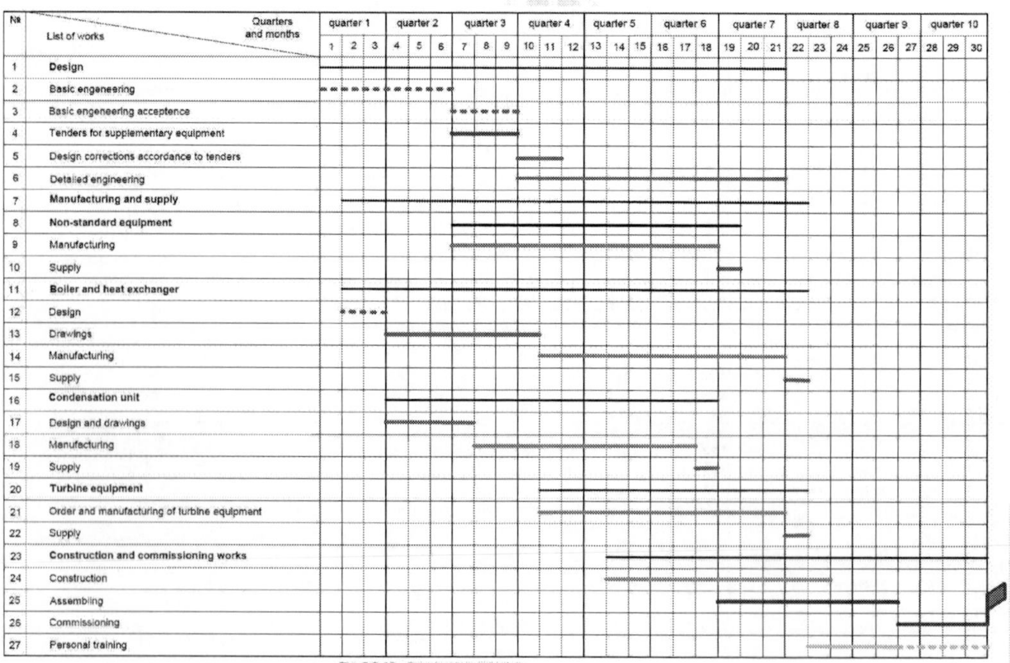

Fig. 5.2.16. Construction schedule

Chairman of the Board VKG Oil AS (left) and
Chief engineer for design of UTT-3000 Mr. Petrov (TTU)
at launch ceremony of UTT-3000 for VKG Oil A.S.

Thank you!

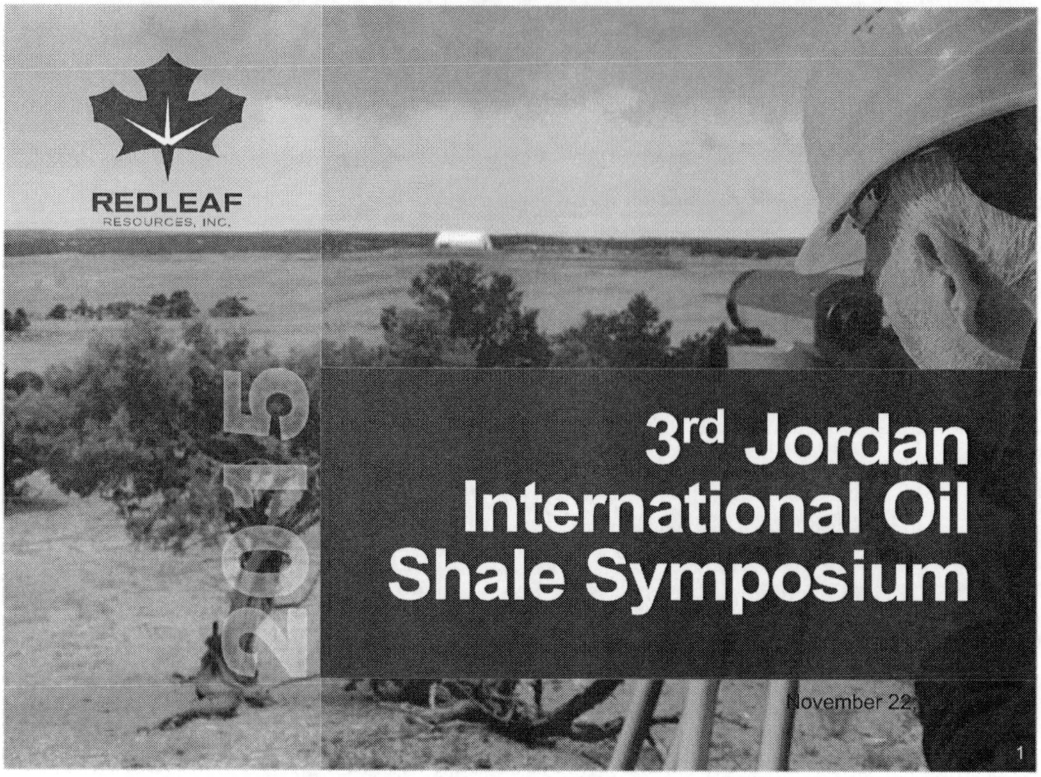

Introduction

- **Red Leaf**
  - Incorporated in Delaware in 2006
- **Concept:**
  - Economically develop the most easily accessible oil shale in the world with a low-cost environmentally responsible approach
- **Original Design:**
  - Utilize clay lined earthen structures as retorts
    - Build in mine pit and reclaim in place
    - Handle shale one time with minimum crushing

Assets

- Expert oil shale team
- EcoShale® technology
  - 14 US patents issued, 15 pending
  - 119 foreign patents issued, 247 pending
  - 6 license agreements
- World class lab facility for oil shale testing
- Utah leases with over 500MMBL of surface recoverable oil
- Red Leaf/TOTAL JV for commercial demonstration project in Utah
  - Final permits received
  - Full funding approved

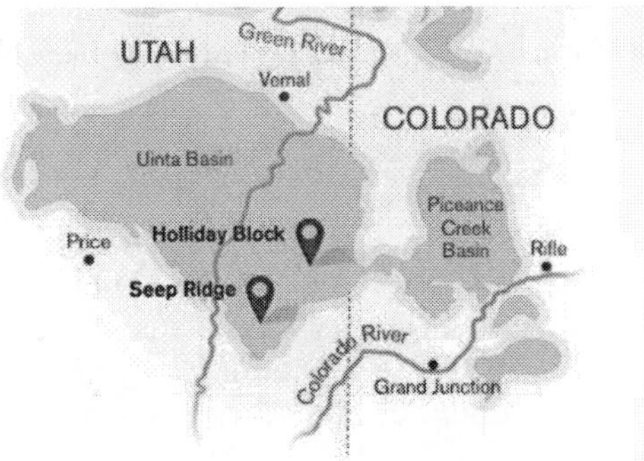

REDLEAF RESOURCES, INC.

Chapter 1:Proof of Concept

- Achievements 2006 – 1Q 2012
  - Patents granted for EcoShale® technology
  - Successful field pilot operation 2008-2009
  - Raised $172MM of equity between 2007 - 2012
  - Entered $400MM joint venture with TOTAL for Utah oil shale projects

REDLEAF RESOURCES, INC.

Chapter 2: Initial EPS

- Achievements 2Q 2012 – 1Q 2015
  - Agreed with TOTAL to construct commercial demonstration project (EPS)
  - Completed detailed engineering for commercial scale first generation capsule and facilities
  - Obtained mining, construction, and groundwater permits
  - Signed Engineering, Procurement, and Construction Management contract with WorleyParsons
  - Mine and construction site fully developed for EPS construction and capsule floor excavated
  - Acquired additional mineral leases in Utah – 100MMB recoverable oil adjacent to existing leases

Chapter 3: Improved Design; Commercial Pre-FEED Study

- 2Q 2015 to current
  - Reduced oil prices led to joint decision with TOTAL to suspend full-scale EPS construction and bring forward design improvements
  - Use time to incorporate technology and cost improvements for improved capsule design:
    - Increase oil production per capsule
    - Simplify capsule heat process
    - Reduce capsule materials and design
    - Simplify capsule construction
  - Conduct Commercial pre-FEED study for Seep Ridge Site
    - Pre-FEED study initiated in Fall 2015
      - First fully integrated plan from mining through oil production
    - Concluded in May 2016

Technology improvements

- Technology improved in the following important areas:
  - Capsule design: lower cost, more robust, reduced risk
    - Direct Heat removed vast majority of piping
    - Removal of piping reduced operating risk
    - Improved capsule constructability with stackers/conveyors
    - Removed inert pipe bedding material – added ore
  - Heat efficiency improved; improved balance of energy
  - CO_2 footprint reduced
  - Liquid yield increased to 85-95% MFA
  - Synthetic membrane qualified for use
  - H_2S sequestration potential shows increasing promise

REDLEAF RESOURCES, INC.

3rd Jordan International Oil Shale Symposium - November 2016 7

Improved Design - Capsule

Initial Design	Improved Design
Indirect, closed loop, flue gas heating	Direct injection of hot working gas
Multiple rows of heating pipe	Single row of heat injection pipe and single row of collection pipe
Significant volume of low grade material for pipe bedding/protection	Greatly reduced volume of low grade material for pipe bedding/protection
400 days for heating and 400 days for cooling	200 days for heating and cooling
Insulation: layers of dry and wetted crushed shale	Same but thinner layers
Bentonite Amended Soil (BAS) for liquid and gas containment	Considering limited Geo-membrane for liquid and gas containment

REDLEAF RESOURCES, INC.

3rd Jordan International Oil Shale Symposium - November 2016 8

Commercial Pre-FEED: Capsule Geometry

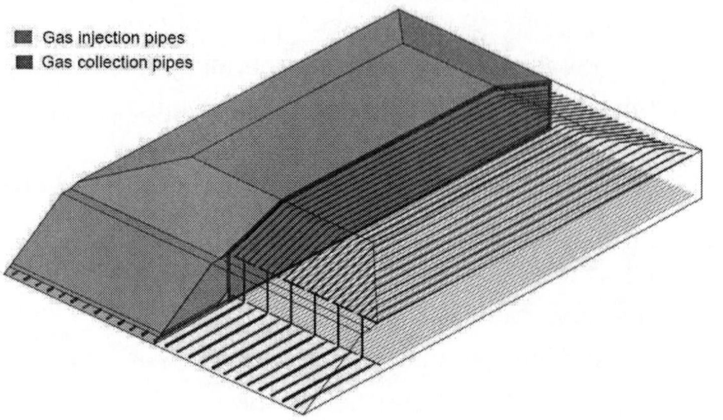

Commercial Pre-FEED Study Results

- Key cost components for 30,000 bbls/day:
  - Capsule cost - $31.8 million for 1 million barrel capsule
  - Oil & gas collection, separation & storage - $1.3 to $4.1 billion for 11,000 – 33,000 bpd
- Pre-FEED commercial project economics:
  - 6% IRR at $80 Oil

- Conclusions:
  - Project technically feasible for capsules and associated process

  - MCC (Capsule cost) – positive
    - Step change improvement in capsule costs as a result of direct heat design change
  - CSS – negative
    - Cost higher than 2014 estimates
  - Project economics, with given Basis of Design assumptions, were marginal

Opportunities for Further Efficiencies

- Further potential improvements identified in pre-FEED improve economics:
  - 10% IRR at $70 oil.
- Increased resource size provides more barrels over which to amortize the CSS
- MFA yield should increase to 90% or above
- H2S sequestration should reduce metallurgy and processing costs
- Original study did not allow for a capsule learning curve
- Economics improved by reducing daily production rate of project

Unquantified possibility for improvement:

- CSS optimization
  - By far the biggest cost reduction opportunity

REDLEAF RESOURCES, INC.

Chapter 4: Path Forward

EPS Engineering and Construction Schedule:

- EPS Basis of Design Study completed in August 2016
- EPS pre-FEED study ongoing
  - Completion expected in spring 2017
- Maintain and amend permits as necessary
- Full-scale construction of improved EPS 2H 2017
- Complete construction in 2H 2018
- Heating/cooling cycle complete in 2019

REDLEAF RESOURCES, INC.

EPS Strategy – Sizing

Current EPS size

- About 850,000 ft^3 of shale
- About 50 ft shale bed height
- Oil production: 17,000 bbl

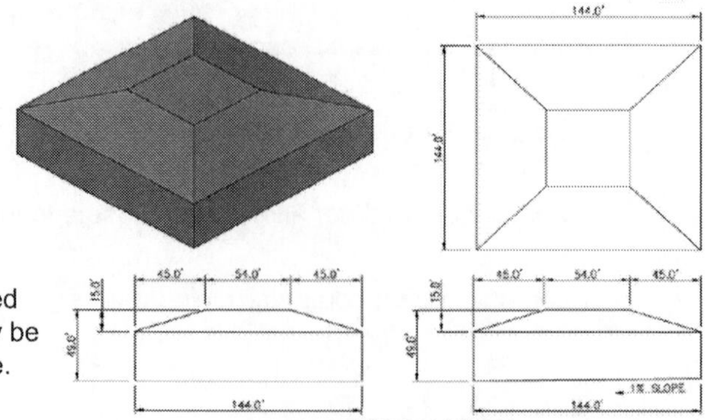

EPS2 CAPSULE SHALE BED
VOLUME = 847,234 ft^3

Sizing and scalability confirmed during BOD and will ultimately be confirmed during FEED phase.

REDLEAF RESOURCES, INC.

Summary

- In Q1 2015, we had a fully engineered design and EPS was under construction
- At TOTAL's request, in Q2 2015, decided to slow construction for two years
- Used the additional time to develop an improved design
  - Switch from indirect heat to direct heat was a step-change
    - Led to simplified design and dramatic reduction in capsule cost
    - Move from up-flow to down-flow was another significant change
    - Direct heat has opened the door to many other improvements
  - Downside of direct heat change is increase in CSS cost
    - Moved complexity from inside capsule to outside capsule
- Schedule:
  - EPS Basis of Design completed in August
  - EPS FEED should be complete in spring of 2017
  - EPS construction in 2017 and 2018
  - EPS Heating and Cooling in 2019
  - Decision on Commercial Project in 2019

REDLEAF RESOURCES, INC.

Reusable Capsules

- Questerre presented us with a problem:
  - Their lease has much thicker overburden and pay zones than our Utah shales
  - Deployment of our Utah capsule design would strand a large majority of the oil in place in Jordan
  - Question: Could our technology be modified to address this problem
- Our Solution: Reusable capsules
  - We had done some work on a reusable idea for Utah but it didn't appear to be a significant advantage over the current approach
  - However, in the Jordan situation, reusable capsules would be a game changer
  - Red Leaf and Questerre are collaborating on the development of a reusable capsule design that could work on their deposit in Jordan
  - Early days, but initial findings point to lower per barrel costs than in Utah

Glossary of Acronyms

- BAS – Bentonite Amended Soil
- BBL or BL – barrels of oil
- BOD – Basis of Design
- BOPD – Barrels of Oil Per Day
- CSS – Collection, Separation and Storage
  - Includes everything outside of the capsule
- EPS – Early Production System (commercial demonstration project)
- FEED – Front End Engineering and Design
- H2S – Hydrogen Sulfide
- IRR – Internal Rate of Return
- MCC – Mining and Capsule Construction
- MM - million
- WTI – West Texas Intermediate (benchmark oil price)

STRICTLY PRIVATE & CONFIDENTIAL

APCO
ATTARAT POWER COMPANY

Oil Shale Power Project and Chinese Financing

2 x 277 MW Oil-Shale Fired Mine Mouth IPP Power Project
at Attarat Um Ghudran, Jordan

22 November 2016

STRICTLY PRIVATE & CONFIDENTIAL

Attarat Power Project : Overview

- Project Company: **Attarat Power Company (APCO)**
- **2 x 277 MW** (gross) **mine-mouth oil shale fired Power Plant**
- Located at Attarat Um Ghudran, 110km south-east of Amman
- 1st **solid fuel** power project in Jordan using local oil shale
- Project comprises of :-
  - Power Plant
  - Fuel & Mixing Yard
  - Oil Shale Mine
  - Water Wells
  - Housing Complex
  - Mine Infrastructure (mine camp, explosive storage etc)
- Signed **30-year Power Purchase Agreement ("PPA")** with National Electric Power Company (NEPCO) and guaranteed by Government of Jordan (GOJ)
- Total Project Cost : **USD 2.1 bn**
- Construction Period : **42 months**
- Target to reach Financial Close **by end of 2016** and commence construction works

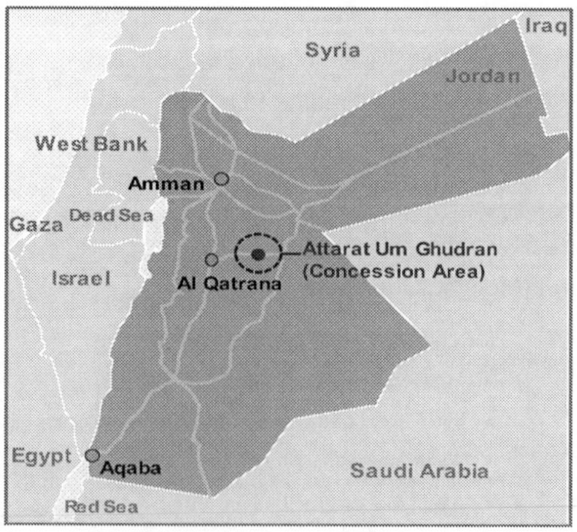

APCO

1

Project Sponsors

YTL POWER	• Malaysian-based international multi-utility group • Listed in Bursa Malaysia with a market capitalization of US$2.7 billion • 5,500 MW of owned and operated CCGT, coal & oil-fired power generation assets in Malaysia, Singapore and Indonesia • Successfully completed five years of O&M contract in Lebanon • 18 mn tonnes quarried annually (by affiliated cement business)
YUDEAN	• Guangzhou-based utility, state-owned Chinese company • Largest power generation enterprise in Guangdong Province with total power generation capacity c. 27,000 MW • Significant coal mining operations in China and overseas (c. 30 Mtpa) • International experience through 20% stake in Intergen (IPP developer) • Ownership: Guangdong Province (76%), China Huaneng (24%)
Enefit	• Largest oil shale-to-energy company worldwide (S&P BBB+; Moody's Baa2) • State-owned enterprise with c 7,500 employees • 2,300 MW of oil-shale fired power generation capacity • 18 mn tonnes of oil-shale mined annually

Significance of the Project

The APCO oil shale power project is a strategic energy project for Jordan:-

- **1st limited recourse project financed oil shale power project in the world**

  All oil shale projects in Estonia and other countries are corporate financed backed by company's balance sheet
- **Largest private sector infrastructure investment in Jordan**

  An inward private investment of more than USD2bn, and it creates more than 5,500 jobs during construction
- **1st baseload power plant in Jordan utilizing indigenous oil shale**

  An important first step towards energy independence for Jordan
- **Biggest mine operations in Jordan in terms of production capacity**

  About 10 millions tonnes oil shale used annually
- **One of the largest private sector overseas power project solely financed and underwritten by Chinese lenders**

  c.USD 1.6 bn of debt underwritten by ICBC and BoC
- **1st time oil shale is commercialized in Jordan and a major breakthrough for the local oil shale industry**

  The success of this project will provide confidence about the oil shale industry in Jordan and will be the cornerstone of future oil shale development in Jordan

STRICTLY PRIVATE & CONFIDENTIAL

Jordan – China Relationship

- Jordan and China have more than 30 years of diplomatic relationship and enjoyed good cooperation
- Jordan was the Honorary Guest at the 2015 China-Arab States Expo held in Yinchuan, China in Sept 2015
- Jordan and China signed MoUs for over USD 7 bn of projects during the visit of His Majesty King Abdullah to the China-Arab States Expo
- China is the largest single source of Jordan's non-oil imports. In 2014, the trade volume between the China and Jordan reached USD 3.6 bn
- Jordan is in a good position to leverage on future Chinese financing for future oil shale or other infrastructure projects

Signing of a strategic partnership pact in Beijing between His Majesty King Abdullah of the Hashemite Kingdom of Jordan and H.E. President Xi Jin Ping of the People's Republic of China
10 September 2015

His Majesty King Abdullah opens the Jordanian pavilion at the 2015 China-Arab States Expo in Yinchuan, China
11 September 2015

4

STRICTLY PRIVATE & CONFIDENTIAL

Advantages of Chinese Financing

- China is encouraging Chinese state owned companies (including commercial banks) to invest outside of China – this is consistent with the One Belt, One Road Policy proposed by the Chinese President Xi Jinping.

Advantages of Chinese Financing are as follows:-

- Chinese banks have huge liquidity and can provide enough volume of debt for mega projects
- Sinosure, the Chinese export credit agency which is part of China Ministry of Finance can provide political and commercial risk insurance cover
- Chinese banks are receptive to oil shale project as long as it fulfils international and local environmental requirements
- Relatively longer financing tenor compared to local and regional banks

There are, however, some challenges and some keys aspects for foreign investors to consider on Chinese financing.

5

General Due Diligence Considerations

- **Relationship of host country with China**
  - ✓ Good relationship between Jordan and China
- **Country's political stability**
  - ✓ Jordan is very stable under H.M. King Abdullah
- **Strong Chinese content**
  - ✓ Chinese EPC Contractor and Chinese investors

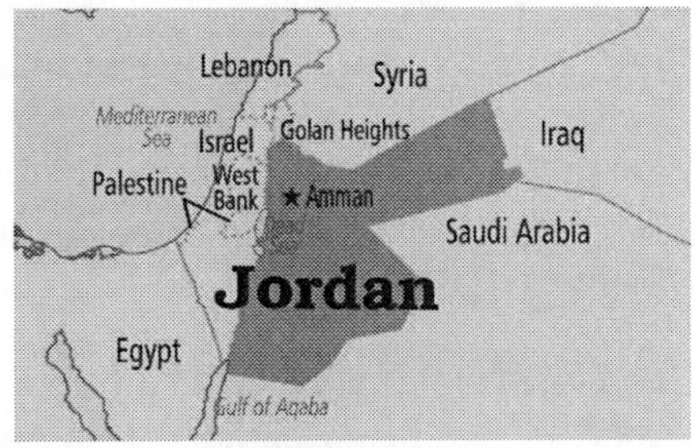

Others:

- Political will on the project
- Population and economic growth
- Legislation and environmental issues
- Energy supply demand
- Offtaker's strength
- Competitiveness in the market

Electricity Consumption of Middle Eastern Countries (kWh per capita)

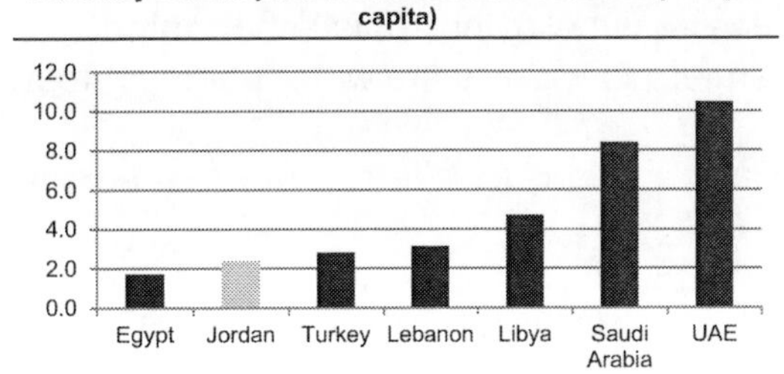

Source: World Bank (2012)

APCO

Oil Shale Resources

- **Oil shale is new to Chinese lenders**
- Resources proven by extensive drilling campaigns
- Oil shale resources certified by independent consultant using international JORC standard
- Detailed mine plan and block modeling
- Bulk sampling and quality testing
- Combustion test on the various oil shale samples

Bulk sample of oil shale at Attarat site

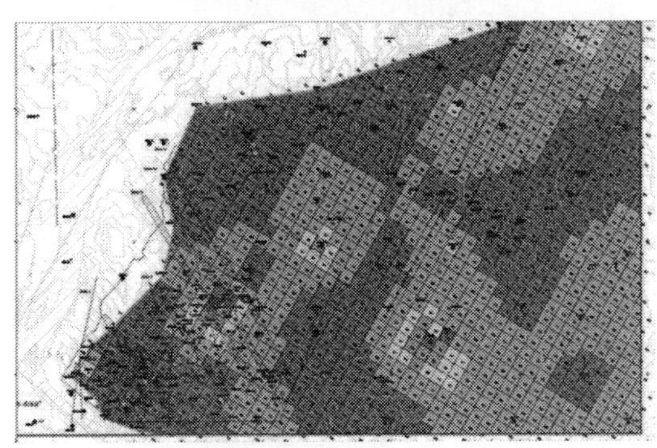

Detailed oil shale quality mapping at Attarat mining site

APCO

Water Availability

- **Concerns of scarcity of water in Jordan**
- Well drilling and testing of various aquifiers
- Longterm pumping tests
- Long term water yield certified by water expert
- Seeing is believing – it is important for lenders to witness the water pumping test
- Back up plans for alternative water supply

Long term water pumping test at Attarat site

STRICTLY PRIVATE & CONFIDENTIAL

Technology

- **Estonia has 100 years of shale experience**
- Select technology and equipment suppliers with track record
- Ability to scale up and identify potential risk and mitigation measures
- Perform combustion tests
- Witness running plant and oil shale mining operation
- Strong EPC Contractor with parent company support

Balti Power Plant in Estonia, 765 MW

Eesti Power Plant in Estonia, 1615 MW

Equity Partners

- Preferable to have a strong Chinese partner
  - *Shares long term values and willing to cooperate with foreign investors*
- Familiar with undertaking international projects
  - *Familiar with international practice and capable to communicate in English*
- Strong financials and relevant technical experience
- Equity partners must be able to demonstrate strong confidence to the lenders
- Advisable not to have too many partners

Signing with our new equity partner Yudean

Financing MoU signing at Yinchuan attended by all equity partners and banks' representatives

APCO

Interesting Photos in China

?

??

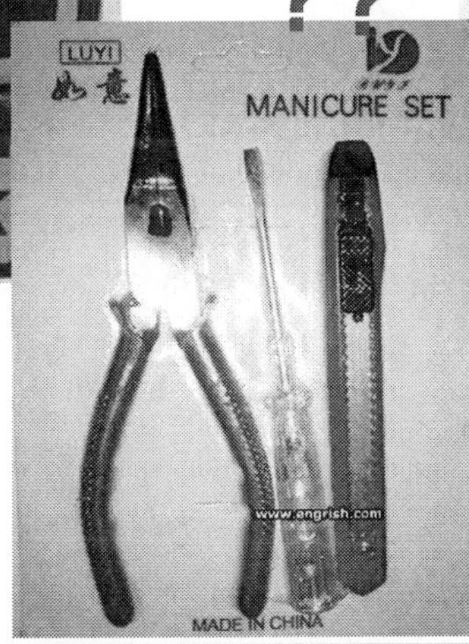

?

APCO

Chinese Culture

- Chinese culture is indirect and vague
- No problem does not mean everything is okay, body language and tone are very important
- Must have a coordinator who speaks and understands Chinese language and culture
- Lenders' internal reports and due diligence are in Chinese (Mandarin)

- Chinese lenders are less market driven and are looking at many projects
- Chinese lenders are less experienced on project financing because Chinese financing for overseas projects are usually done on a G-to-G level
- Project progress requires continuous monitoring and follow up, and to provide guidance and assistance to their due diligence work
- Most importantly, must have the 2Ps – patience and perseverance.

APCO
ATTARAT POWER COMPANY

END OF PRESENTATION

Thank You شكرا 谢谢

Third Jordan International Oil Shale Symposium - 2016

"Enhancement of Oil Shale Viability and Economics"
Case Study: Retorting and Direct Burning Projects
Eng. Munther Bsieso

AL QAMAR

Oil Shale Composition

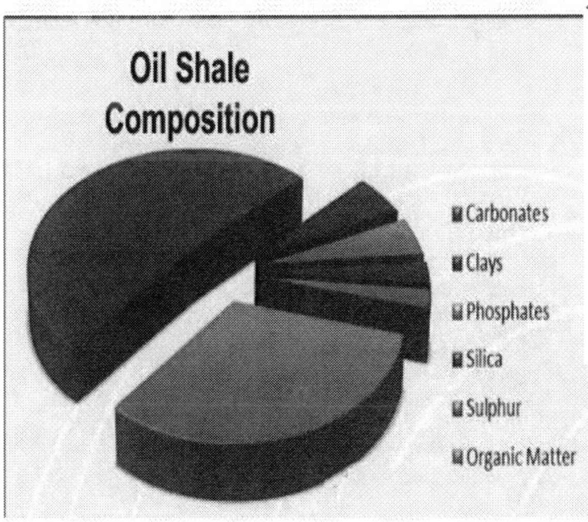

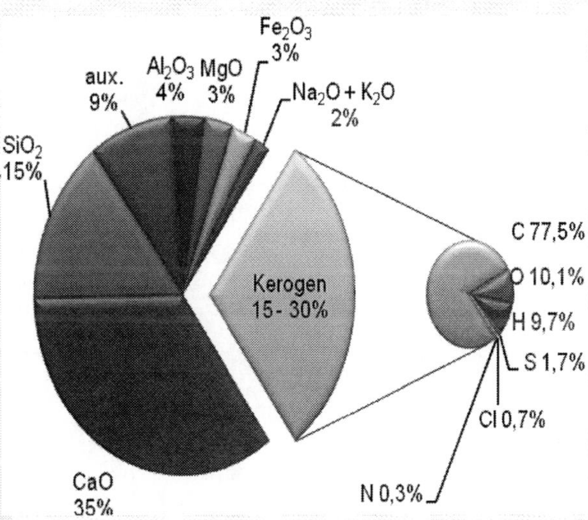

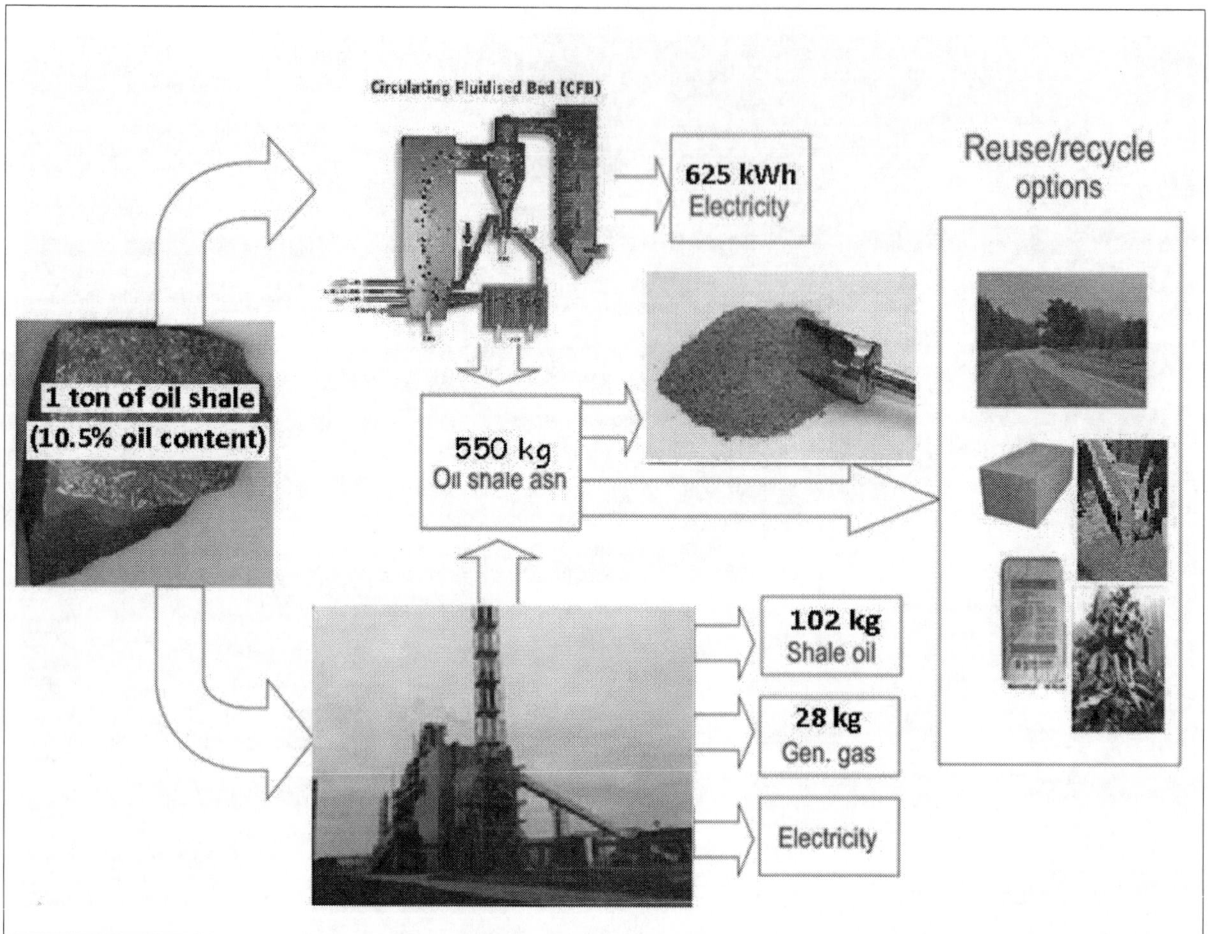
Circulating Fluidised Bed (CFB)
625 kWh
Electricity
Reuse/recycle
options
1 ton of oil shale
(10.5% oil content)
550 kg
Oil shale ash
102 kg
Shale oil
28 kg
Gen. gas
Electricity

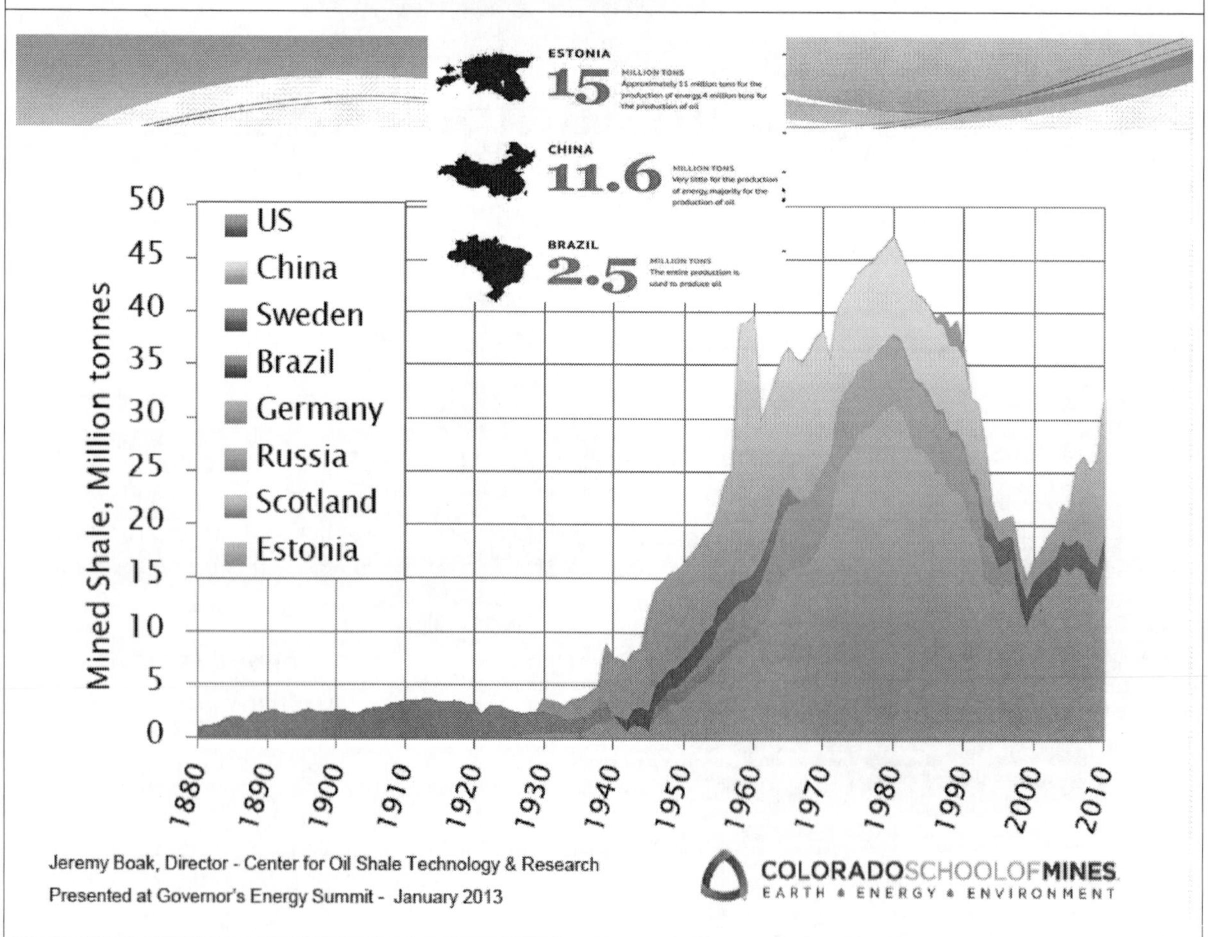
ESTONIA
15
MILLION TONS
CHINA
11.6
MILLION TONS
BRAZIL
2.5
MILLION TONS
The entire production is
used to produce oil
Mined Shale, Million tonnes
50
45
40
35
30
25
20
15
10
5
0
US
China
Sweden
Brazil
Germany
Russia
Scotland
Estonia
1880
1890
1900
1910
1920
1930
1940
1950
1960
1970
1980
1990
2000
2010
Jeremy Boak, Director - Center for Oil Shale Technology & Research
Presented at Governor's Energy Summit - January 2013
COLORADOSCHOOLOFMINES.
EARTH • ENERGY • ENVIRONMENT

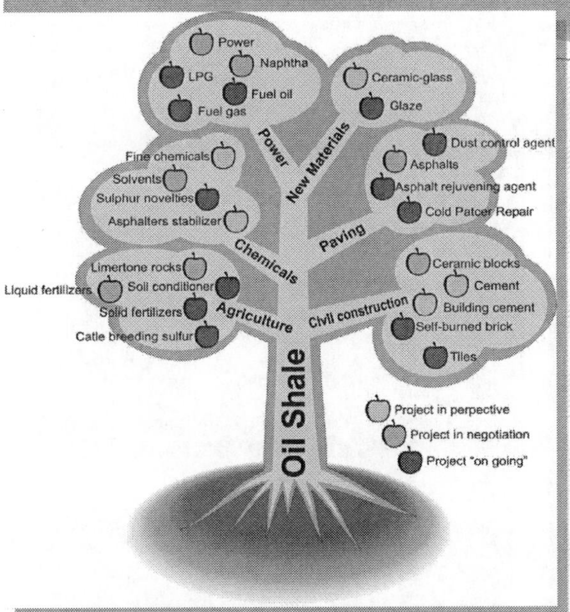

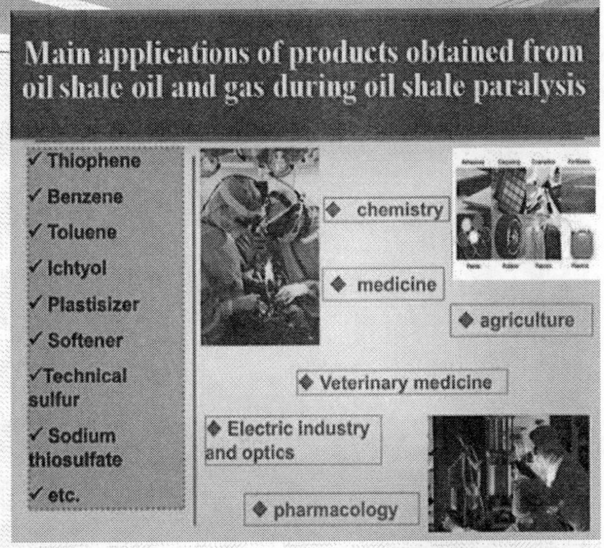

- Additional line of petrochemical equipment will be required with cost about US$ 5 million for each UTT-3000. Output products are: thiophene at selling price of 50,000 $/ton. Thiophene group products yield one ton per 1000 tons oil shale or 1000 tons per year from each UTT-3000.

In-Place Resources

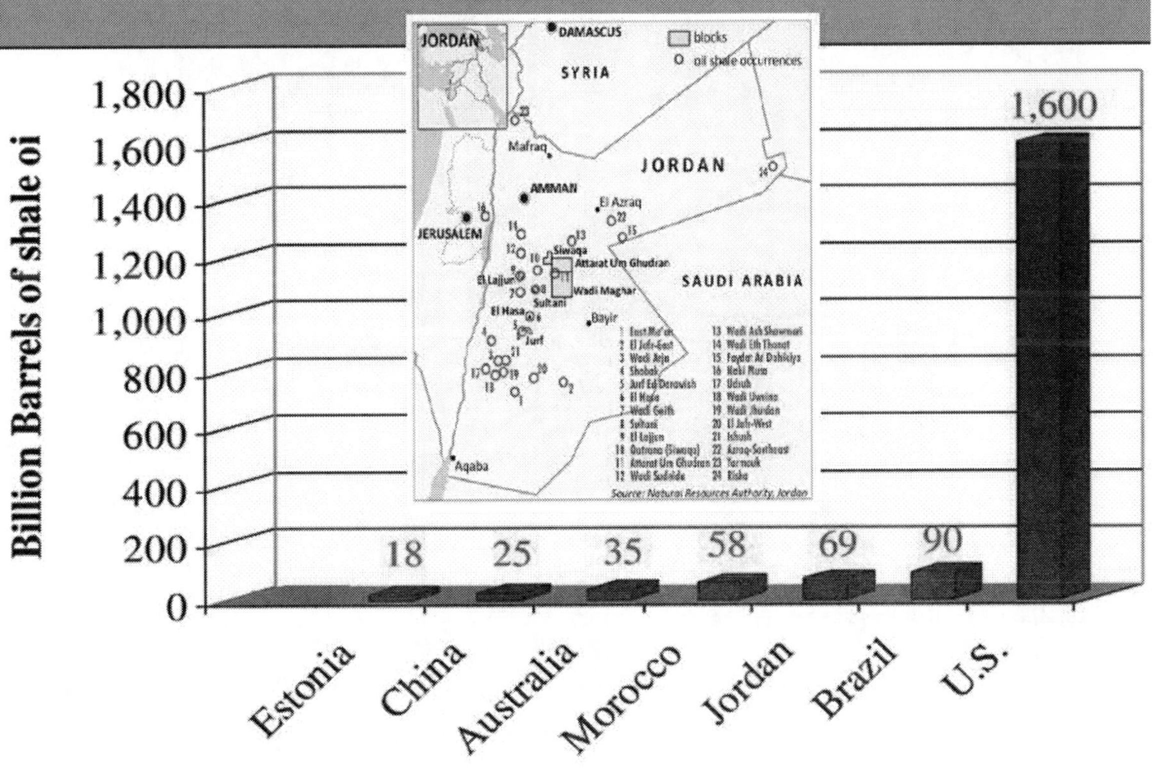

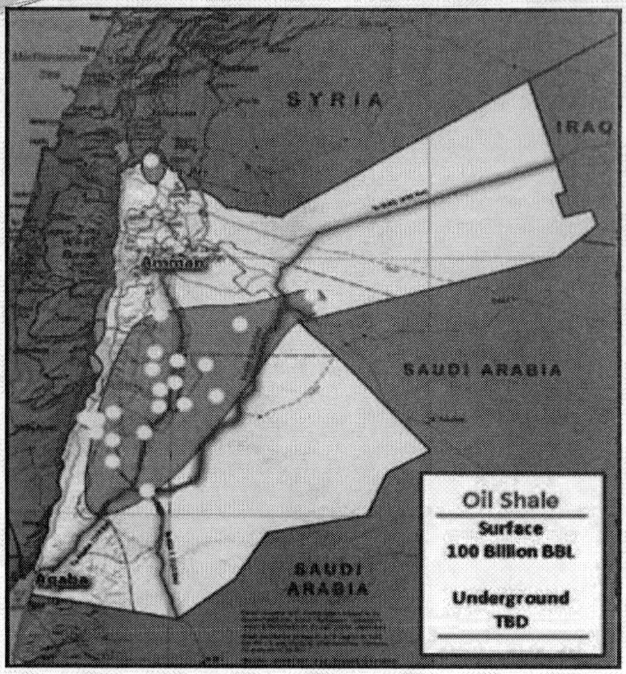

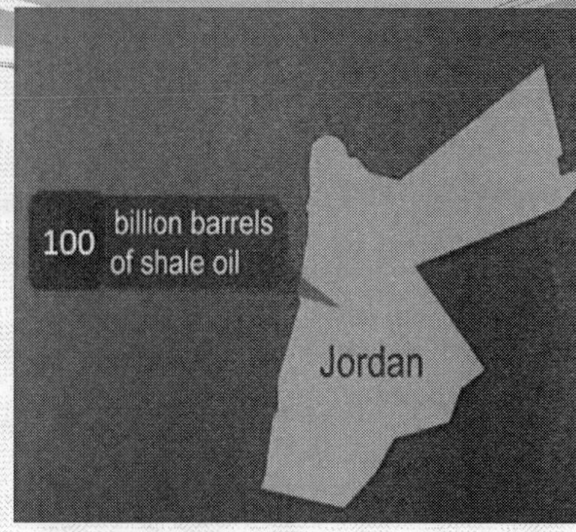

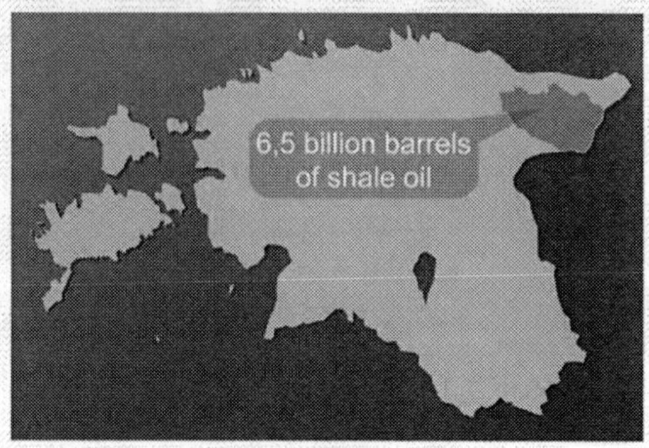

Oil Shale Characteristics

Brazil	U.S.A	Euorope	China	Estonia	Morocco	Jordan	
1340	1800	1986	1500	2000	1100	1800	C.V (k cal/kg
7	9.7	13	7	20	7	12	oil content (%)
2.6	17.3	20	12	19	11	19	Organic Carbon(%)
1.9	1.7	2.7	2	2.5	2	2	H2 (%)
2.7	2.1	11.5	7	7.5	9.6	1.8	O2 (%)
0.6	1.3	0.2	0.38	0.2	_	0.34	N2 (%)
0.6	0.65	6	0.3	1.55	1.2	3.2	S (%)
5.3	0.7	12.7	13	11.5	10	2.5	H2O (%)
79.8	68.3	45	71	44	64	55	Ash

"it is so essential for the Government to speed up the search for other sources of energy other than importing oil, and in front of that is to complete the imple-mentation of oil shale projects and any other sources of energy"

Quoted from the Royal speech of His Majesty King Abdullah directed to the Prime Minister on April 13th, 2006

Jordan, China Sign Agreements Worth $1.7 billion For Oil Shale Deal

Jordan faces significant challenges to meet future demand for energy. These challenges are caused by rising demand for foreign oil and other petroleum products.

Jordan's economy is highly dependent on imported oil and gas .

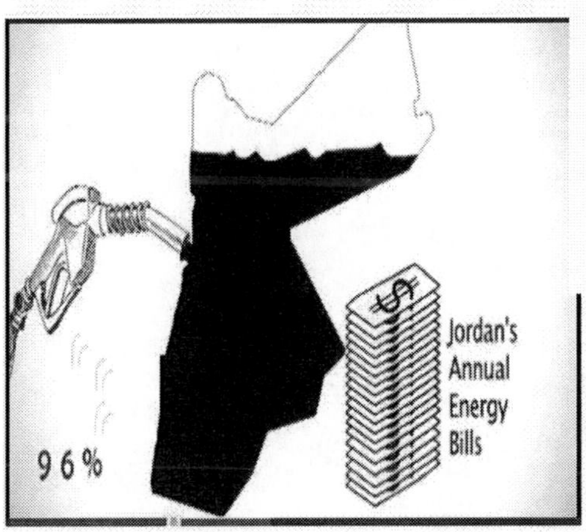

Energy Master Plan in Jordan - 2020

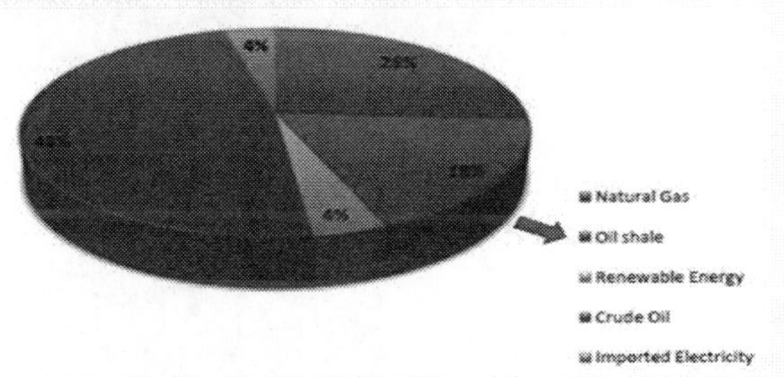

Jordan's Electricity Supply by Fuel Type

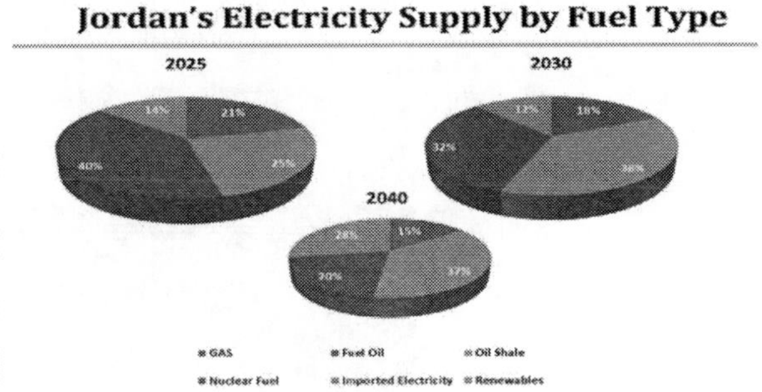

Phosphate

توقيع مذكرة تفاهم بين الفوسفات والكرك الدولية للبترول لشراء منتجاتها من الوقود الثقيل

The deal will enable the JPMC to extract raw phosphate found under oil shale rocks

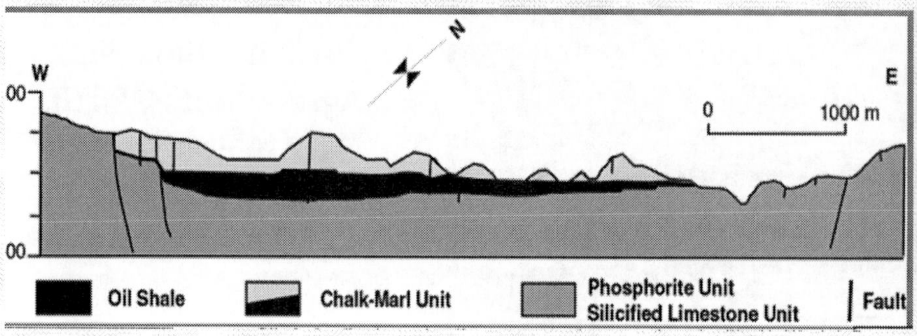

Marble

Limestone

	Layers										PPM				H.H.V		
	down to up	thickness m	MgO %	SiO2 %	CaO %	P2O5 %	S * %	P2O5 * %	Al2O3 %	Fe2O3 %	V	Pb	Cr	Cd	Kcal/Kg	Ash %	Thick.%
1	A	8	0.83	43.25	38.39	2.87	6.15	2	2.46	1.27	288	3.9	337	41.5	2343.2	69.73	36%
2	B	6	1.19	18.12	38.93	5.04	4.97	3.32	2.42	1.3	393	4.1	336	92.4	2724.8	65.8	27.50%
3	C	4.5	1.11	19.21	25.32	3.8	2.25	2.79	2.72	1.27	328	2.3	175	80	1876.6	73.4	20.5%
4	D	3.5	0.78	13.25	41.62	5.68	1.97	3.92	2.66	1.34	277	1.6	357	20.7	1908.9	68.94	16.0%
	Avgerage		0.98	26.61	36.38	4.11	4.36	2.83	2.53	1.29	323	3.26	306.7	60.1	2283.0	69.3	100%

In general, oil yield and/or the calorific value of Jordanian oil shales are relatively high, approximately 10 % by weight and about 5–6 MJ kg^{-1}, respectively [15]. Based on recent experimental investigations, the average oil yield is around 100 litres per tonne of shale, depending on the pyrolysis conditions and type of shale [16].

Besides the organic matter, Jordanian oil shale deposits contain metals, such as uranium, molybdenum, vanadium, chromium, cobalt and nickel, in low concentrations, as well as aluminium and iron in higher concentrations [17]. These metals have strategic significance in the international market. A preliminary estimate is that rare metals worth at least 30 US$ per tonne

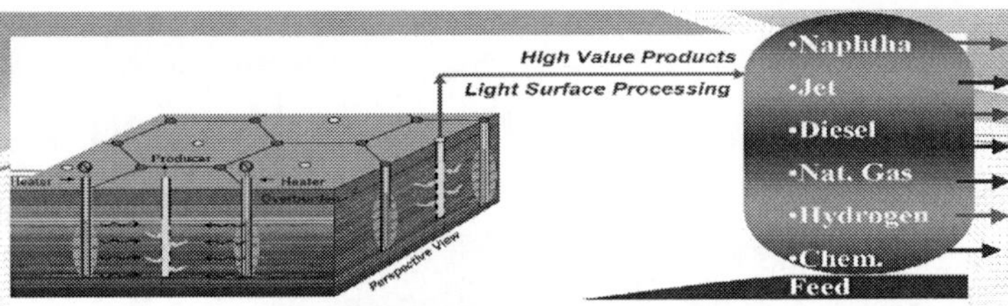

Jordanian Oil Shale VS Estonian Oil Shale Key Figures

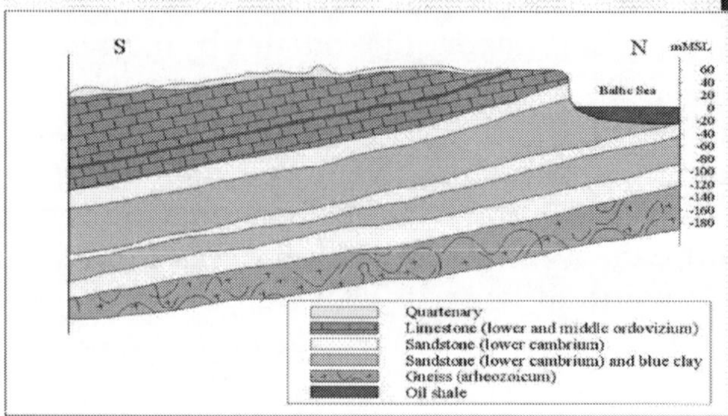

	Estonia	Jordan
Oil Shale Layers (meters)	3	50 - 75
Stripping Ratio	10:1	1:1
Mining & Crushing Costs ($/T)	35	7.22
Oil Shale Prices - Taxes ($/T)	9.5 (Prices)	2.1 (Taxes)
Rate of Pollution Charges for ash land filling ($/T)	3.3	0
Rate of Pollution Charges for Particulates ($/T)	160.1	0
Oil Shale Reserves (billion barrels)	6.5	100

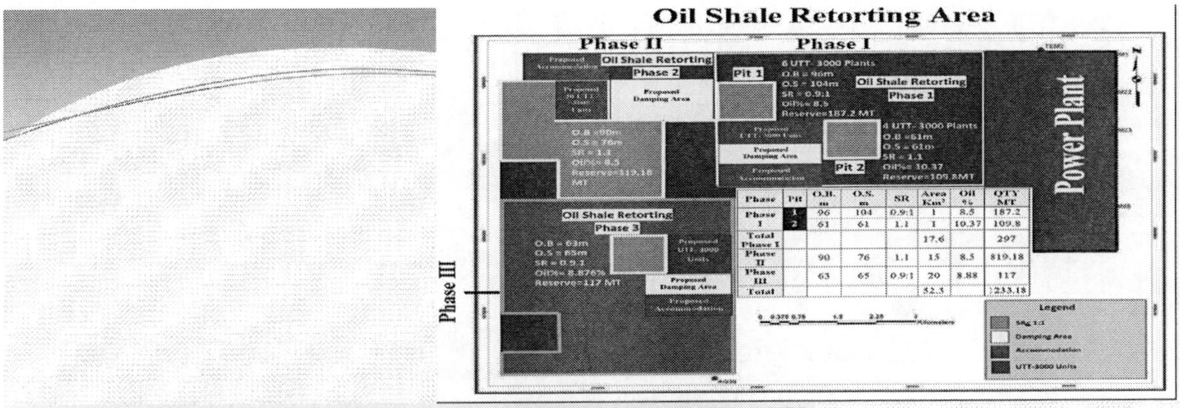

Phase	Pit	O.B. m	O.S. m	SR	Area Km²	Oil %	QTY MT
Phase I	1	96	104	0.9:1	1	8.5	187.2
	2	61	61	1.1	1	10.37	109.8
Total Phase I					17.6		297
Phase II		90	76	1.1	15	8.5	819.18
Phase III		63	65	0.9:1	20	8.88	117
Total					52.3		1233.18

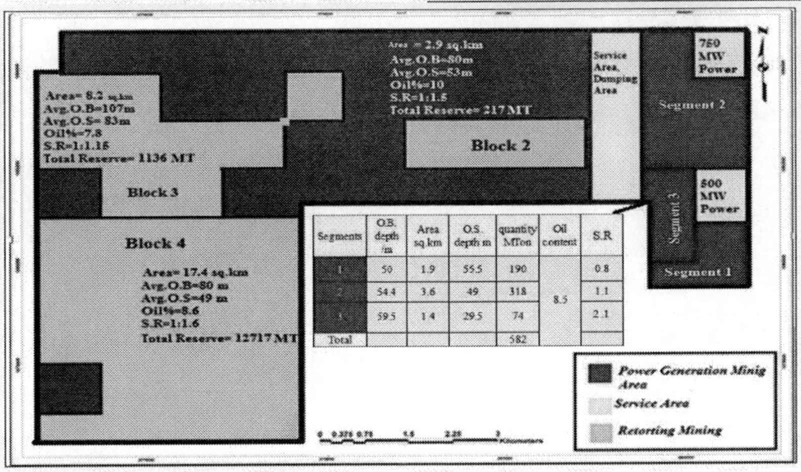

Segments	O.B. depth m	Area sq.km	O.S. depth m	quantity MTon	Oil content	S.R
1	50	1.9	55.5	190	8.5	0.8
2	54.4	3.6	49	318		1.1
3	59.5	1.4	29.5	74		2.1
Total				582		

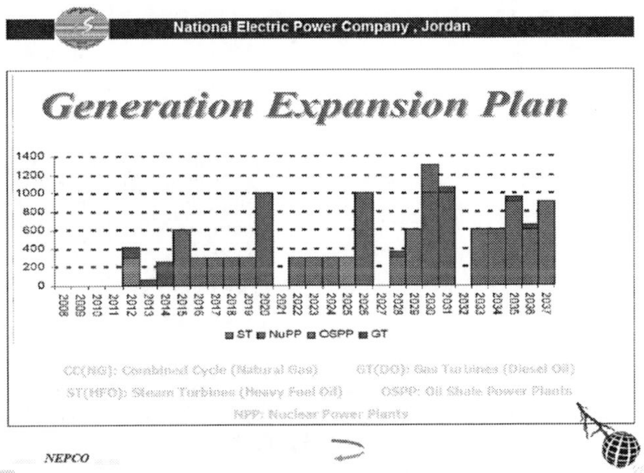

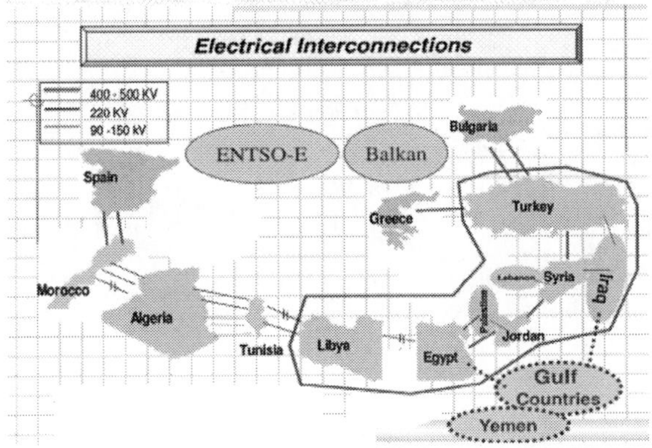

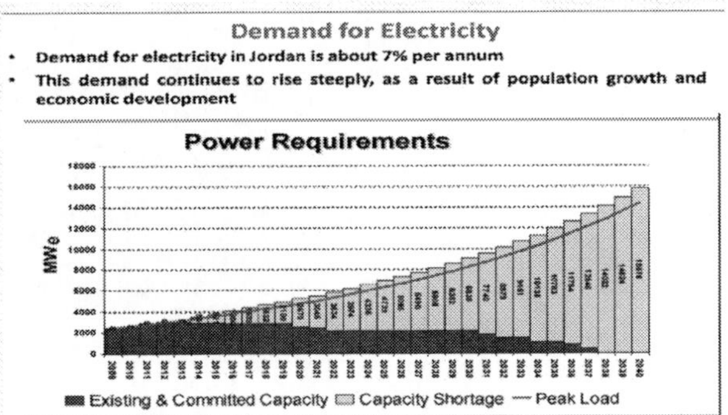

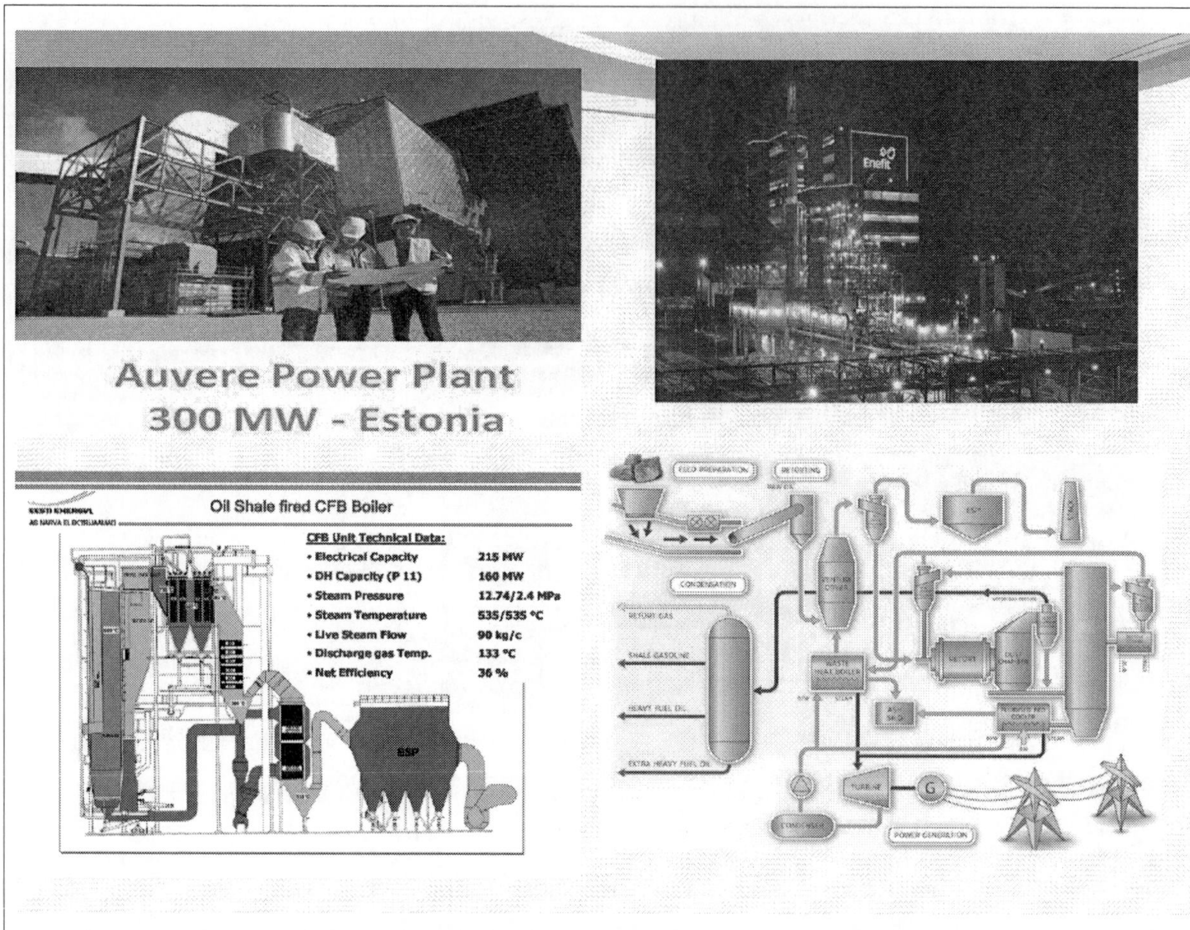
Auvere Power Plant
300 MW - Estonia
Enefit
Oil Shale fired CFB Boiler
CFB Unit Technical Data:
• Electrical Capacity 215 MW
• DH Capacity (P 11) 160 MW
• Steam Pressure 12.74/2.4 MPa
• Steam Temperature 535/535 °C
• Live Steam Flow 90 kg/c
• Discharge gas Temp. 133 °C
• Net Efficiency 36 %
ESP
FEED PREPARATION
RETORTING
CONDENSATION
RETORT GAS
SHALE GASOLINE
HEAVY FUEL OIL
EXTRA HEAVY FUEL OIL
WASTE HEAT BOILER
RETORT
TURBINE
G
CONDENSER
POWER GENERATION

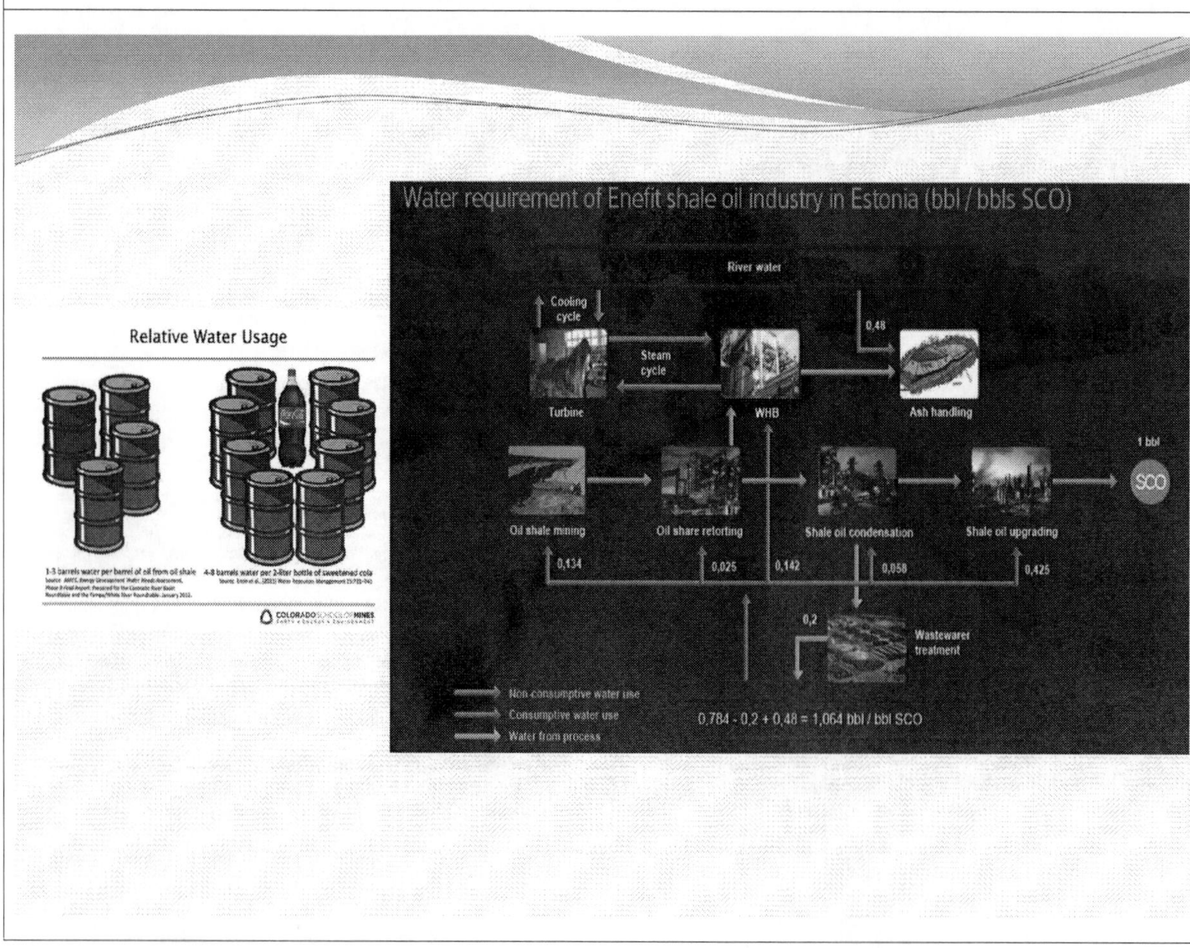
Relative Water Usage
1-3 barrels water per barrel of oil from oil shale
4-8 barrels water per 2-liter bottle of sweetened cola
COLORADO SCHOOL OF MINES
Water requirement of Enefit shale oil industry in Estonia (bbl / bbls SCO)
River water
Cooling cycle
Steam cycle
0,48
Turbine
WHB
Ash handling
1 bbl
SCO
Oil shale mining
Oil share retorting
Shale oil condensation
Shale oil upgrading
0,134
0,025
0,142
0,058
0,425
0,2
Wastewarer treatment
Non-consumptive water use
Consumptive water use
Water from process
0,784 - 0,2 + 0,48 = 1,064 bbl / bbl SCO

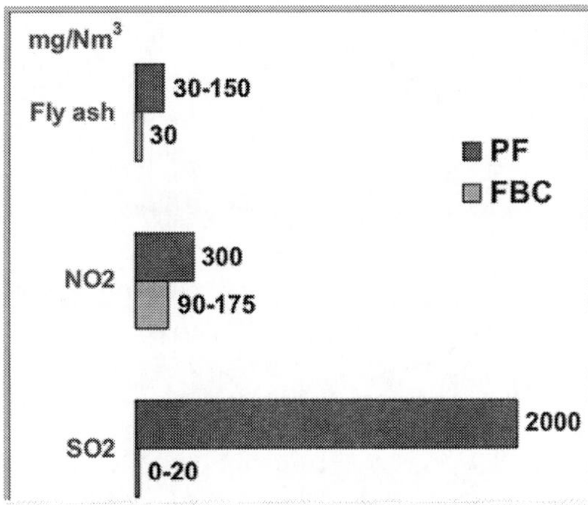

Subject of Impact	Source	Contaminants	Note, concentration
Atmosphere	Chimneys	Dust	Meets to EU norms
		SO2	Meets to EU norms
		CO	Meets to EU norms
		CO2	Meets to EU norms
		Nox	None
		Hydrocarbon	Within allowed limits for refineries

Emissions

Dust < 30 mg/Nm^3
NO_x < 200 mg/Nm^3
SO_2 < 50 mg/Nm^3
CO_2 2.1 t/t shale oil

In Jordan, One Tone – 0.89 Barrel

Material Balance	Mass Balance (kg/t oil shale)	UTT Balance (t/hr)	UTT Balance (t/Day)
Input			
Raw oil shale	1000	139	3336
Air	1243	172.8	4146.9
Water	495	68.8	1650.9
Total input	2738	380.6	9133.8
Output			
Flue gas	1424	197.89	4749.48
Ash	717	99.65	2391.68
Steam	495	68.79	2391.68
Shale oil	**102**	14.24	341.71
Gas	**28**	3.94	94.49
Total output	2766	384.51	9228.33

130 kg / 0.92 = 141.3 liters

141.3 / 159 = 0.89 bbl

UTT unit capacity is 3336 ton/day

Daily oil yield of one unit of UTT-3000 is 3336 X 0.89 = 2963 bbl/day

Economics

Key figures	
Capital costs for 1 x UTT-3000 unit	~ $ 100,000,000
Construction time	2 years
Raw oil shale processing	1,000,000 t / year
Energy consumption	5 MW
Water consumption	10 cu.m / hour
Energy produced	~ 160,000 MWh / year
Oil produced	~ 120,000 t / year ~ 780,000 bbl / year
Ash produced (can be used for cement production)	~ 500,000 t / year

160000MWh per year = 280000bbl of oil equivalent

Capital Cost Calculation (UTT-3000)

- Unit Capital Cost = \$100 million
- C.R. Factor (8%, 25 yrs) = 0.094
- Annual Investment = \$ 9.4 million
- Oil Yield is 3336 X 0.89 = 2969 bbl/day
- Annual Production=0.975 million bbl
- Investment cost=\$ 9.4 million/0.975 million bbl

\$ 9.64 per barrel

- Cost of mining, as reported by one local mining company is \$6.5 per cubic meter or \$3.6/ton
- SR 1:1 (remove one ton of overburden and one ton of oil shale, cost is \$7.2 per ton.
- The crushing and transportation costs are \$1.00 per ton.
- Total mining cost is \$8.2 per ton.
- **Mining cost per barrel is \$8.2 /0.89 = \$9.2**

- Annual O&M Cost is 3% of Capital Cost

\$100 m X 0.03 = \$ 3 million per year

3/0.97 (annual oil production) = \$3.08 per barrel

- UTT Royalty (3 days revenue) = \$0.545 /bbl

(Assuming selling oil price \$ 60/bbl)

- Miscellaneous = \$2.5 /bbl
- Risk factor = \$1.0 /bbl

TOTAL PRODUCTION COST before taxes is

\$ 26 /bbl

Taxes = \$6 /bbl

Total Production Cost = \$32/bbl

Oil Shale Retorting Projects are Based on Royalty and Petroleum Tax Regime

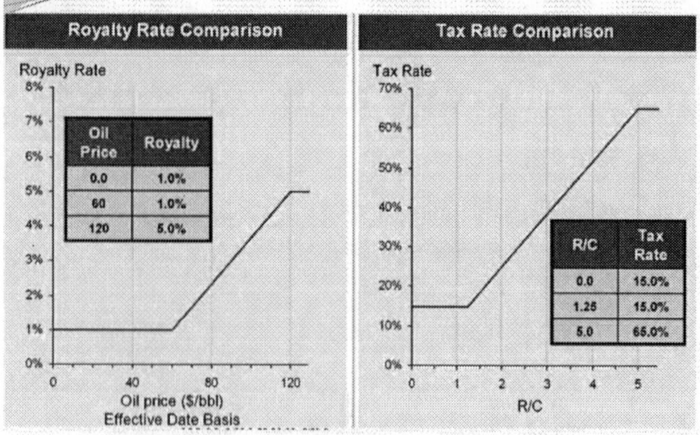

Assumptions:

Production Cost (before Taxes) = 26

Selling Price 60

Profit before taxes= 34

- Petroleum Taxes
- R factor =Revenue/Cost = 34/26= 1.31
- Petroleum Taxes percentage = (13.3 X R/C) – 1.66
- Petroleum Taxes percentage = (13.3 X 1.31) – 1.66 = 15.76%
- Petroleum Taxes = (15.76 X \$34/b)/100 =\$5.36
- Royalty Taxes Formula = (Oil price/15)-3
- Royalty Taxes (%) = (60/15)-3 = 1.0%
- Royalty Taxes (value) = (1.0X \$60/b) / 100 = \$0.6
- Total Taxes = 5.36 + 0.6= \$5.96/b

R/C	Petroleum Tax Rate %
R/C < 1.25	15
1.25 = or < R/C < 5	13.33 X R/C - 1.66
5= or <R/C	65

Oil Price (\$/Bbl)	Royalty %
0 < O.P < 60	1
60 = or < O.P < 120	(O.P/15) - 3
120= or <O.P	5

C: Cost

R: Revenue

O.P: Oil Price

UTT-3000 Retorting Plant		
Retort max. capacity of oil shale	tons per day	3336
Produced oil equivalent per year	Thousand barrels	973.837
Mining Costs (Crushing & Transportation)	US$/ton	8.220
Plant Capital Cost	Million US$	100
Operation & Maintenance Cost	% of capital cost	3
Cost Breakdown per Barrel		
Investment cost	US$/ barrel	9.620
Operation & Maintenance Cost	US$/ barrel	3.081
Mining Costs	US$/ barrel	9.250
Miscellaneous	US$/ barrel	2.500
Risk	US$/ barrel	1.000
UTT Royalty (3days revenue)	US$/barrel	0.545
Cost per Barrel (before Tax)	**US$/ barrel**	**25.996**
Suggested Oil Price	**US$/ barrel**	**60**
Total Taxes per Barrel	US$/ barrel	5.965
Barrel Cost After Tax	**US$/ barrel**	**31.960**
Net Profit	**Million US$**	**31.246**
Internal Rate of Return (Project)		**27.4%**

Production Cost Breakdown ($/bbl)

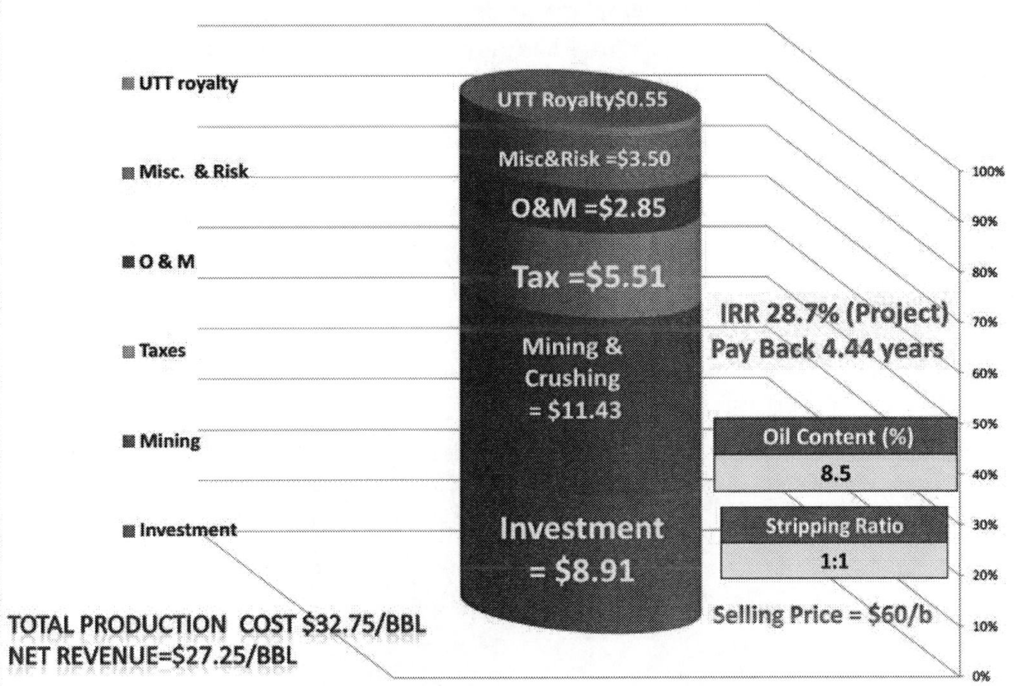

Selling Price $50/Barrel

Retort max. capacity of oil shale	tons per day	3336
Produced oil equivalent per year	Thousand barrels	973.837
Mining Costs (Crushing & Transportation)	US$/ton	8.220
Plant Capital Cost	Million US$	100
Operation & Maintenance Cost	% of capital cost	3
Cost Breakdown per Barrel		
Investment cost	US$/ barrel	9.620
Operation & Maintenance Cost	US$/ barrel	3.081
Mining Costs	US$/ barrel	9.250
Miscellaneous	US$/ barrel	2.500
Risk	US$/ barrel	1.000
UTT Royalty (3days revenue)	US$/barrel	0.455
Cost per Barrel (before Tax)	**US$/ barrel**	**25.905**
Suggested Oil Price	**US$/ barrel**	**50**
Total Taxes per Barrel	US$/ barrel	4.114
Barrel Cost After Tax	US$/ barrel	30.019
Total Taxes	Million US$	4.007
Net Profit	**Million US$**	**23.309**
Internal Rate of Return (Project)		**20.9%**

Two units UTT-3000 were put into operation in 1987 at the Power Plant at city Narva, Estonia (in the time of former USSR).
It demonstrate successful experience of thermal processing of Baltic oil shale as well as of waste automobile tires, solid and liquid organic waste in a mixture with oil shale.

The result is the obtaining of high calorific liquid and gaseous fuels as well as the very valuable chemical products.

The production cost of one barrel of shale oil constitutes from 14 to 17 US$ that allows to construct the competitive industrial units practically with unlimited number.

One ton of Lajjun oil shale can produce 625 kWh of electricity

Oil shale specific consumption per kWh $= \frac{Heat\ Rate}{Calarofic\ Value} = \frac{2389}{1500} = 1.6$ **Kg /kWh.**

Heat Rate $= (860 \frac{Kcal}{KWH} \div 0.36) \approx 2389$

Each one ton (1000 kg) / 1.6 = 625 kWh

Capital Cost Rated 300 MW
$ 2273 per KW installed.

Lahmeyer Results – CFB options

100 MW

Project Cost Summary	Estimated Cost US$
I Specialized Equipment	136,917,850
II Other Equipment	46,867,410
III Civil	7,712,665
IV Mechanical	36,818,826
V Electrical Assembly & Wiring	4,626,401
VI Buildings & Structures	5,434,492
VII Engineering & Plant Startup	12,509,934
VIII Contractor's Soft & Miscellaneous Costs	34,448,285
Contractor's Price	285,335,853
Cost per kW	2,857.8

Criteria	100 MW	200 MW	300 MW
Capital Cost (US$M)	285	461	681
Capital Cost (US$/kW)	2858	2305	2273
Power Sales Costs	As per cash flow analysis	As per cash flow analysis	As per cash flow analysis

ALSTOM total cost of power plant Rated 600 MW - Euros 900M ($1750 per installed KW)

Estonian Oil Shale Electricity Production

In 2003, the Unit 11 of Balti Power Plant was reconstructed (Foster Wheeler) to use the circulated fluidized bed combustion (CFBC) technology, which is more efficient and environmental-friendly (lower SO2 and CO2 emissions) than PC technology.

On 14 January 2011, Narva Power signed a contract with the French power engineering company ALSTOM for construction of the new power plant next to existing Eesti Power Plant. The €950 million contract foresees construction of two 300 MW oil shale-fired units. The units will utilize circulating fluidized bed boiler technology.

According to the contract, a first unit of 300 MW will cost €540 million. There is an option for a second unit of 300 MW for €410 million. The first unit will be completed by 2015. A decision on construction of the second unit will be made in 2012

(Cap. Cost X CRF) + Fuel Cost + O&M Cost / Annual Electricity Output

Assumptions

Power plant capacity = 100 MW

Capital cost = $2500 / kW installed

Capital Recovery Factor (8%, 25 yrs) = 0.094

Stripping ratio = 1:1

Oil Shale Heating Value = 1500 k cal/kg

Mining , Crushing and Tax Costs = $9.82 per ton

O&M costs = 3% of Capital cost

Fuel Costs = 9.82 X 1.6 = 1.56 cents / kWh

Investment Costs =($250 m X 0.094) /700800 MWh = 3.35 cents / kWh

O&M Costs = $ 250 m X 0.03 /700800 MWh = 1.07 cents / kWh

Total Production Costs = 6 cents / kWh

Item	Unit	Value	
Plant capacity	Mega Watts	500	
Efficiency	%	36	
Load Factor	%	80	
Annual Power Production	MWh	3,504,000	
Oil Shale Calorific Value	Kcal/Kg	1500	
Stripping Ratio		1	
Oil Shale Mining, Crushig and Taxes	US$/ton	9.82	
Oil Shale Specific Consumption	Kg/KWh	1.593	
Annual Oil Shale Consumption	Million tons	5.583	
Project Lifetime	Years	25	
Discount Rate	%	8	
Capital Recovery Factor		0.094	
Construction Period	Years	3	
Capital Investment	US$/KW	2500	
Total Plant Capital Cost	Million US$	1250	
Operation And Maintenance Cost	% of capital cost	3	
Total Annual Cost	**Million US$**	**209.425**	
Of which			%
Investment Cost	Million US$	117.098	55.914
Fuel Cost (Mining, Crushing and taxes)	Million US$	54.827	26.180
Operation And Maintenance Cost	Million US$	37.500	17.906

Total Production Cost Sent Out	**c/KWh**	**5.977**
Of which		
Investment Cost	c/KWh	3.342
Fuel Cost (Mining, Crushing and taxes)	c/KWh	1.565
Operation And Maintenance Cost	c/KWh	1.070
Power Sale Price to NEPCO	c/KWh	13.5
Net Revenue	c/KWh	7.523
	Profits	
Project Annual Net Profit	Million US$	263.615
Income Tax (24%)	Million US$	63.268
Net Profit After Tax	Million US$	**200.347**
	Internal Rate of Return	
Shareholders	Project pre tax	Project post tax
24%	18%	14%
Payback period (Project Post Tax)	**Years**	**7.39**

المحطة	الوحدة	نظام (CFB)
القدرة	م.و	300
معامل القدرة	%	85
الكفاءة	%	36
الطاقة المولدة	ج.و.س /سنة	2,234
استهلاك الوقود	غم /ك.و.س	2,000
العمر الاقتصادي	سنة	25
سعر الخصم	%	8
سعر الوقود	دولار /طن مكافئ	0
تكلفة الوقود	سنت / ك.و.س	0.40
مدة الإنشاء	سنة	3
تكلفة المعدات	دولار /ك.و	1,400
تكاليف الإنشاء	مليون دولار	443
الفائدة اثناء التركيب	مليون دولار	54.61
التكلفة الرأسمالية	مليون دولار	498
التكلفة الرأسمالية	دولار /ك.و	1,660
معامل استرداد رأس المال		0.09
التكلفة الرأسمالية	سنت / ك.و.س	2.089
التشغيل والصيانة	سنت / ك.و.س	0.63
تكلفة التوليد الإجمالية	سنت / ك.و.س	3.12

NEPCO's Sensitivity Analysis (CC: Combined Cycle NG, ST; Steam Cycle HFO: OSPP: Oil Shale Power Plant: Nupp: Nuclear Power Plant Uranium Fuel)

Type:	Unit	CC	ST	OSPP	NuPP
Capacity:	MW	380	300	300	400
Capital Cost:	US$/KW	550	900	1650	2300
Life:	Years	25	30	30	40
Construction Time:	Years	2.5	3	3	5.5
Capacity Factor:	%	85%	85%	85%	85%
IDC:	%	8.82%	10.69%	10.69%	20.65%
IDC:	US$/KW	49	96	176	475
Total Capital Cost	US$/KW	599	996	1826	2775
Total Capital Cost	MUS$	227	299	548	1110
Discount Rate:	% p.a	8%	8%	8%	8%
Annual Capital Cost	US$/KW	$56	$88	$162	$233
Annual Capital Cost	MUS$	$21	$27	$49	$93
Fixed O&M Cost	US$/KW	2.4	6.6	7.2	72
Total Annual Cost	US$/KW	$58	$95	$169	$305
Efficiency:	%	46%	38%	37%	40%
Fuel Type:		NG	HFO	OS	Nu
fuel Consumption:		7414	233	1571	
Fuel Price:		4	316.9	10	
Fuel Cost:	US$/KWH	0.030	0.074	0.016	0.008
Variable O&M Cost	US$/KWH	0.0007	0.0008	0.0009	0.0014
Total Variable Cost	US$/KWH	0.030	0.075	0.017	0.009
Levelised Annual Cost	US$/KWH	0.007852	0.012771	0.022755	0.040922
Generation Cost:	**Cents/KWH**	3.8	8.8	3.9	5.0
Generation Cost:	**Fils/KWH**	27.1	62.1	27.9	35.7

Type:	Unit	CC	ST	OSPP	NuPP
Generation Cost:	**Cents/KWH**	**3.8**	**8.8**	**3.9**	**5.0**
Generation Cost:	**Fils/KWH**	**27.1**	**62.1**	**27.9**	**35.7**

Extreme Case

Item	Unit	Value
Plant capacity	Mega Watts	500
Efficiency	%	34
Load Factor	%	70
Oil Shale Calorific Value	Kcal/Kg	1000
Stripping Ratio		4
Oil Shale Mining, Crushing and Taxes	US$/ton	20.65
Oil Shale Specific Consumption	Kg/KWh	2.531
Construction Period	Years	3
Capital Investment	US$/KW	3000
Operation And Maintenance Cost	% of capital cost	4
Total Production Cost Sent Out	**c/KWh**	**11.766**
Of which		
Investment Cost	c/KWh	4.583
Fuel Cost (Mining, Crushing and taxes)	c/KWh	5.226
Operation And Maintenance Cost	c/KWh	1.957
Power Sale Price to NEPCO	c/KWh	13.5
Net Revenue	c/KWh	1.734

- NEPCO has reached agreement with APCO including a 30 year Power Purchase Agreement ("PPA") under which NEPCO agrees to purchase the entire dependable capacity and net energy output of the Plant.
- The price per kilowatt hour that the government will buy from the plant will be levelised, ranging from a minimum of 78 fills per kilowatt hour to a maximum of 99 fills per kilowatt hour.

Ministry of Energy and Mineral Resources
The Hashemite Kingdom of Jordan

Home • Tenders • Useful Links • Complaints and Suggestions • Careers

Search عربي

A+ A A-

The Ministry | Legislations | Projects | E-Services | Media Center | Open Data | Contact Us

Projects Mineral Resources

Projects

Oil and Gas

Petroleum and Oil Shale Projects

Petroleum Exploration Blocks in Jordan

Direct combustion to produce electricity Projects

Attarat Energy Company (Estonian / Malaysian / Jordan):

Oil shale direct combustion agreements were signed with Al Attartat Power Company/ the Estonian/ Malaysian /Jordanian company, with a capacity of 470 MW in October 2014,on the basis of the electricity tariff sold to the government from the company at the value of 95 fils per kilowatt.

Ministry of Energy and Mineral Resources and the National Electricity Power Company expect that the company will reach the financial close at the end of the current year knowing that the financial investment expected for the project is estimated at 1. 9 billion US dollars.

NEPCO has stated that The cost of electricity production in 2014 totaled 146 fills/kWh (21 cents)

NEPCO Cost Recovery Under Different International Oil Price Assumptions

	$45/bbl	$70/bbl	$99/bbl	$115/bbl
NEPCO average cost (fils/kWh) 1/	87	116	153	169

1/ Computed as average cost-recovery bulk supply tariff assuming 2014 fuel mix. Estimate cost-recovery in 2014: 146 fils/kWh

- Oil shale is a potentially promising mineral resource but not because of its organic content which is pretty high. Economically more interesting aspect are the phosphate , limestone, marble and heavy metals.
- Ash has also potential economic value to be used in various applications including building materials.

THANK YOU

Eng. Munther Bsieso
Technical Manager
079 96 97 859
bsieso@gmail.com

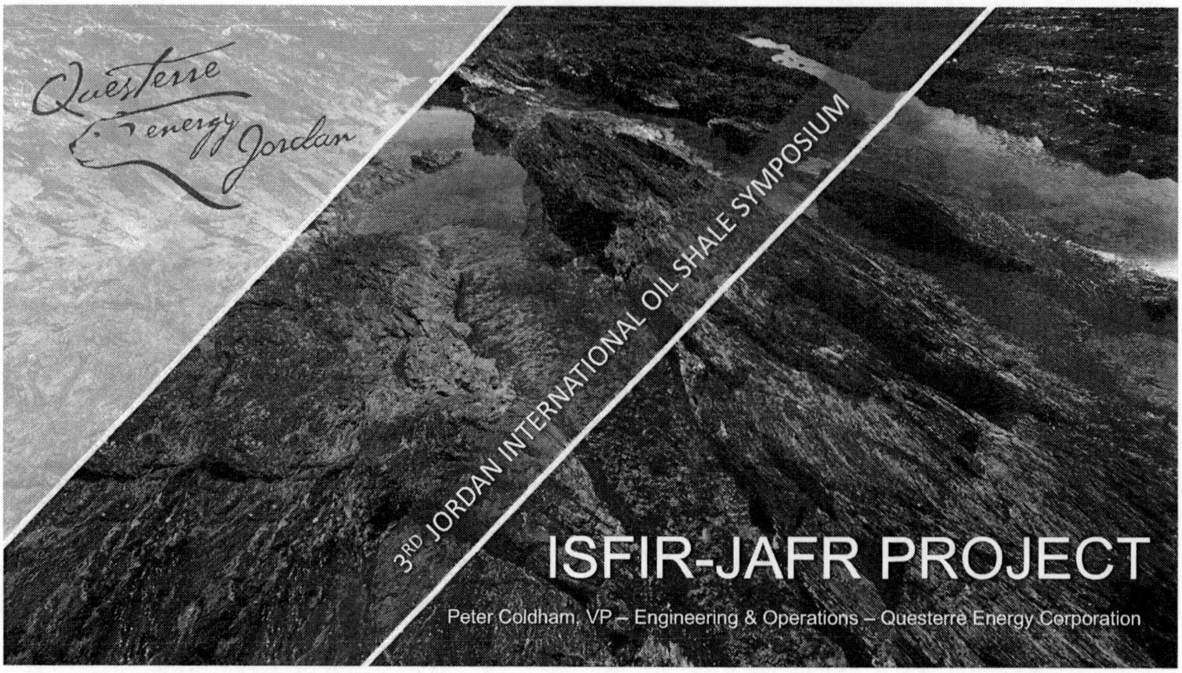
Questerre energy Jordan
3RD JORDAN INTERNATIONAL OIL SHALE SYMPOSIUM
ISFIR-JAFR PROJECT
Peter Coldham, VP – Engineering & Operations – Questerre Energy Corporation

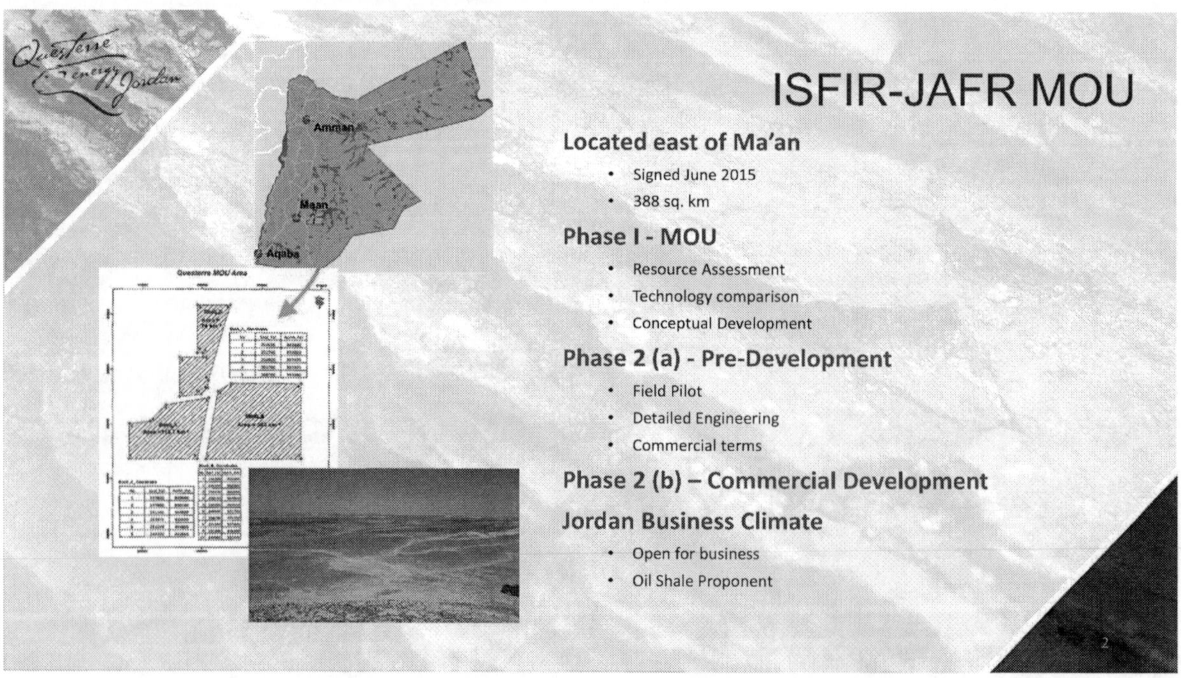
Questerre energy Jordan
Amman
Maan
Aqaba
Questerre MOU Area
ISFIR-JAFR MOU
Located east of Ma'an
• Signed June 2015
• 388 sq. km
Phase I - MOU
• Resource Assessment
• Technology comparison
• Conceptual Development
Phase 2 (a) - Pre-Development
• Field Pilot
• Detailed Engineering
• Commercial terms
Phase 2 (b) – Commercial Development
Jordan Business Climate
• Open for business
• Oil Shale Proponent
2

GEOLOGY

MCM Oil Shale

- Porous marl (chalk) / limestone
  - main rock forming components - carbonate, kerogen, minor clay, silica
  - pyrite in voids
- Estimate 45% porosity, 25% free water, 10-12% kerogen
- Subcrop on block
- Low compressive strength

Overburden

- URC formation
  - argillaceous marls and occasional cherts
  - low compressive strength

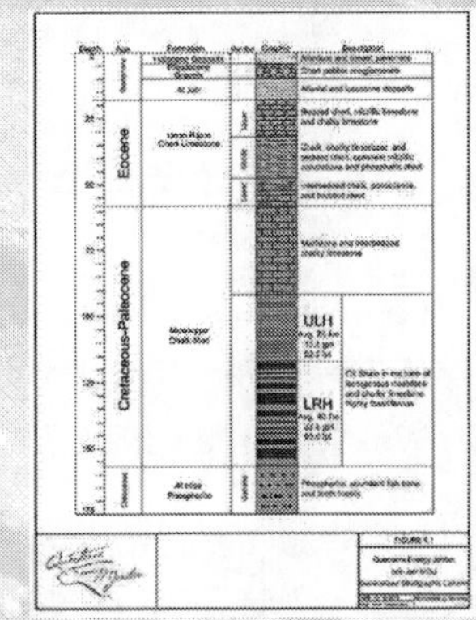

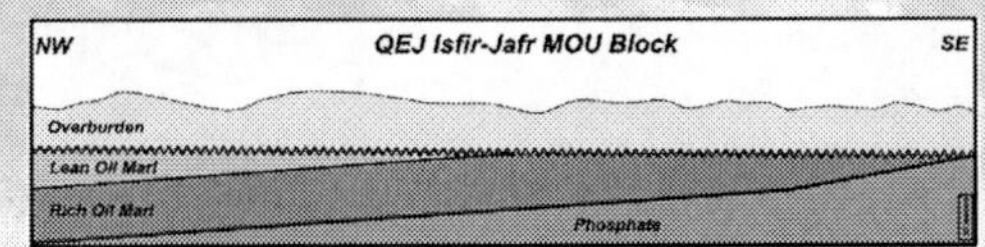

3

RESOURCE ASSESSMENT

Independent evaluation by Millcreek Mining Group

PIIP Calculations

- 2:1 Strip ratio cutoff
- 15 gpt cutoff
- MFA calibrated to ASTM
- Mean strip Ratio 1.2 : 1
- Area 213 km2
- Mean thickness 50.7 m
- Mean oil Yield 22.6 gpt
- **8.6 Billion Barrels**

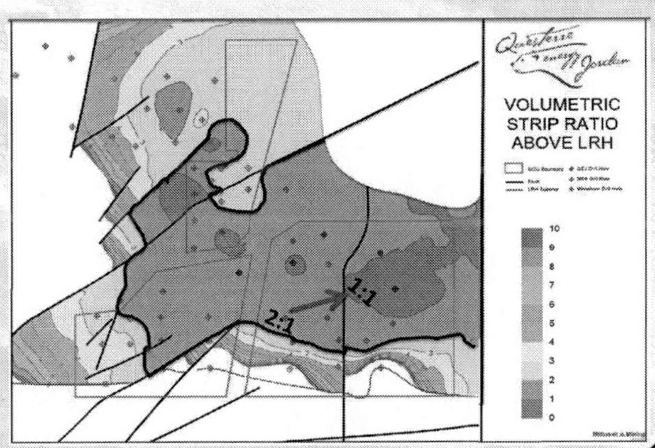

4

TECHNOLOGY

Initial Technologies

- Alberta Taciuk Process (ATP)
- Enefit 280
- EcoShale

Compare on basis of 20,000 BPD raw oil

Why look at a variety of technologies?

- Ore properties unknown
- Applicability for commercial development on this ore
- Efficiency

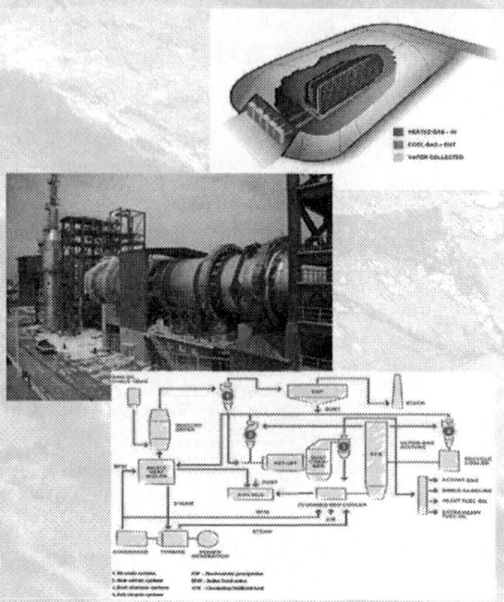

5

EVALUATION PROCESS

Sequential evaluation is key

Ore and Overburden Characterization

Retort Assessments

- From Suppliers – incl. Capex and Opex
  - Retorting Units - Yields and Product Quality
  - Product Separation
  - Ash Handling

Common Studies

- Mining
- Feed Processing
- Upgrading
  - Hydrotreating – SCO, Refined Products
  - Sulphur
- Marketing
- Utilities
  - Electrical Power
  - Water
  - Roads and Buildings

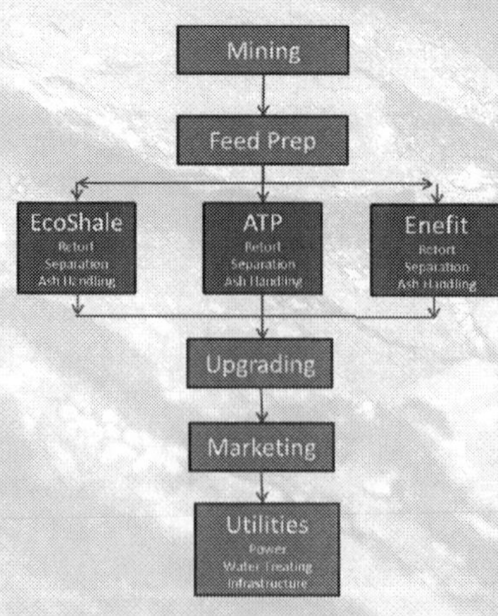

6

EVALUATION UPDATE

Resource Characterization

- Integrated the ministry data
- Used 5 new cores to gather key data on areas of concern
  - free water, mechanical properties, sulphur
- Developed correlations for legacy data

Mining study completed

- Fixed facilities and capsule versions

Retort studies

- Enefit and UMATAQ evaluated core samples
- EcoShale delayed because of redesign to down flow
  - Currently working on reusable capsule concept

Upgrading study completed

- Options for SCO or finished products

Feed Prep study completed

RFP for Marketing and Utilities

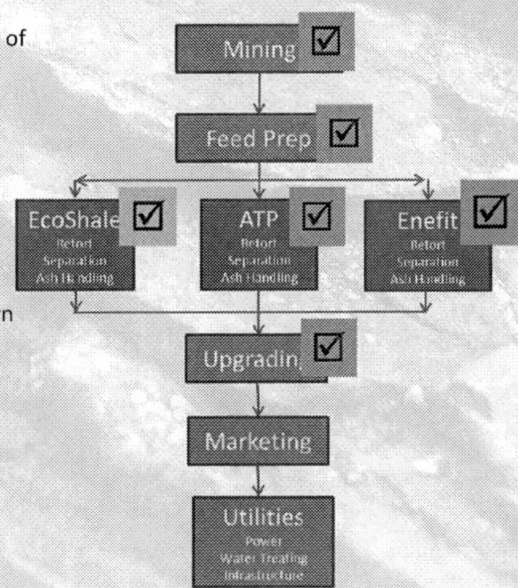

7

ATP

Scoping Level Study Goals

- Additional resource and product characterization
- Study to review:
  - indications of potential operational problems
  - identify suitability of ATP process and quantify cross battery limit stream flows
  - review size, number of units, CAPEX and OPEX

Results

- No real operational problems
- Particle attrition to be evaluated at larger scale
- Ore requires pre-drying prior to ATP plant
- Ash leachate analysis results below instrument detectability
- 3 ATP trains required – indicative CAPEX and OPEX defined
- Quantified mass flows and rates for power, water, etc
- Raw oil requires hydrotreating and sulphur removal

Figure 9 – Laboratory Facilities, UMATAC R&D Centre, Calgary

Figure 14 – UMATAC Batch Retort Testing

8

ENEFIT

Scoping Level Study Goals

- Potential operational problems
- Suitability of Enefit process
- Quantify stream flows, units, CAPEX and OPEX

Results

- No technical obstacles
  - Requires scale up of retort unit size, and changes to the flue gas and dry sections
- 6 Enefit units required
- Mass flows, utilities, CAPEX and OPEX defined
- Additional testing proposed to further define crushing requirements, particle attrition, drying, etc

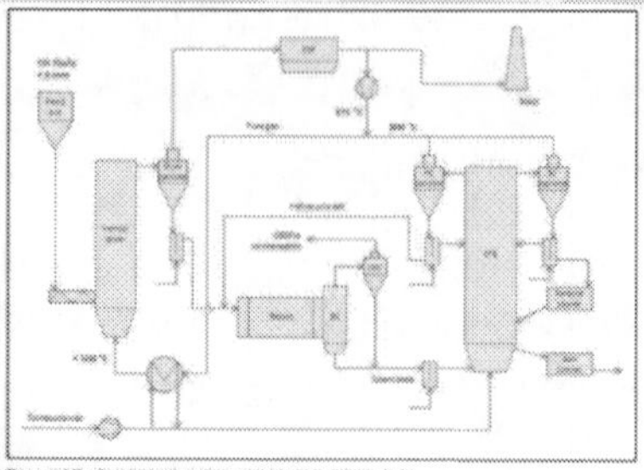

Figure 4-1 Enefit solid heat carrier process for Isfir-Jafr oil shale

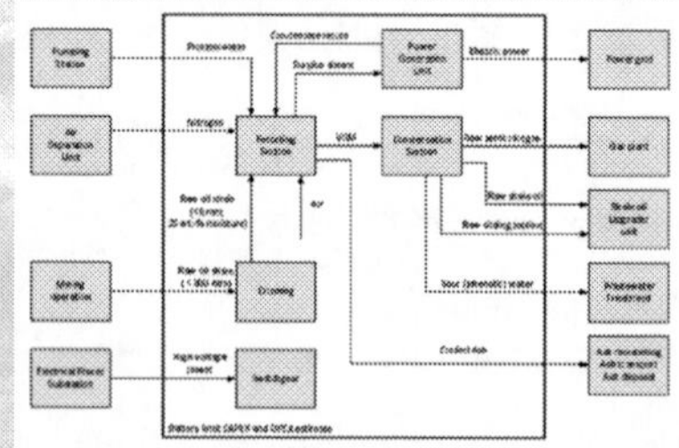

Figure 6-1: Battery limits for CAPEX and OPEX estimate.

9

ECOSHALE

Process redesign to down-flow

- Increases heating efficiency, eliminates reflux product, increases yields, etc.

Single use capsules not effective for Isfir-Jafr deposit. Concerns:

- Sterilizing resources
- Available construction materials
- Compaction (gas envelope integrity)
- Cost

Developing a reusable down-flow capsule – steel construction

- Simple construction, recovers free water, avoids sterilizing resources, avoids settling issues, no gas membrane/insulation issues
- RLR is completing preliminary design ideas and scoping cost calculations to provide line -of-sight to commercial project economics
- Will required lab work to complete study
  - Lab currently busy with EPS II FEED work
- Promising adaptation of EcoShale technology for this ore body

10

NEXT STEPS

Finish studies

- Marketing, Utilities
- EcoShale lab work and evaluation
- Integrate data and build detailed economic models

Further drilling program **(as required)**

- Collect additional geotechnical data
- Further testing as required

Review technology selection for Isfir-Jafr

Conceptual Development Plan

Pilot Program Plan

11

JOSCO

JORDAN OIL SHALE COMPANY

HSSE and Operational Excellence at JOSCO Operations

Definitions and Cautionary Note

The companies in which Royal Dutch Shell plc directly and indirectly owns investments are separate entities. In this presentation "Shell", "Shell group" and "Royal Dutch Shell" are sometimes used for convenience where references are made to Royal Dutch Shell plc and its subsidiaries in general. Likewise, the words "we", "us" and "our" are also used to refer to subsidiaries in general or to those who work for them. These expressions are also used where no useful purpose is served by identifying the particular company or companies. ''Subsidiaries'', "Shell subsidiaries" and "Shell companies" as used in this presentation refer to companies over which Royal Dutch Shell plc either directly or indirectly has control. Companies over which Shell has joint control are generally referred to "joint ventures" and companies over which Shell has significant influence but neither control nor joint control are referred to as "associates". In this presentation, joint ventures and associates may also be referred to as "equity-accounted investments". The term "Shell interest" is used for convenience to indicate the direct and/or indirect ownership interest held by Shell in a venture, partnership or company, after exclusion of all third-party interest.

This presentation contains forward-looking statements concerning the financial condition, results of operations and businesses of Royal Dutch Shell. All statements other than statements of historical fact are, or may be deemed to be, forward-looking statements. Forward-looking statements are statements of future expectations that are based on management's current expectations and assumptions and involve known and unknown risks and uncertainties that could cause actual results, performance or events to differ materially from those expressed or implied in these statements.. There are a number of factors that could affect the future operations of Royal Dutch Shell and could cause those results to differ materially from those expressed in the forward-looking statements included in this presentation, including (without limitation): (a) price fluctuations in crude oil and natural gas; (b) changes in demand for Shell's products; (c) currency fluctuations; (d) drilling and production results; (e) reserves estimates; (f) loss of market share and industry competition; (g) environmental and physical risks; (h) risks associated with the identification of suitable potential acquisition properties and targets, and successful negotiation and completion of such transactions; (i) the risk of doing business in developing countries and countries subject to international sanctions; (j) legislative, fiscal and regulatory developments including regulatory measures addressing climate change; (k) economic and financial market conditions in various countries and regions; (l) political risks, including the risks of expropriation and renegotiation of the terms of contracts with governmental entities, delays or advancements in the approval of projects and delays in the reimbursement for shared costs; and (m) changes in trading conditions. All forward-looking statements contained in this presentation are expressly qualified in their entirety by the cautionary statements contained or referred to in this section. Readers should not place undue reliance on forward-looking statements. Additional risk factors that may affect future results are contained in Royal Dutch Shell's 20-F for the year ended December 31, 2015 (available at www.shell.com/investor and www.sec.gov). These risk factors also expressly qualify all forward looking statements contained in this presentation and should be considered by the reader. Each forward-looking statement speaks only as of the date of this presentation, **22 Nov 2016**. Neither Royal Dutch Shell plc nor any of its subsidiaries undertake any obligation to publicly update or revise any forward-looking statement as a result of new information, future events or other information. In light of these risks, results could differ materially from those stated, implied or inferred from the forward-looking statements contained in this presentation.

We may have used certain terms, such as resources, in this presentation that United States Securities and Exchange Commission (SEC) strictly prohibits us from including in our filings with the SEC. U.S. Investors are urged to consider closely the disclosure in our Form 20-F, File No 1-32575, available on the SEC website www.sec.gov. You can also obtain these forms from the SEC by calling 1-800-SEC-0330.

Agenda

- Shell HSSE Control Framework System
- Operational Excellence system
- Overview of JOSCO Operations.
- JOSCO Drilling Campaign
- Jordan Field Experiment Construction Phase
- Jordan Field Experiment Operations phase
- JOSCO HSSE and Operational Excellence

HSSE Control Framework

- HSSE: refers to Health, Safety, Security and Environment
- HSSE Control Framework: Set of manuals, specification and standards for HSSE management
- Developed by Shell International and adopted by ventures under operational influence

Shell HSSE & SP Control Framework

Commitment & Policy, Standards, Manuals, Specifications and Glossary

HSSE MS

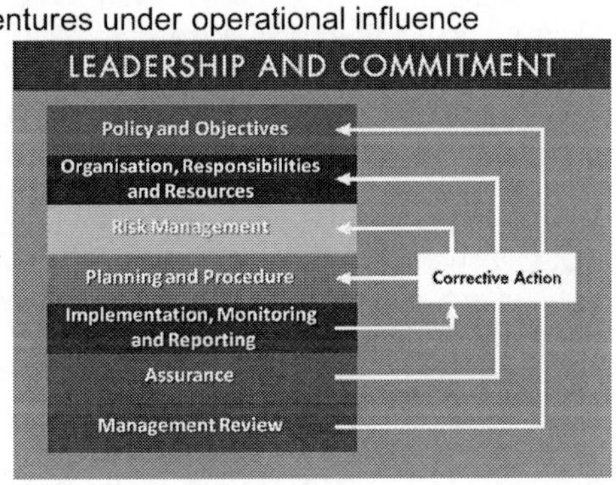

Manuals

- Health
- Personal Safety
- Security
- Process safety
- Transport
- Contractor HSSE Management
- Environment
- Projects
- Product stewardship
- Social Performance

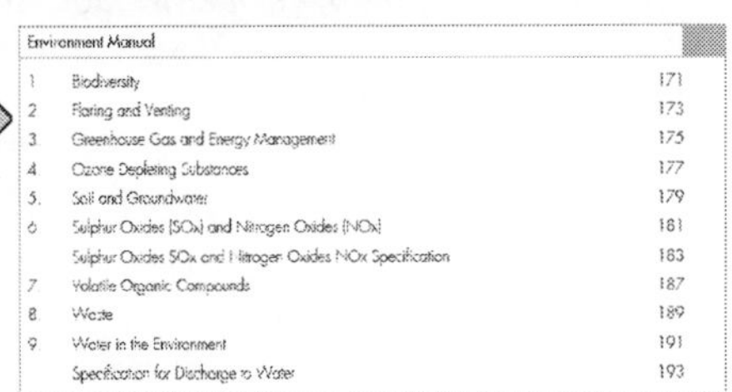

Operational Excellence Elements

- Developed by Shell International to achieve operational excellence at operated assets
- Covers 4 areas related to People, Process, Plant and Production
- Operational excellence includes 28 elements, referred to as “Blades”
- Purpose of these elements to avoid safety incidents, maintain asset integrity, achieve high process performance and meet production targets.

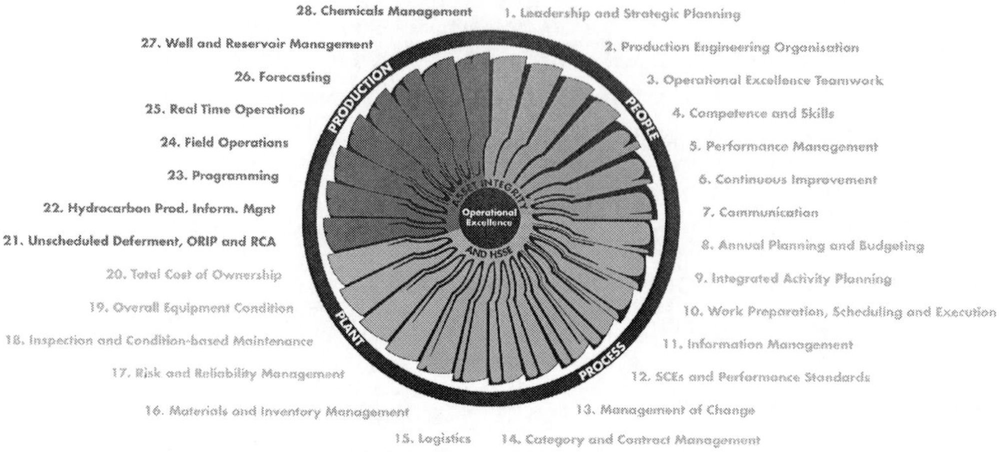

Jordan Oil Shale Operations- Overview

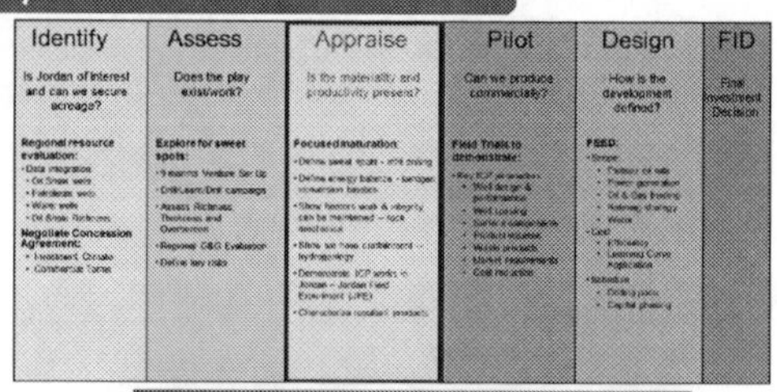

Introduction:
- Development of the oil shale reserves requires significant field operations
- These operation embed high risk activities with Health, Safety and Environmental (HSSE) risks.
- Operational Excellence and Safe Operations required controls be put in place to address the HSSE risks.
- JOSCO operational model is “Own & Operate”, proved to be effective in terms of HSSE performance, affordability, raising local capabilities, and long-term sustainability.

Drilling campaign (2009 – 2015)
- Drilling more than 340 well across Jordan.
- Hydrogeological testing campaign for reservoir containment and aquifer yield
- Wireline operations for logging drilled wells
- Exploration campaigns by JOSCO Geological department
- Lab Operations (Mechanical and chemical testing, Fisher assay analysis)

Jordan Oil Shale Operations- Overview

Jordan pilot facility, Jordan Field Experiment- JFE (2012 – Now)

- Construction of the pilot facility JFE
- Drilling and completion of unconventional wells
- Commissioning and Start Up of the JFE facility
- Sour operations of the JFE (H2S 50+%)
- Lab Operations (Gas Chromatography analysis)

Supporting activities

- Logistics operations to supportt field operations
- Equipment support, mechanical workshop activities (mechanical, electrical and welding)
- Management of field suppliers and contractors
- Warehousing and storage of field equipment

JOSCO drilling Campaign

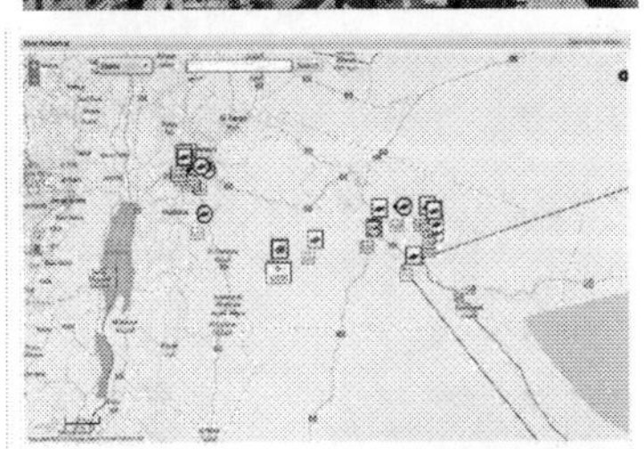

- 2 drilling rigs/camps, 1 Hydrogeology testing unit/camp, 2 wireline units, and geochemistry lab for sample analysis.
- Drilled more than 340 wells, Exploration & Appraise phases.
- JOSCO lab has analyzed safely the equivalent of 3+ km of core samples

Focusing on road transport safety and worksite hazard management

- Road transport safety: Biggest cause of death in the Oil and Gas industry
- Vehicle specifications, Defensive Driving training, Daily vehicles inspections, Use of IVMS system to monitor and improve driving performance, etc.

- Worksite hazard management: personal safety risks
- Use of Permit to Work system, Internal Audit system as mean of quality assurance, Continuous training to crew, Use of Personal protective equipment, Management of Change system, etc

JFE Construction and Commissioning Phase

Focusing on Personal safety during the Construction and drilling phase of JFE

- Construction included Civil, Mechanical, Electrical and Instrumentation works of the Facility
- Drilling and completions of complex wells.
- Main Contractor for Construction Phase CCC
- Several international contractors for drilling and completions e.g. Scientific Drilling, SCLB.
- Main risks related to: Road transport, Worksite Hazard Management and contractor management.
- Significant HSE training and Supervision was put in place with JOSCO Contractor to improve HSE competence.
- Construction and drilling campaigns completed safely at the JFE more than 300k hrs completed without recordable incidents.

JFE Operations Personal and Process Safety Focus

Focusing on Personal and process safety during the Operations phase

- Personal Safety focus included
  - Developing operational procedures for management of personal safety in Sour environment.
  - Carry out theoretical and practical H2S training.
  - Carry out emergency response drills.
  - Several types of alarm systems and gas testing installed in the facility and wells area.

- Process Safety focuses on maintaining containment included
  - Developing Performance standards for Safety Critical equipment.
  - Continuous monitoring of JFE wells and facility for any potential loss of containment.
  - Use of Nitrogen for purging purposes and pressure measurements.

Operational Excellence at JFE Project

1. Challenges:

Sour Process: With very high H_2S composition 50+% in the production stream, and unexpected operational challenges

Novel ICP Technology: Unconventional development applied for the first time with Jordanian oil shale

Local O&G Services: Oil and gas know-how and capabilities are not available in-country

2. Actions:

Staff Development & Training: Raised the qualifications and capabilities of JOSCo employees to operate a sour facility

Applied MoC System: Assured that all process, procedural and organizational changes were properly managed

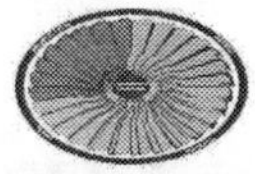

Proactive O&M Strategy: Developed a reliable M&I schedule, and involvement with P&T Houston for critical operational challenges

3. Results:

Achieved 1000+ GZD: no LTI's, no process safety incidents, no leaks and proved full containment of oil shale reservoir

100 % Uptime Facility: 0% downtime as a result of full compliance to JFE M&I system, and robust monitoring and protection systems

ICP Technology Proves To Be Successful: Creating business opportunities for technology commercialization

JOSCO HSSE and Operation Excellence 2010 -2015

- More than 3.1 million exposure hours includes (JOSCO Exploration, Wells staff and contractors).
- JOSCO logistics achieved the prestigious Shell 2014 UI Impact award: more than 5 million km driven and more than 300 rig moves executed safely.
- JOSCO drilled more than 340 wells through the Exploration and Appraise phases.
- JOSCO achieved 100% up time at JFE operations with no loss of containment incidents.

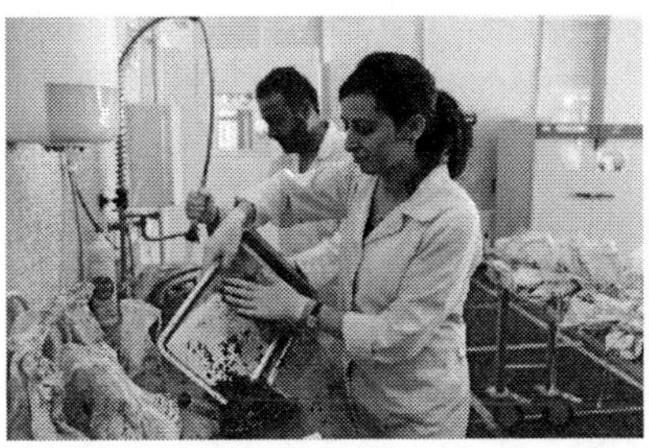

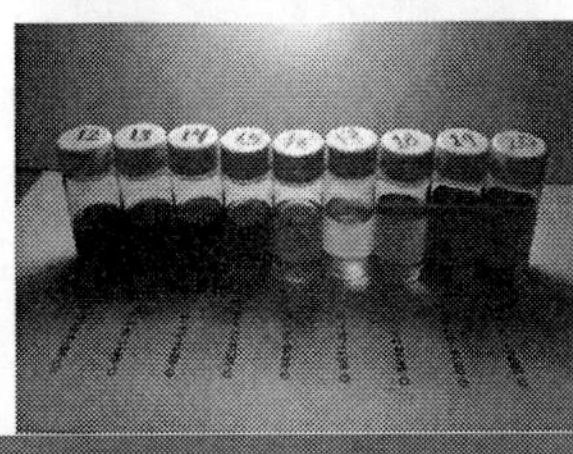

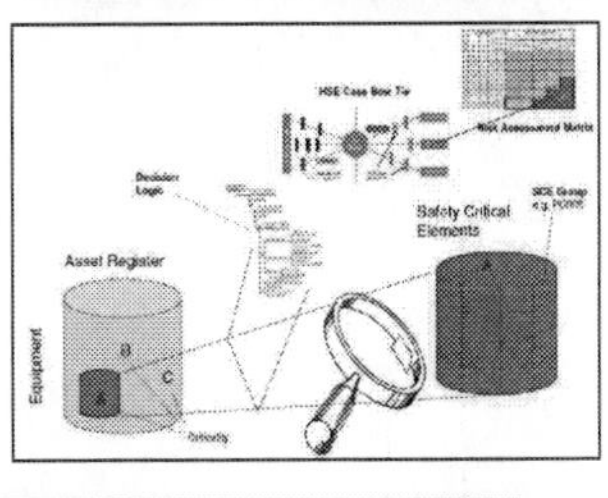

Thank you.

Al-Qamer for Energy and Infrastructure Company

Oil Shale Project

Outline

- **Part I** Al Qamer Company Profile.
- **Part II** Exploration Program.

Introduction

Al Qamer for Energy & Infrastructure ltd (QEI) is a company registered at the Ministry of Industry and Trade as Limited Liability Company under the law of Hashemite Kingdom of Jordan with registration number of (23147).dated 28-10-2010. QEI signed a Memorandum of Understanding (MOU) with GoJ represented by Ministry of Energy & Mineral Resources (MEMR) to explore and exploit oil shale resources at Attarat Umm Ghudran area.

Al Qamer company engaged in development of sectors such as: oil shale commercial venture through the establishing of an oil shale industry capable of profitably producing a shale oil substitute including petrochemical industries, power generation through direct combustion, solar and wind energy and any other industrial minerals.

Location and Access

- The MOU area is located in the central part of Jordan, about 110 km south of Amman, it covers 64.3 km^2

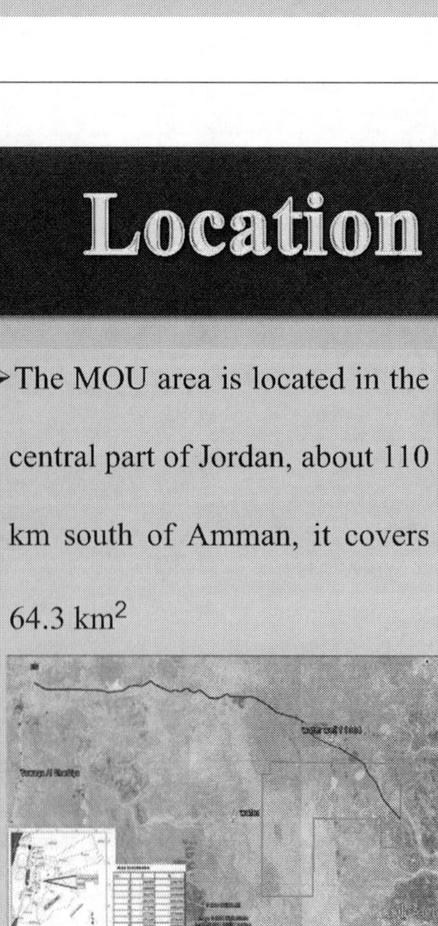

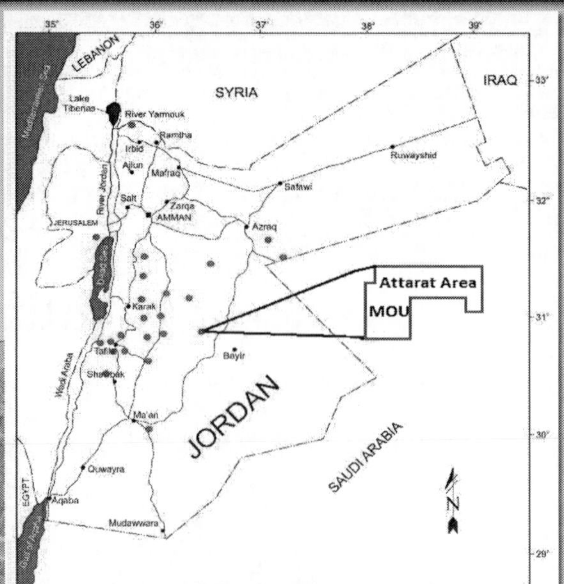

Oil shale occurrences in Jordan including the MOU area

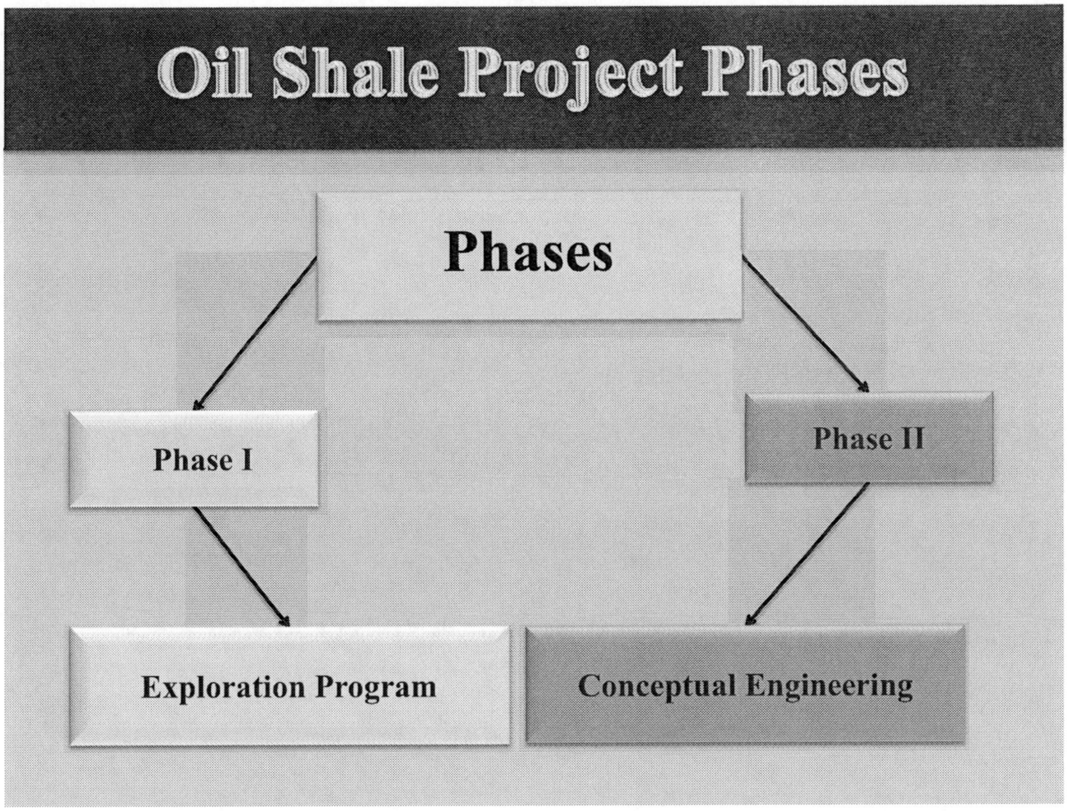

Part II

Exploration Program

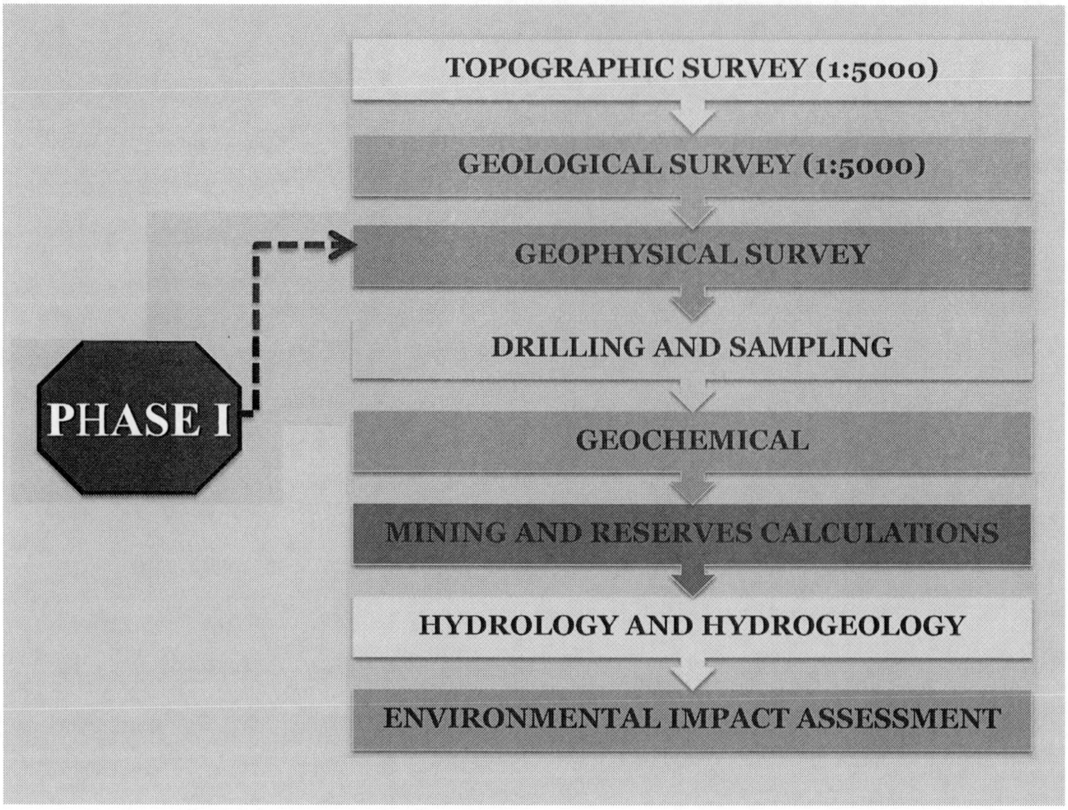

Topographic Survey

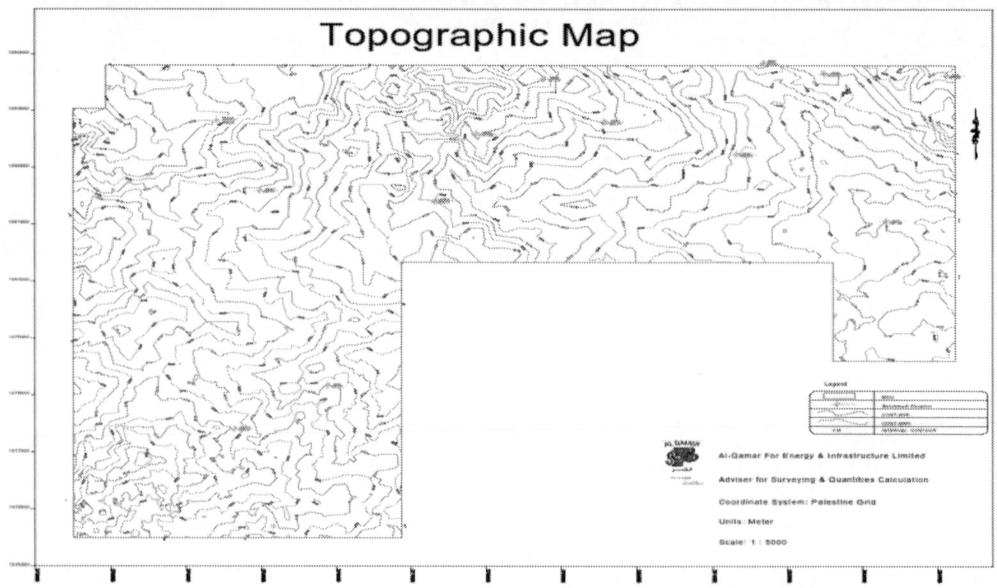

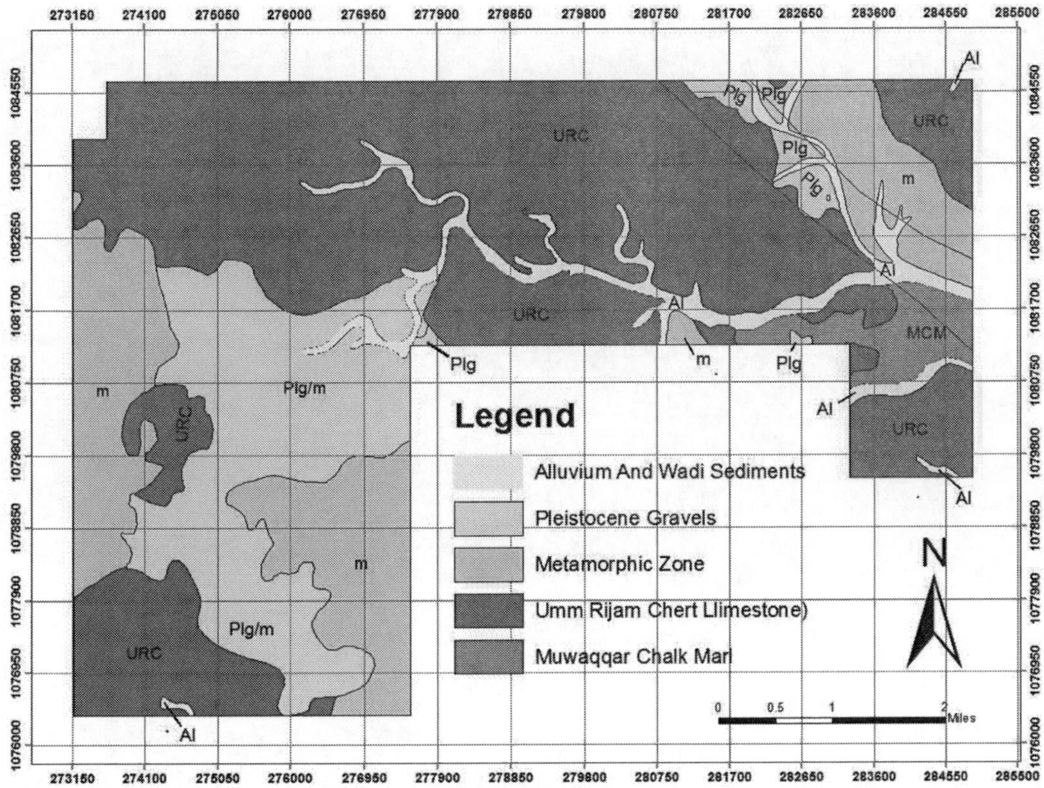
Legend
Alluvium And Wadi Sediments
Pleistocene Gravels
Metamorphic Zone
Umm Rijam Chert Llimestone)
Muwaqqar Chalk Marl
N
URC
MCM
Plg
Plg/m
m
Al
0 0.5 1 2 Miles

Field Activity
Canon

Field Activity Geophysical Survey

Geophysical Survey

The main goals of this study was to determine:

- subsurface layering;
- layer thickness (overburden and oil shale);
- Water table and
- Structural Geology

Geophysical Survey

- Geophysical survey activities stipulated in the MOU using Time Domain Electromagnetic (TDEM) method, and the implementation completed of seventy five (75) TDEM points.
- Main Results

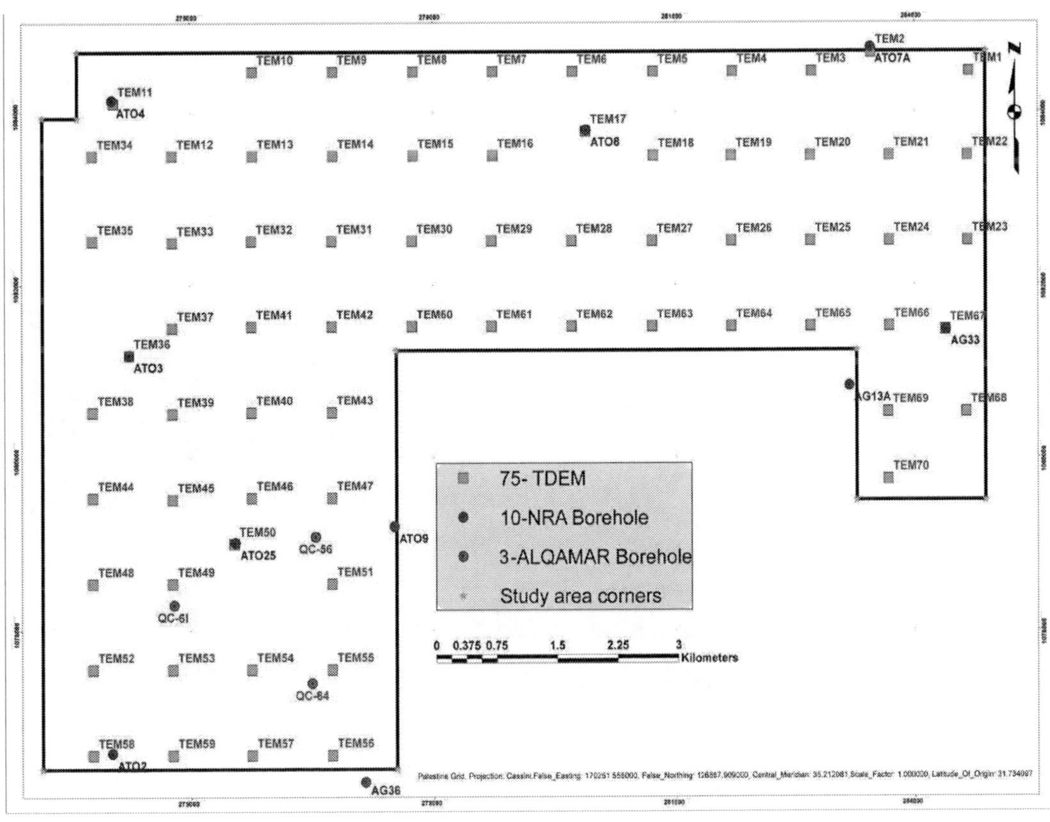

Geophysical Survey

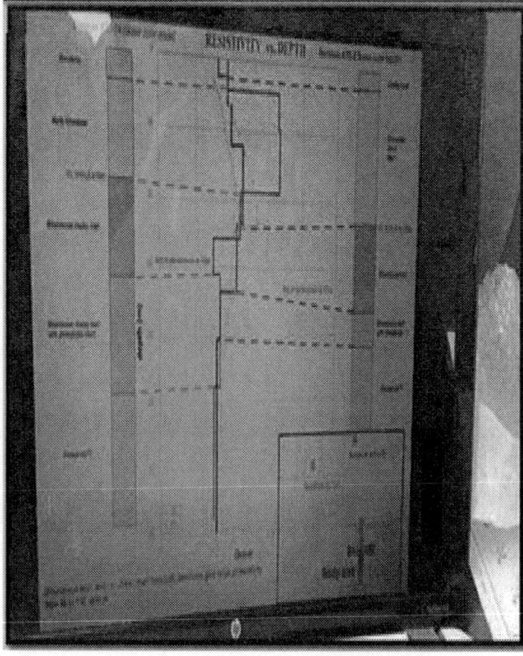

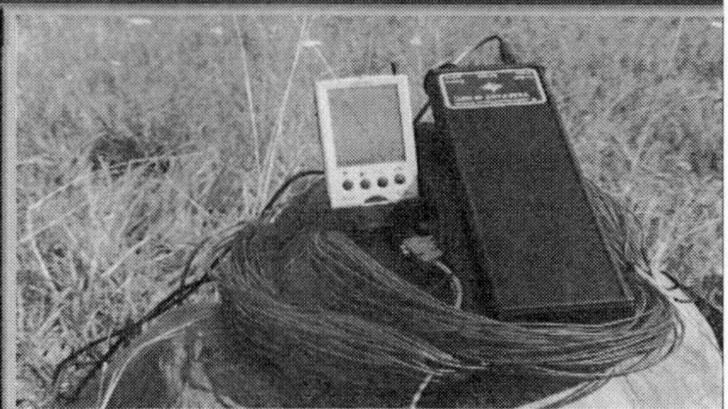

Exploration Work

Al Qamer for Energy & Infrastructure Ltd has Started implementation of the exploration program and field work activities, in Attarat um Ghudran to implement of MoU Requirements.

Drilling

1. Boreholes will be drilled using percussion drilling for the overburden and core drilling for the oil shale. 40 boreholes will be drilled during this phase.
2. 5 boreholes core drilling for OB & OS
3. Completion drilling of 30 boreholes and other expected to be completed by the end of the next month.
4. Boreholes should be drilled up to reach the Al Hasa Phosphorite Formation (AHP) that located beneath the oil shale layer.

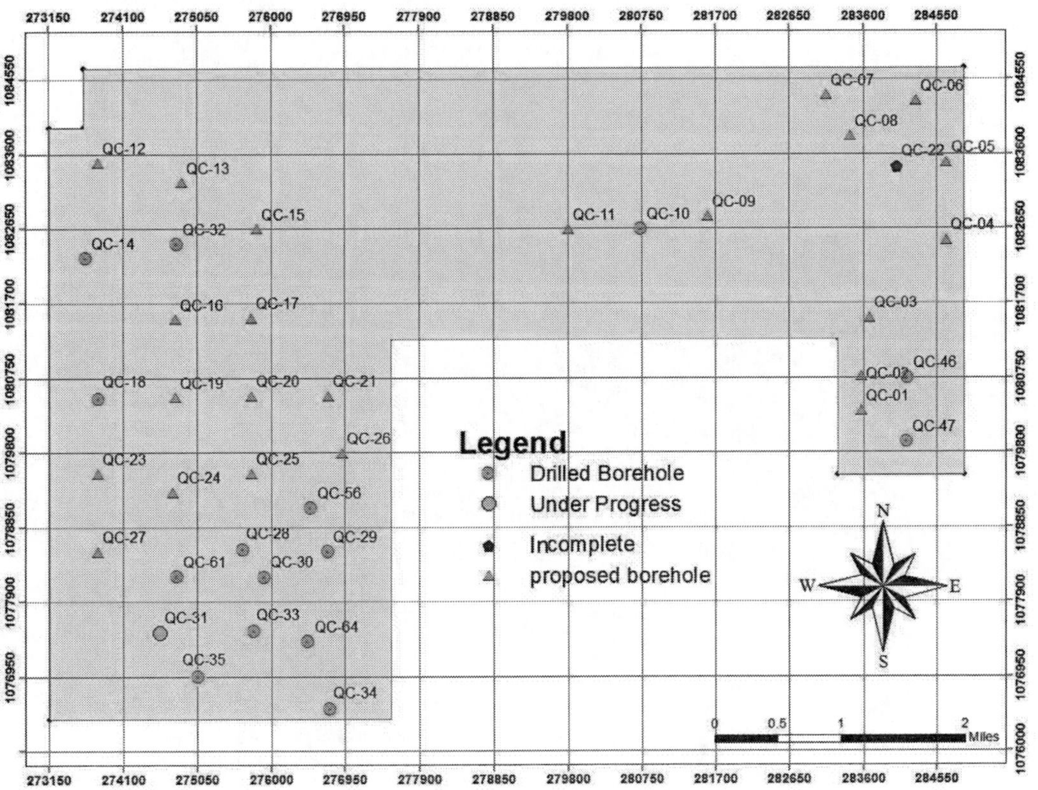

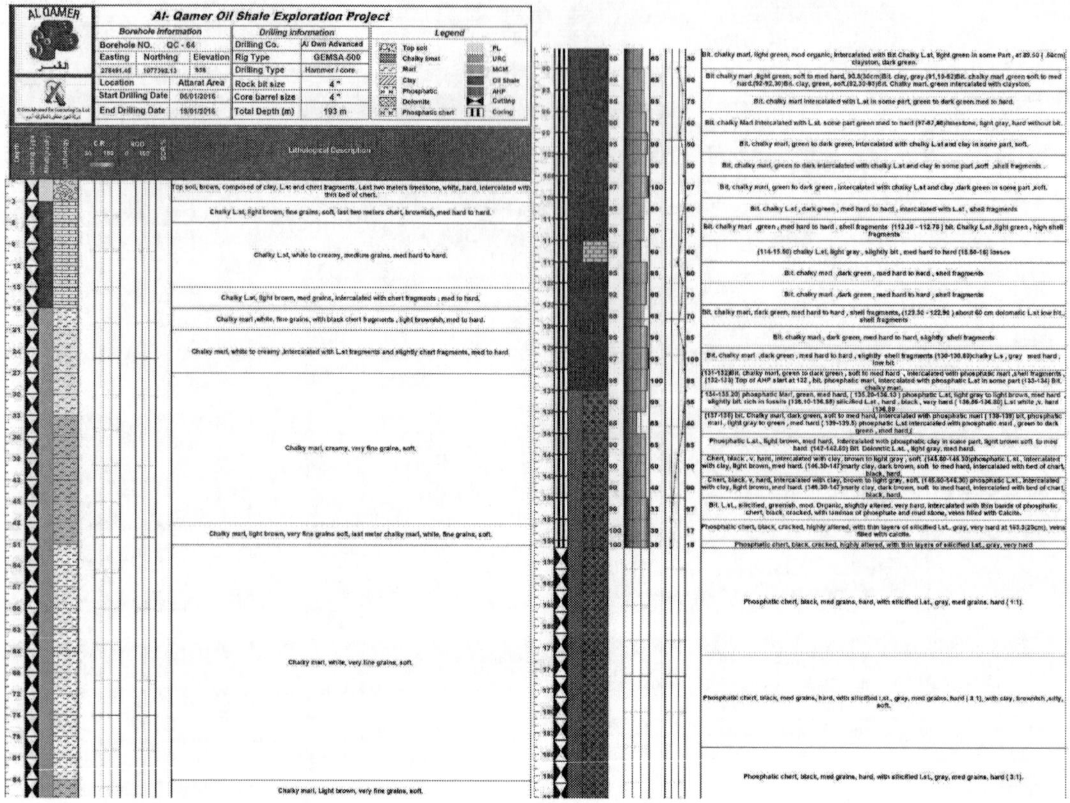

Geochemistry & Mineralogy

- Representative oil shale core samples (**1- 3 meters length**).
- Chemical analysis for sample will be conducted in the MEMR laboratories for the following tests:

XRF	**Calorific Value**
XRD	**Sulfur Content**
Fischer Assay Analysis	**Oil Shale Trace Elements**
Oil Content	**Specific Gravity and Bulk Density measurements**
Spent Shale	**Calcium carbonate**
Organic Carbon	

Mining

- The mining plan including: type of mining, slope stability, infrastructure requirements, mine layout, initial estimates of mining labor, equipment and materials requirements and costs.

Hydrology and Hydrogeology

- **Groundwater resources**
  - ❖ Determination of the depth of groundwater in the MOU area;
- **Water resources**
  - ❖ Estimation of water resources (Quality & Quantity);
  - ❖ Hydrological study includes runoff volume, infiltration and water budget calculations.
  - ❖ Determination of best location to drill water wells in the MOU area to support project life cycle.

Environmental Impact Assessment (EIA)

The main objectives of EIA study are:

- ❖Identify the key environmental issues;
- ❖Identify the legal requirements and framework for the project through its life;
- ❖Identify the relevant component studies to establish the relevant baseline for the project area;
- ❖Finalize the terms of references (
- ❖Prepared final EIA study including Environmental Management Plan (EMP) that should be approved by MoE.

Environmental Impact Assessment (EIA)

The EIA is used to evaluate different impacts of the different project phase activities on:

Public Health

Water Resources

Biodiversity

Socio-Economics

Occupational Health and Safety

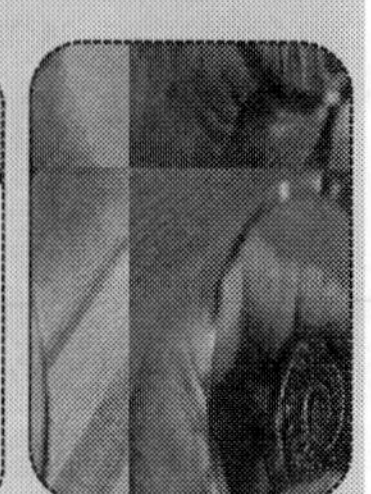

Archaeology

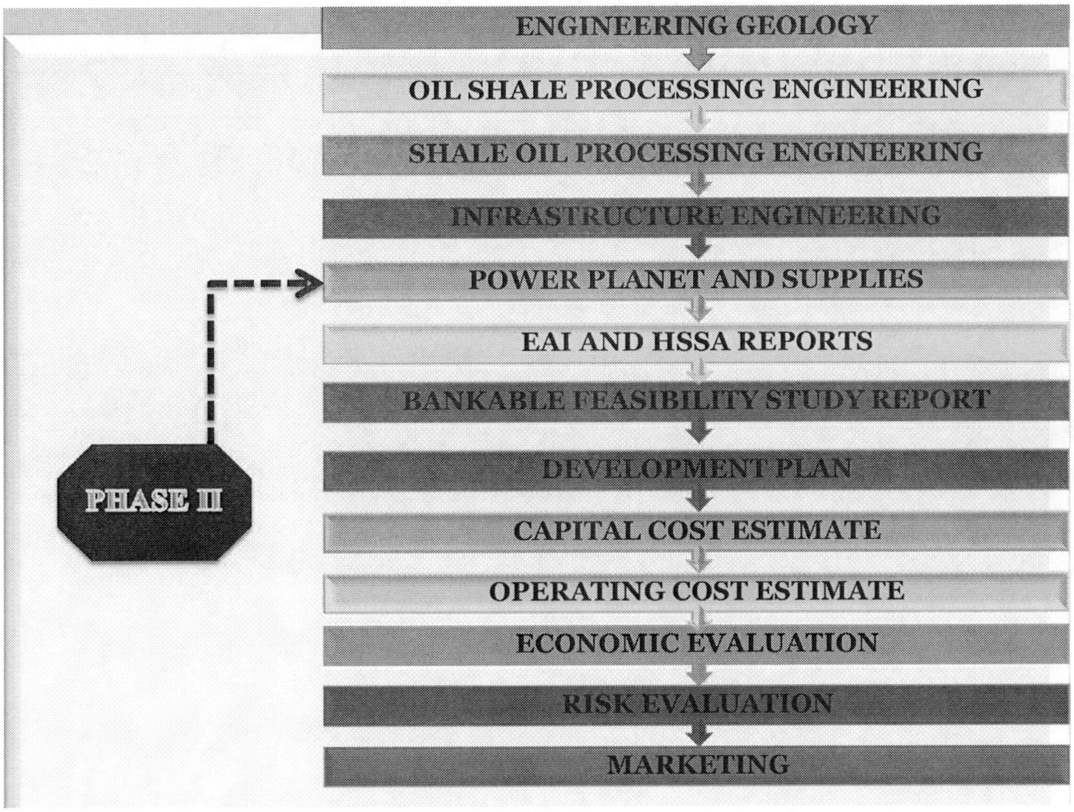

Implementation of Phase II Program

- **QEI** signed Services Agreement With TTU Company in order to achieve the requirements of phase II including the main topics: -
- Processing;
- Infrastructure;
- Hydrology (Unite Water Usage);
- Project Development Plan;
- Capital and Operation Cost Estimates;
- Economic and Risk Evaluations;
- Marketing and
- Upgrading Study
- **All these activities expected to be completed by the mid of 2017.**

Conceptual Engineering Work Program (Phase II)

The main objectives of the Conceptual Engineering program are:

- ❖ Identify and consider all the technical, legal, health, safety and environmental requirements and associated project risk factors to mitigate it;
- ❖ Define all of the necessary operations and facilities to be conducted and infrastructure required to support commercial scale implementation and operation of the project;.
- ❖ Estimate capital and operating costs and overall project economics and project development plan .

Infrastructure Design and Site Plan

➢ The main facilities design and site plan will include:-

- ❖ Description of existing site infrastructure;
- ❖ Identification and description of all roads requirement;
- ❖ Infrastructure and facilities for mining;
- ❖ Galoter and power generation constructions and operations;
- ❖ Product recovery and storage;
- ❖ Materials handling & equipment;
- ❖ Water supply and drainage system, Storm (Flood) water drainage, communications;
- ❖ Overview of power availability, regional unit power costs, housing accommodations and allied facilities and
- ❖ Estimated costs of all infrastructure and facilities design.

AL QAMER
Thank You
القمر

WORLD
ENERGY
COUNCIL

The Role of Oil Shale in the Broader Energy Environment

Zulandi van der Westhuizen | 21 November 2016

Some interesting developments

WORLD
ENERGY
COUNCIL

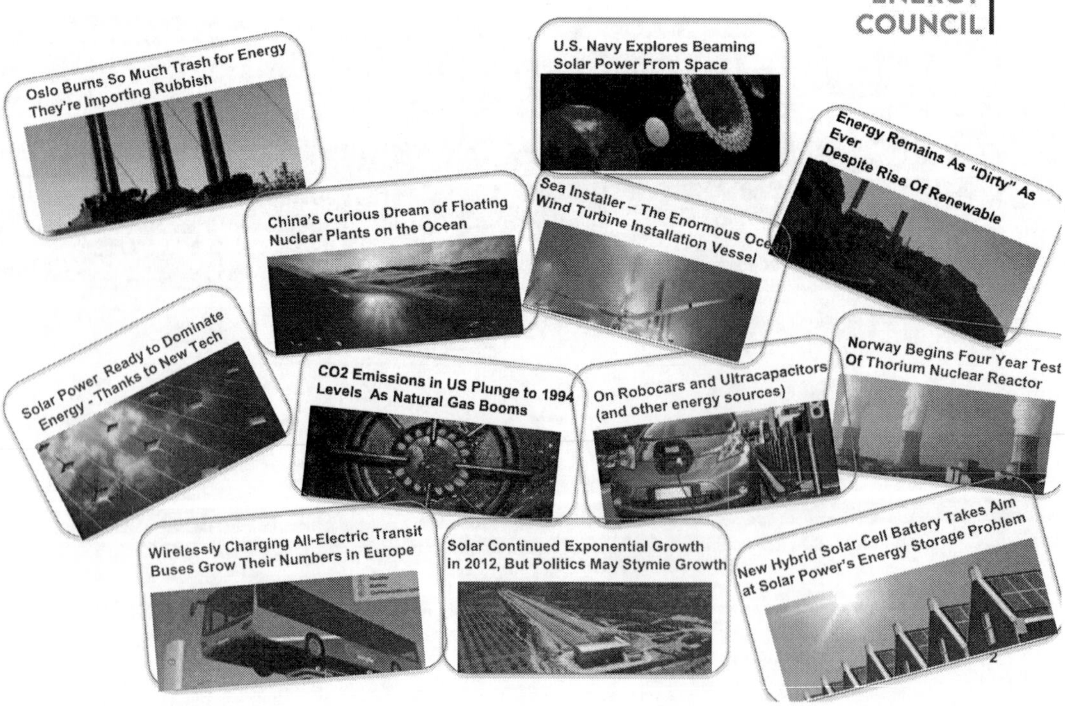

2

Energy context: Resources Trends

WORLD ENERGY COUNCIL

- **Cheaper, smaller, closer**
  General trend of smaller generation across technologies and resources
  Energy sources are moving closer to consumers
  Costs are coming down for most energy resources and technologies

- **Subsidies and support**
  Regulatory support toward clean energy is starting to change with some markets reducing and withdrawing support for renewables.
  Subsidies for fossil fuels are decreasing in some countries.

- **Competitive auctions**
  The major policy trend is the continued move towards competitive tendering of financial support in auctions. This move radically changes the development process and risk profile of the individual projects.

 3

Energy context: Resources Trends

WORLD ENERGY COUNCIL

- **Biggest growth occurring in emerging and new markets**
  China and India continue to lead and growing electricity demand is expected to outstrip average regional growth. China remained the world's largest growth market for energy for a fifteenth consecutive year.

- **Energy players**
  For renewable energy, on the one hand there is a trend of consolidation, but on the other hand, also new players specially from China and India.
  Continued trend of upstream oil and gas companies diversifying revenue streams through adding new technologies and downstream P&U assets.

 4

'NIMBY' Not In My Back Yard – a global phenomena

WORLD ENERGY COUNCIL

- Boosting **energy security**
- Foster **economic development**

However → Social & environmental resistance

Energy globally: What has changed?

WORLD ENERGY COUNCIL

- The past 15 years have seen unprecedented change in the consumption of energy resources
- Unexpected high growth in the renewables market.
- Growth of unconventional resources
- Improvements in technology evolution for all the resources
- Decrease in energy prices
- Decoupling of economic growth and GHG emissions
- More diversified energy mix achieved
- Growth in community ownership and evolution of micro grids

Globally: An outlook over the past 15 years

WORLD ENERGY COUNCIL

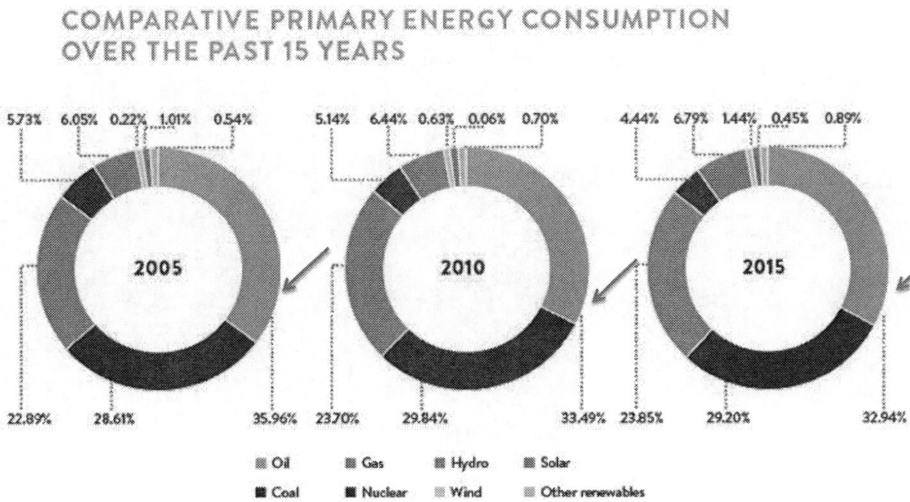

Which energy source to choose?

WORLD ENERGY COUNCIL

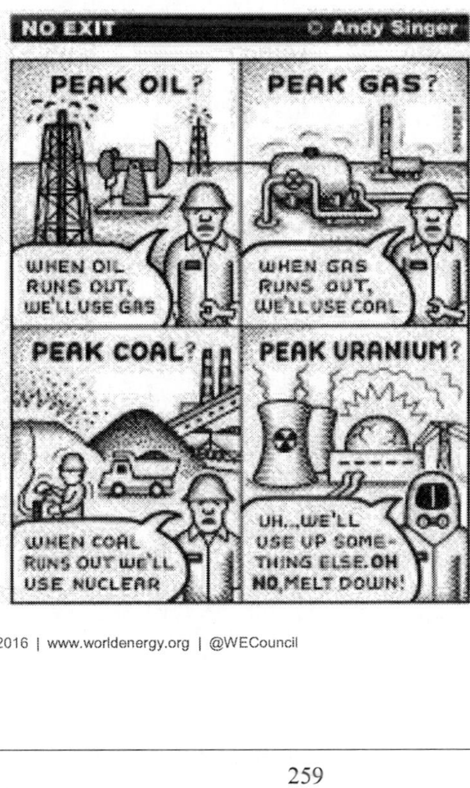

Which energy source to choose?

WORLD ENERGY COUNCIL

"I've worked out how to keep our solar panels generating electricity during winter."

9

No perfect and no 'one-size-fits-all' solutions

WORLD ENERGY COUNCIL

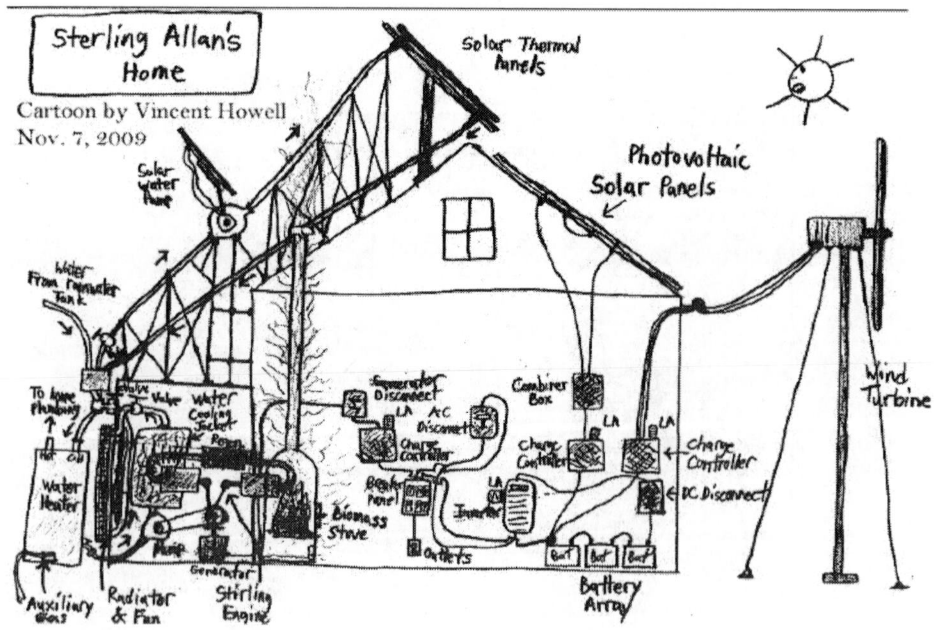

10

Factors impacting the oil market and Jordan on the Issues Monitor

WORLD ENERGY COUNCIL

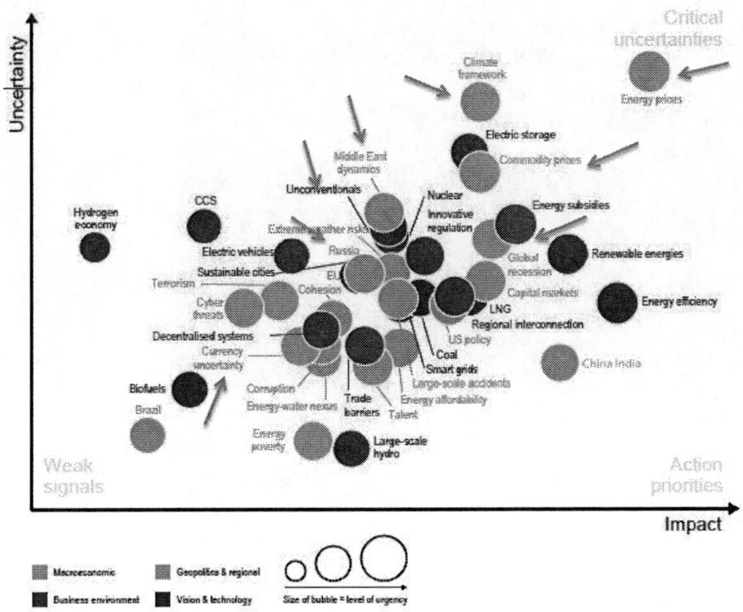

Critical uncertainties and action priorities MENA 2016

WORLD ENERGY COUNCIL

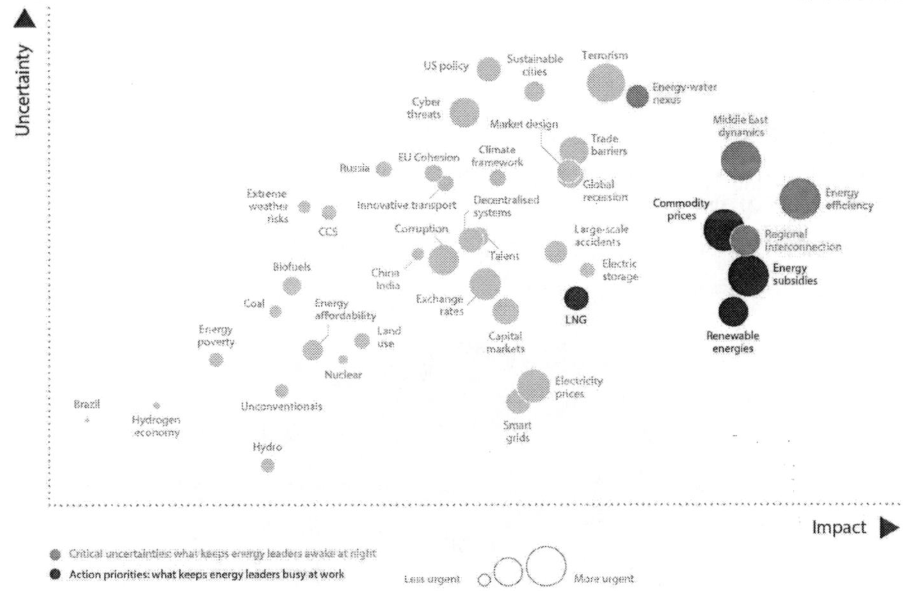

WORLD
ENERGY
COUNCIL

"This next song is about oil. Then again, isn't everything?"

Some quick facts

WORLD
ENERGY
COUNCIL

- Oil remains the world's leading fuel, accounting for 32.9% of total global energy consumption.
- Emerging economies now account for 58.1% of global energy consumption and global demand for liquid hydrocarbons will continue to grow
- Oil is the most widely traded physical commodity in the world
- Transport remains the biggest user and source of demand growth

Some quick facts

WORLD ENERGY COUNCIL

It is a political commodity:
centre stage in international affairs

Peak oil has ceased to be of concern thanks to unconventionals

Industry experiences its worst downturn (oil prices plunged by 73%)

Global challenges

WORLD ENERGY COUNCIL

The four biggest challenges in the global energy industry as a whole, are:

- technology,
- geopolitics,
- the environment and
- the world's growing population

The rise of unconventionals

WORLD ENERGY COUNCIL

- Enormous potential of unconventional oil resources that can be developed across a number of countries.
- Opportunity for many nations to become more self-sufficient, for developing countries to have better access to energy, and for enormous economic benefits to flow on from the resulting reduction in energy poverty.

 17

Oil – Economics and markets

WORLD ENERGY COUNCIL

How the global economy and the oil industry are correlated?

- Oil is of crucial importance to the world economy, the value of oil production corresponded to 4.8% of the global GDP in 2013
- Oil exports generated 12% of global trade
- Indispensability, large fluctuations in the real price of oil and the uncertainty about the future developments of oil price, are also factors that makes oil important for the global economy

 18

Oil – Economics and markets

WORLD ENERGY COUNCIL

The most important events that have influenced the price of oil:

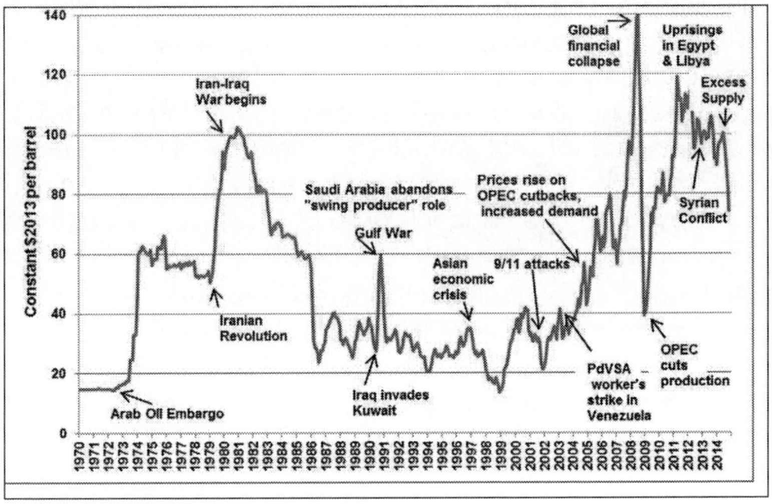

19

Jordan economic indicators

WORLD ENERGY COUNCIL

Long-term economic & social development	1980	1990	2000	2014*
GDP per capita (US$)	1714	1239	1774	4855
Inflation (%)	11.1	16.2	0.7	2.9
Population (mn)	2.28	3.36	4.77	7.38
Urban population (% of total)	60.0	73.3	79.8	83.5
Life expectancy (years)	66.2	69.9	71.8	74.1
Source : Oxford Economics & World Bank				

Structural economic indicators	1990	1995	2000	2014*
Current account (US$ million)	-227	-259	28	-2608
Trade balance (US$ million)	-1237	-1518	-2175	-11966
FDI (US$ million)	38	13	913	1760
Debt service (US$ million)	628	600	844	1396
Debt service (% of exports)	24.4	16.7	20.1	8.5
External debt (% of GDP)	200.3	113.8	130.8	67.7
Oil production (000 bpd)	0.3	0.0	0.0	0.0
Oil consumption (000 bpd)	66.2	85.4	101.1	134.0
Source : Oxford Economics / World Bank / EIA				

Structure of GDP by output	2014
Agriculture	3.8%
Industry	29.8%
Services	66.4%
Source : World Bank	

Economic outlook for Jordan

WORLD ENERGY COUNCIL

- The economy continues to show resilience despite the numerous external shocks. The conflicts in Syria and Iraq have undermined activity through disruption to trade routes, plunging tourism, lower business confidence and the fiscal costs related to hosting some 1.3m refugees. In addition, the disruption to gas imports has added to the strains on the fiscal and external deficits. In Q1, GDP still grew by 2.3% y/y, albeit the weakest for four quarters, benefitting from very low oil prices. Financial and business services (5.1% y/y) and transport and communications (3%) sectors were notably strong.
- Notwithstanding a continuing difficult regional environment (though we assume some improvement next year), we expect growth to average close to 3% this year and then 3.5% in 2017. Economic confidence is being boosted by the progress made in strengthening the fiscal and external positions.

 21

Economic outlook for Jordan

WORLD ENERGY COUNCIL

- In addition, the economy is likely to receive a boost from the IMF's approval of a three-year programme which aims to reduce public debt and drive inclusive growth higher. Some US$723m has been approved, including US$72m for immediate disbursement, with the increased prospect of donor support from the international community in line with promises made at the London Conference in February.

- Low global oil and food prices, coupled with a dollar peg and weak demand, have resulted in inflation falling into negative territory between February 2015 and August 2016. But we expect inflation to turn positive in the next few months, averaging 2.5% in 2017 on the back of higher oil prices and stronger growth, after -0.6% this year.

 22

Jordan energy resources

WORLD ENERGY COUNCIL

- Jordan, unlike its immediate neighbours, does not possess significant energy resources. As of January 2014, the *Oil & Gas Journal* estimated Jordan's proved oil reserves at just 1 million barrels and its proved natural gas reserves at slightly more than 200 billion cubic feet (Bcf).
- Oil shale resources have the potential to increase Jordan's reserves significantly As a result of its lack of significant energy resources, Jordan relies heavily on imports of crude oil, petroleum products, and natural gas to meet domestic energy demand. Government sources indicate that energy imports meet more than 90% of Jordan's energy demand, and those imports account for more than 40% of the country's budget.
- Government statistics indicate that in 2012 the use of crude oil and oil products accounted for approximately 88% of Jordan's total primary energy demand, but that domestic sources of oil and natural gas met less than 2% of that demand.

 23

Geopolitical outlook and impact

WORLD ENERGY COUNCIL

- Political risk to supply will remain extraordinarily elevated in the next few years, both on the upside and the downside, after years of chronic disruptions in the oil-rich Middle East and North Africa region.
- Lower oil prices may indeed heighten the risk of political disturbances in countries where social spending requires high oil export and fiscal revenues and buffers are insufficient to make up for the shortfall, while territorial gains and activism by increasingly aggressive terrorist groups also pose a threat to supply, not least in Iraq.
- For producer countries, lower export and fiscal revenues provide an incentive to maximise output and stimulate production growth, in a bid to make up in volume for per-barrel losses. Down cycles typically lead producer countries to tone down resource-nationalistic policies and thus can in some ways at least ease above-ground hurdles to supply.

© World Energy Council 2016 | www.worldenergy.org | @WECouncil 24

Carbon emissions in Asia 1994

WORLD ENERGY COUNCIL

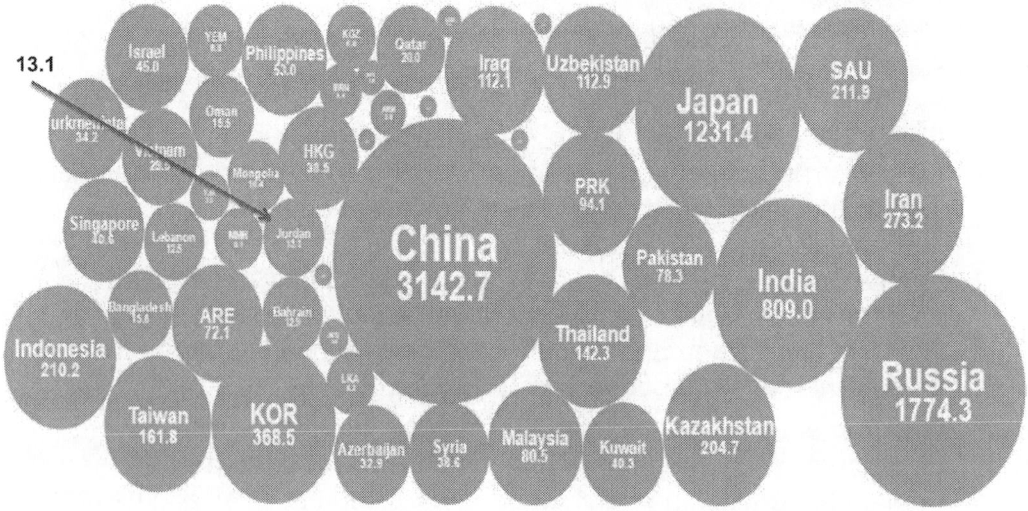

Source: British Gas

Per capita, measured in million metric tonnes

Carbon emissions in Asia 2004

WORLD ENERGY

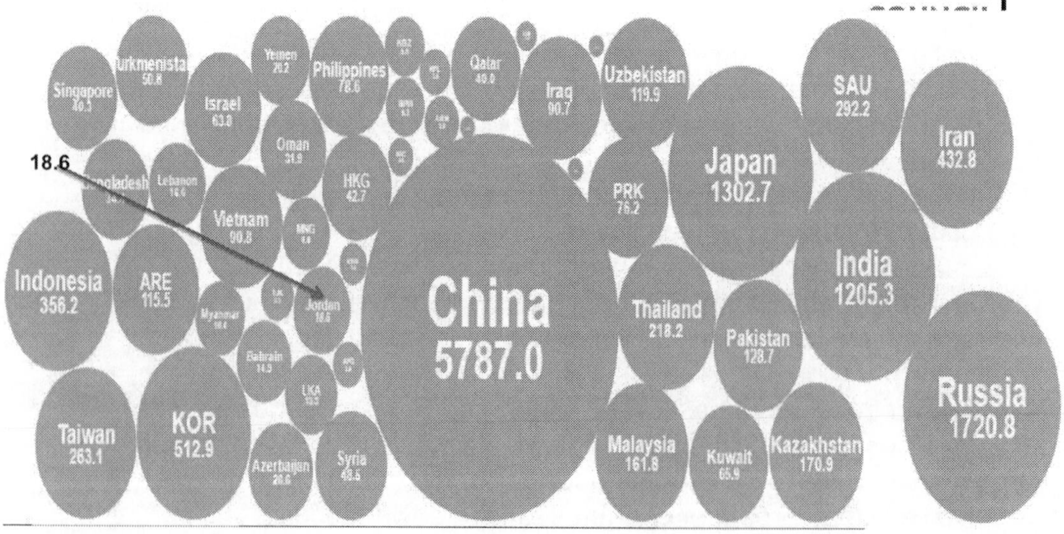

Source: British Gas

Per capita, measured in million metric tonnes

Carbon emissions in Asia 2014

WORLD ENERGY COUNCIL

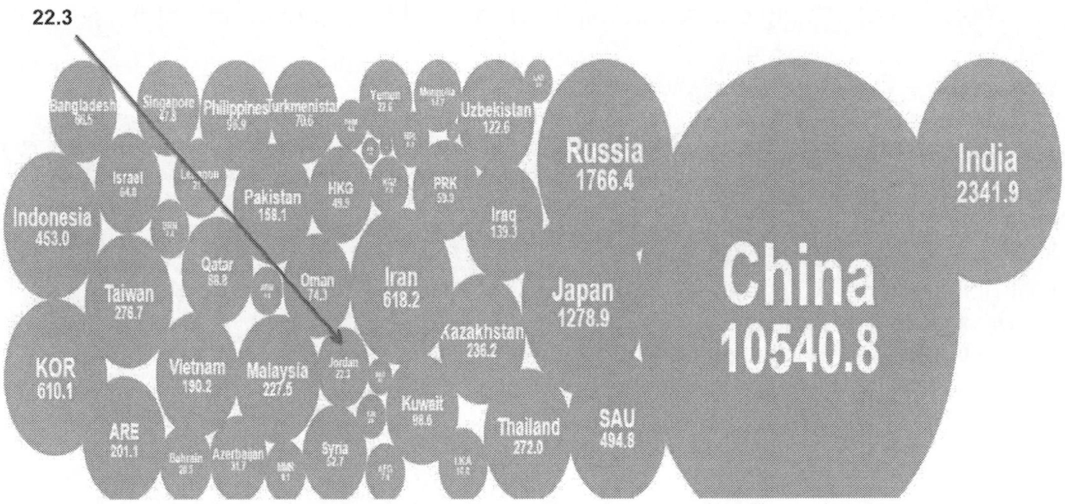

Source: British Gas

Per capita, measured in million metric tonnes

 27

Potential role of oil shale

WORLD ENERGY COUNCIL

- **Sharp rebound in oil and food prices or a further slowdown in global or regional growth** – these are particular concerns for Jordan, given its fiscal and external vulnerabilities and the openness of its economy. Higher than expected oil prices would drive up the cost of electricity and the level of fuel subsidies. Disruption to fuel imports is another risk.
- **Delay in implementing reforms including its energy strategy** – this strategy is aimed at meeting rising power demand and reducing reliance on costly and unreliable imports. It includes higher tariffs (while protecting the poor), diversifying energy sources, including nuclear power and raising energy efficiency.
- **Population growth** - the hosting of 1.3m, mainly Syrian, refugees has put immense pressure on the economy and above all on the public finances since 2012, given inadequate international aid.

 28

Message 1: Costs are reducing

WORLD ENERGY COUNCIL

New systems will boost oil recovery and extend the life of the field while reducing operating costs and minimising safety and environmental risks

New technologies include:

- Improved exploration data ('Big Data')
- Robotic technologies that allow workers to remotely control oil wells and make visual and ultrasonic inspections
- Uses of applied acoustics to deliver live drilling and completion data
- The use of lasers to drill is close to becoming a reality
- Automation
- Nanotechnology

 29

Message 2: Inventories and consumption is growing

WORLD ENERGY COUNCIL

- Global demand for liquid hydrocarbons will continue to grow
- The growth of population and the consumer class in Asia will support oil demand increase
- Modern methods of evaluation of shale oil reserves allow considerable uncertainty
- The main increase comes from transportation in developing countries
- Increase of oil production in North America won't lead to a global oil price collapse

 30

Message 3: Prices remain volatile

WORLD ENERGY COUNCIL

- Very hard to predict price:
  for 2017 ~ 50 USD/bbl
  for last 150 years ~ 35 USD/bbl

- Price has been declining due to oversupply and lower economic growth
- Cost of unconventionals is between 40-100 USD/bbl
- The growing cost of reserve replacement, the balancing role of OPEC and the depreciation of the US dollar will help to support the current levels of oil prices in the long term
- The decrease in US gasoline imports and the commissioning of new highly effective oil refineries in the Middle East and Asia will continue to have a long-term negative effect on European producers

Price Outlook

WORLD ENERGY COUNCIL

- Neither sudden geopolitical developments, nor OPEC decisions, nor any supply side discontinuity has driven the recent price collapse, the market is already rebalancing
- The main driver of price changes has been the gradually building up OPEC spare capacity and the emergence of non-OPEC production, especially US Light Tight Oil (LTO)
- On the demand side, the reducing oil intensity of Chinese demand has been a contributing factor
- The market has re-balanced very quickly to the new price levels, countries like India and Indonesia have absorbed the surplus

Outlook

WORLD ENERGY COUNCIL

- Despite the temporary price drop, the fundamentals of the oil industry remain strong. Price fluctuations seen of late have been neither unexpected nor unprecedented.
- Neither sudden geopolitical developments, nor OPEC decisions, nor any supply side discontinuity has driven the recent price collapse, the market is already rebalancing.
- The main driver of price changes has been the gradually building up OPEC spare capacity and the emergence of non-OPEC production, especially US Light Tight Oil (LTO).
- Substitution of oil in the transport sector is not yet imminent and is not expected to reach more than 5% for the next five years.
- New and increased use of technologies such as high-pressure, high-temperature (HPHT) drilling; multi-stage fracking; development in Flow Assurance for mature fields; greater sophistication in well simulation techniques, reservoirs modelling; 3-D seismic technologies, EOR developments and many more are having a positive impact on safety and E&P possibilities.

33

Opportunities

WORLD ENERGY COUNCIL

Increasing demands for increased **quality and reliability** of energy services

→ Industrial growth
→ Economic and price stability
→ Increased resilience
→ Economic development purposes
→ FDI

Bilateral trade

New demand for electricity

Attracting domestic markets

Evolving energy mix and market dynamics

34

WORLD
ENERGY
COUNCIL

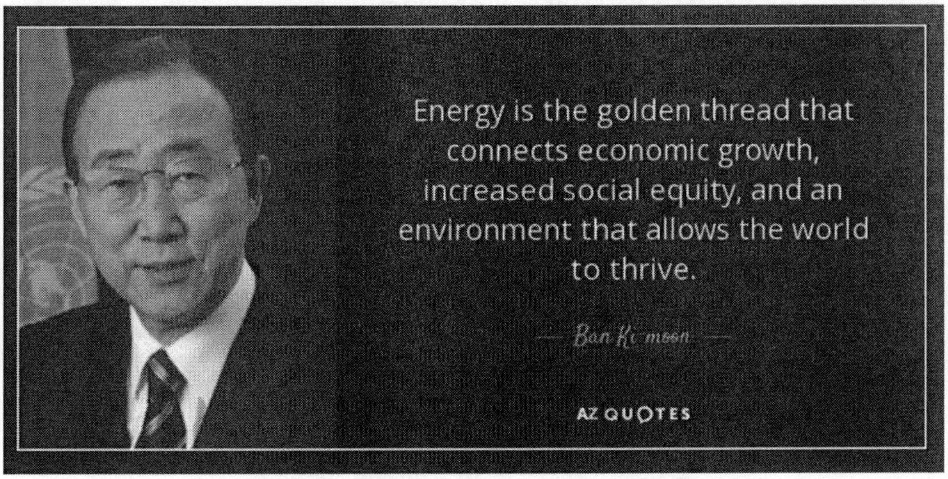

 35

WORLD
ENERGY
COUNCIL

Thank you

Zulandi van der Westhuizen
westhuizen@worldenergy.org

www.worldenergy.org | @WECouncil

INGRAIN

3rd Jordan International Oil Shale Symposium 2016

36th CSM Oil Shale Symposium
COLORADOSCHOOLOFMINES
EARTH • ENERGY • ENVIRONMENT

3rd Jordanian International Oil Shale Symposium & The 36th Colorado School of Mines Symposium
Converting Challenges into Opportunities in Oil Shale Industry

Variations of Shale Rock Properties from Different Formations in the Middle East

Moustafa Dernaika, Safouh Koronfol, Osama Al Jallad, Joel Walls and Gary Sinclair
Ingrain Inc.

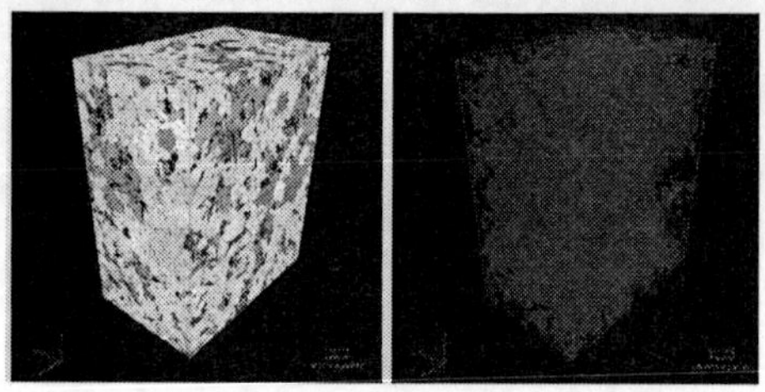

King Hussein Bin Talal Convention Center, Dead Sea, Jordan, 23 Nov 2016, 12.10pm-12.30

INGRAIN

3rd Jordan International Oil Shale Symposium 2016

36th CSM Oil Shale Symposium
COLORADOSCHOOLOFMINES
EARTH • ENERGY • ENVIRONMENT

Presentation Outline

- Challenges in Unconventional Shale Reservoirs
- Introduction:
  - Geology of the Source Rocks
  - Study objectives
- Workflow and Methodology:
  - Core Characterization and Plug Selection.
  - High-Resolution SEM Imaging
  - Segmentation and Creation of 3D rock model
  - Quantifying the Pore volume and connectivity.
- Results
- Key Findings and Summary

INGRAIN

3rd Jordan International Oil Shale Symposium 2016

36th CSM Oil Shale Symposium

COLORADO SCHOOL OF MINES
EARTH • ENERGY • ENVIRONMENT

Challenges in Shale Source Rocks

- Shale comprises a variety of rocks that composed of extremely fine-grained particles.
- Shale formations are very complex and heterogeneous geologically, chemically and petrophysically at core and plug scales.
- It has very low porosity and permeability and small PTSD.

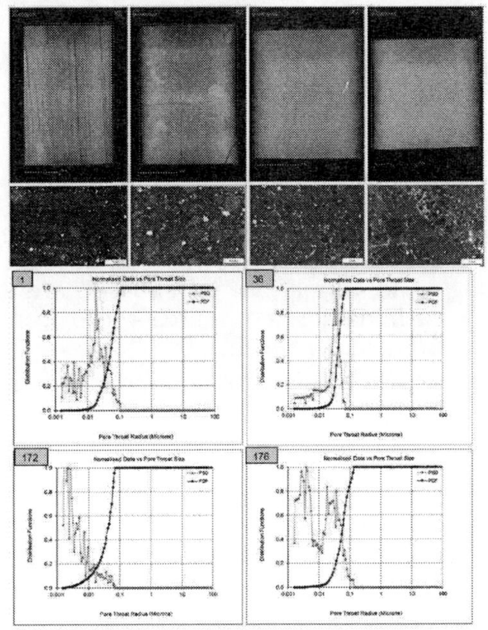

INGRAIN

3rd Jordan International Oil Shale Symposium 2016

36th CSM Oil Shale Symposium

COLORADO SCHOOL OF MINES
EARTH • ENERGY • ENVIRONMENT

Challenges in Shale Source Rocks

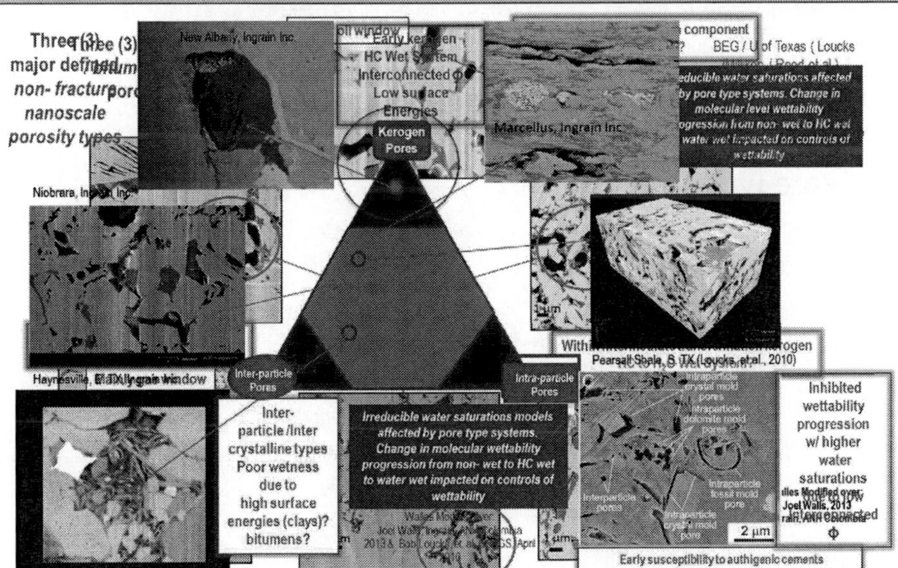

Shale are also very complex at nano- scale since the pores have different porosity types that are detected within kerogen and rock matrix.

INGRAIN

3rd Jordan International Oil Shale Symposium | 2016

36th CSM Oil Shale Symposium

COLORADO SCHOOL OF MINES
EARTH • ENERGY • ENVIRONMENT

Introduction: Geology of the Source Rocks

- More than one-thousand data sets from 14 wells and three different formations in the Middle East were analyzed.

- Formation "A" is a siliciclastic shale source rock of lower Silurian age while Formation "B" and "C" are a carbonate shale source rocks from Cretaceous epoch.
- They are very tight with low porosity and permeability.

- These source rocks are vertically and laterally heterogeneous because of depositional variability and diagenetic alteration through space and time.

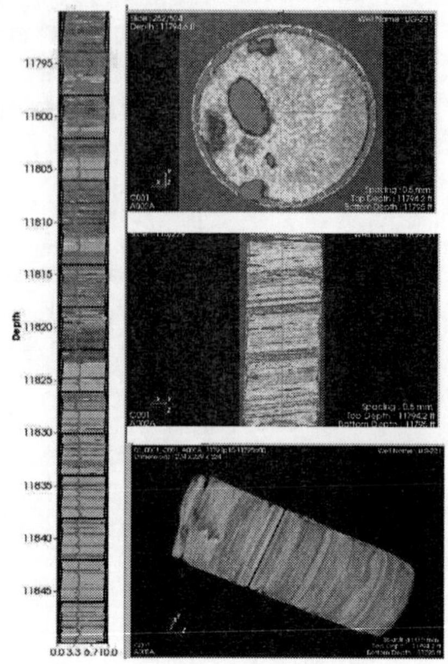

INGRAIN

3rd Jordan International Oil Shale Symposium | 2016

36th CSM Oil Shale Symposium

COLORADO SCHOOL OF MINES
EARTH • ENERGY • ENVIRONMENT

Study Objectives

- Detailed understanding of three complex unconventional source rocks characteristics including:
  - Pore types.
  - Pore volumes.
  - Pore connectivity.
  - Organic Matter Content.
  - Brittleness Index.
  - Mineralogy and Clay Content.

INGRAIN

3rd Jordan International Oil Shale Symposium 2016

36th CSM Oil Shale Symposium

COLORADO SCHOOL OF MINES — EARTH • ENERGY • ENVIRONMENT

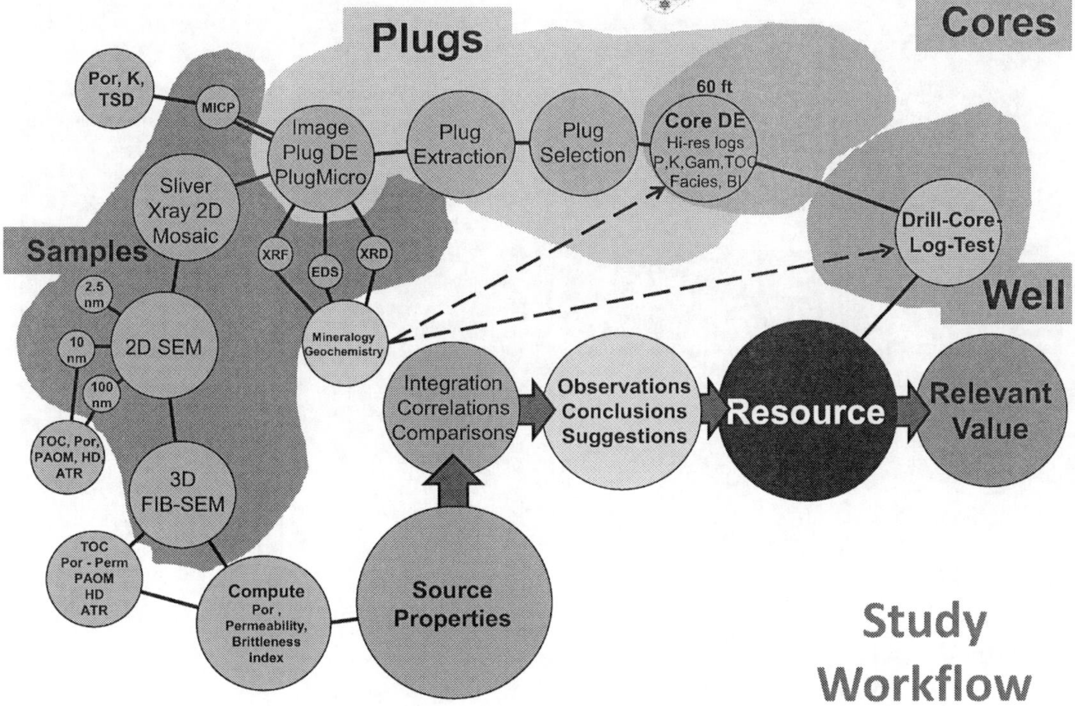

Study Workflow

INGRAIN

3rd Jordan International Oil Shale Symposium 2016

36th CSM Oil Shale Symposium

COLORADO SCHOOL OF MINES — EARTH • ENERGY • ENVIRONMENT

Core DE Scanning

- The core sections was fully imaged in 3D at a resolution of 0.5 mm per vertical slice by using DE HR CT scanner.
- Consequently, high resolution imageRhob and Zeff values were computed for each vertical slice.

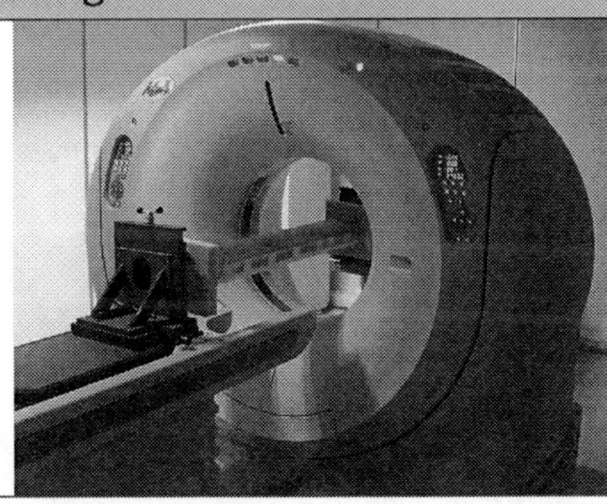

$$\mu(E) = \rho Z^{3.8}\,\alpha(E) + \rho\beta(E)$$

2 unknowns

Beer's law: $I = I_o e^{-\mu h}$

Attenuation Coefficient; dependent on

- Bulk Density (BD)
- Atomic Number (Z)

Scan core at two different energy levels:

$$\mu(E) = \rho Z^{3.8}\,\alpha(E) + \rho\beta(E)$$

$$\mu(E) = \rho Z^{3.8}\,\alpha(E) + \rho\beta(E)$$

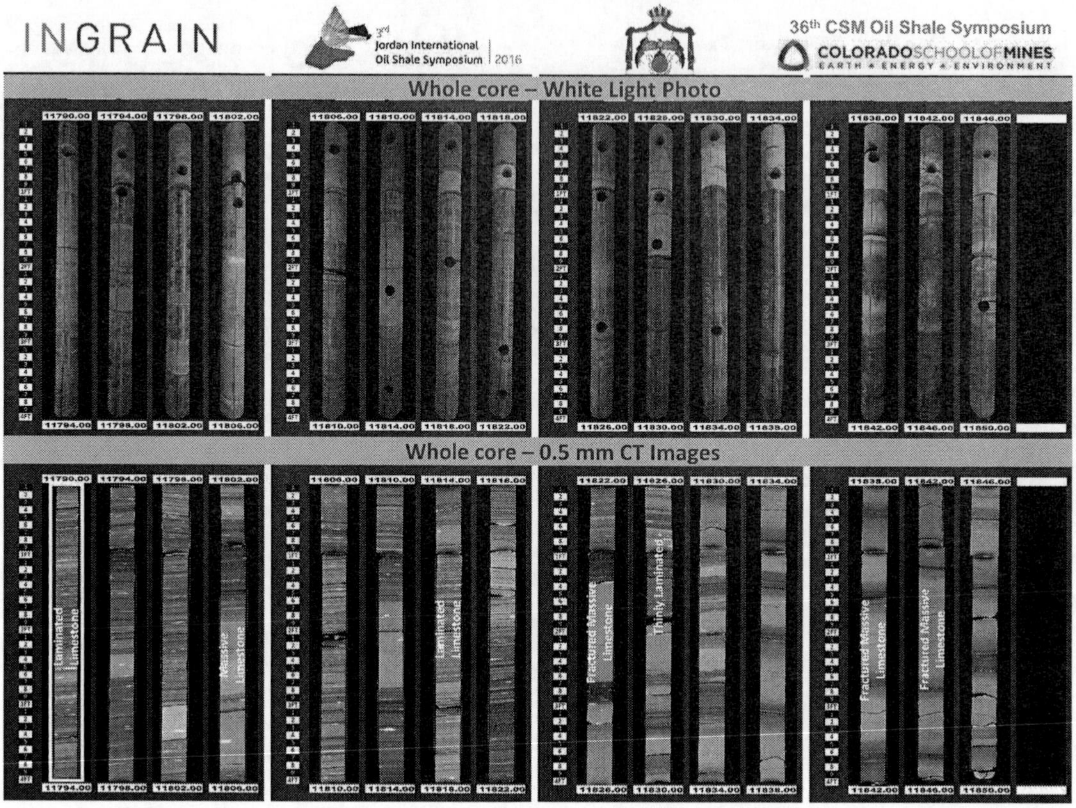
INGRAIN
3rd Jordan International Oil Shale Symposium 2016
36th CSM Oil Shale Symposium
COLORADO SCHOOL OF MINES
EARTH • ENERGY • ENVIRONMENT
Whole core – White Light Photo
11790.00 11794.00 11798.00 11802.00
11806.00 11810.00 11814.00 11818.00
11822.00 11826.00 11830.00 11834.00
11838.00 11842.00 11846.00
11794.00 11798.00 11802.00 11806.00
11810.00 11814.00 11818.00 11822.00
11826.00 11830.00 11834.00 11838.00
11842.00 11846.00 11850.00
Whole core – 0.5 mm CT Images
Laminated Limestone
Massive Limestone
Laminated Limestone
Fractured Massive Limestone
Thinly Laminated
Fractured Massive Limestone
Fractured Massive Limestone

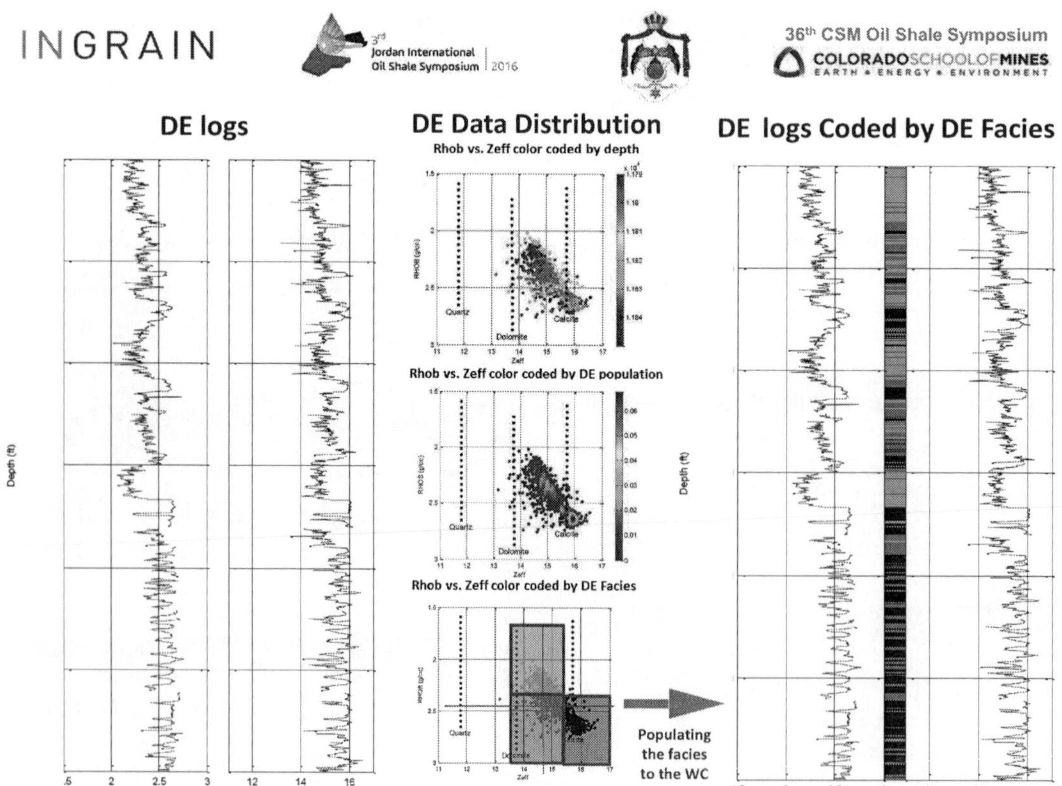
INGRAIN
3rd Jordan International Oil Shale Symposium 2016
36th CSM Oil Shale Symposium
COLORADO SCHOOL OF MINES
EARTH • ENERGY • ENVIRONMENT
DE logs
Depth (ft)
RHOB (g/cc)
Zeff
DE Data Distribution
Rhob vs. Zeff color coded by depth
Rhob vs. Zeff color coded by DE population
Rhob vs. Zeff color coded by DE Facies
Quartz
Dolomite
Calcite
Populating the facies to the WC
DE logs Coded by DE Facies
Depth (ft)
RHOB (g/cc)
Zeff

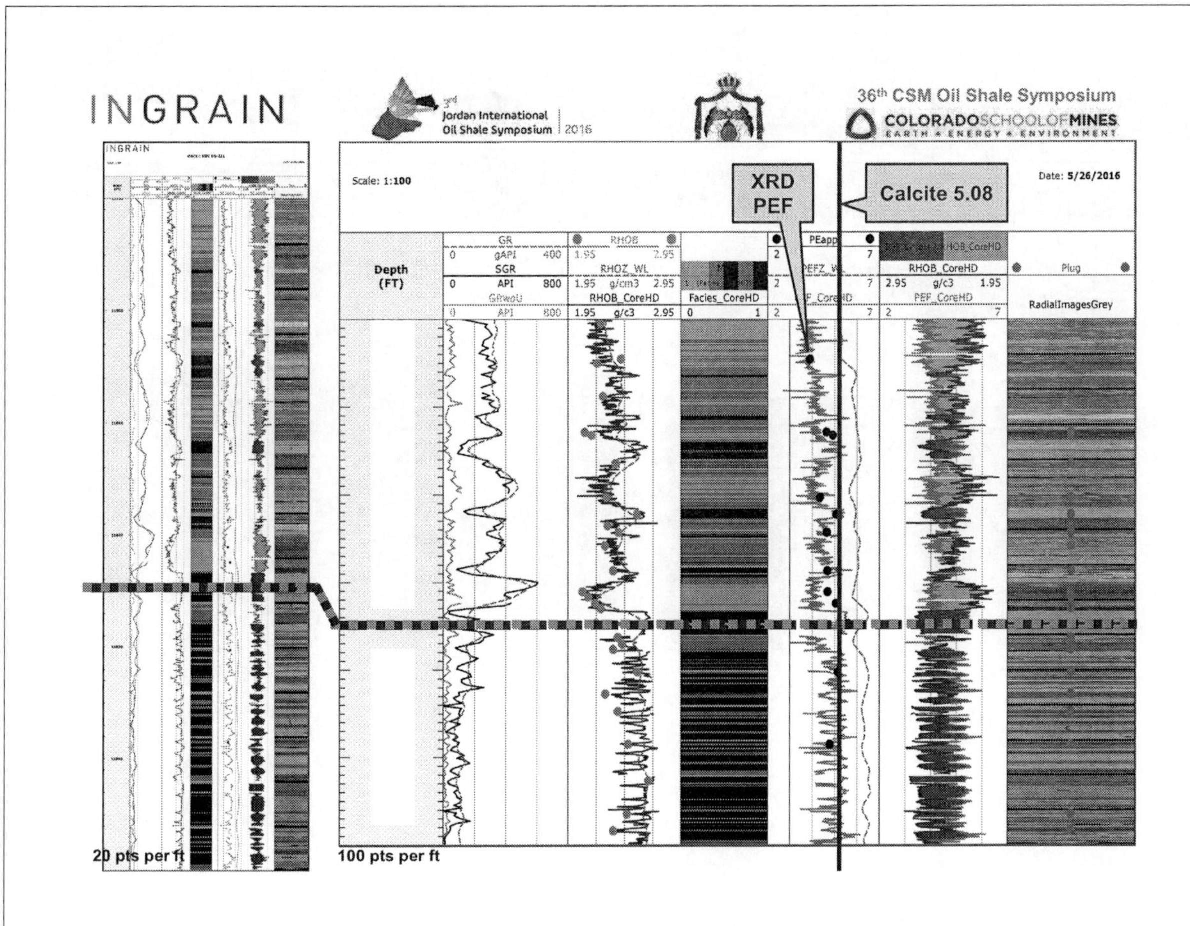
INGRAIN
3rd Jordan International Oil Shale Symposium 2016
36th CSM Oil Shale Symposium
COLORADO SCHOOL OF MINES
EARTH • ENERGY • ENVIRONMENT
Scale: 1:100
XRD PEF
Calcite 5.08
Date: 5/26/2016
Depth (FT)
GR
gAPI
SGR
API
RHOB
RHOZ_WL
g/cm3
RHOB_CoreHD
g/c3
Facies_CoreHD
PEapp
PEFZ_WL
PEF_CoreHD
RHOB_CoreHD
Plug
RadialImagesGrey
20 pts per ft
100 pts per ft

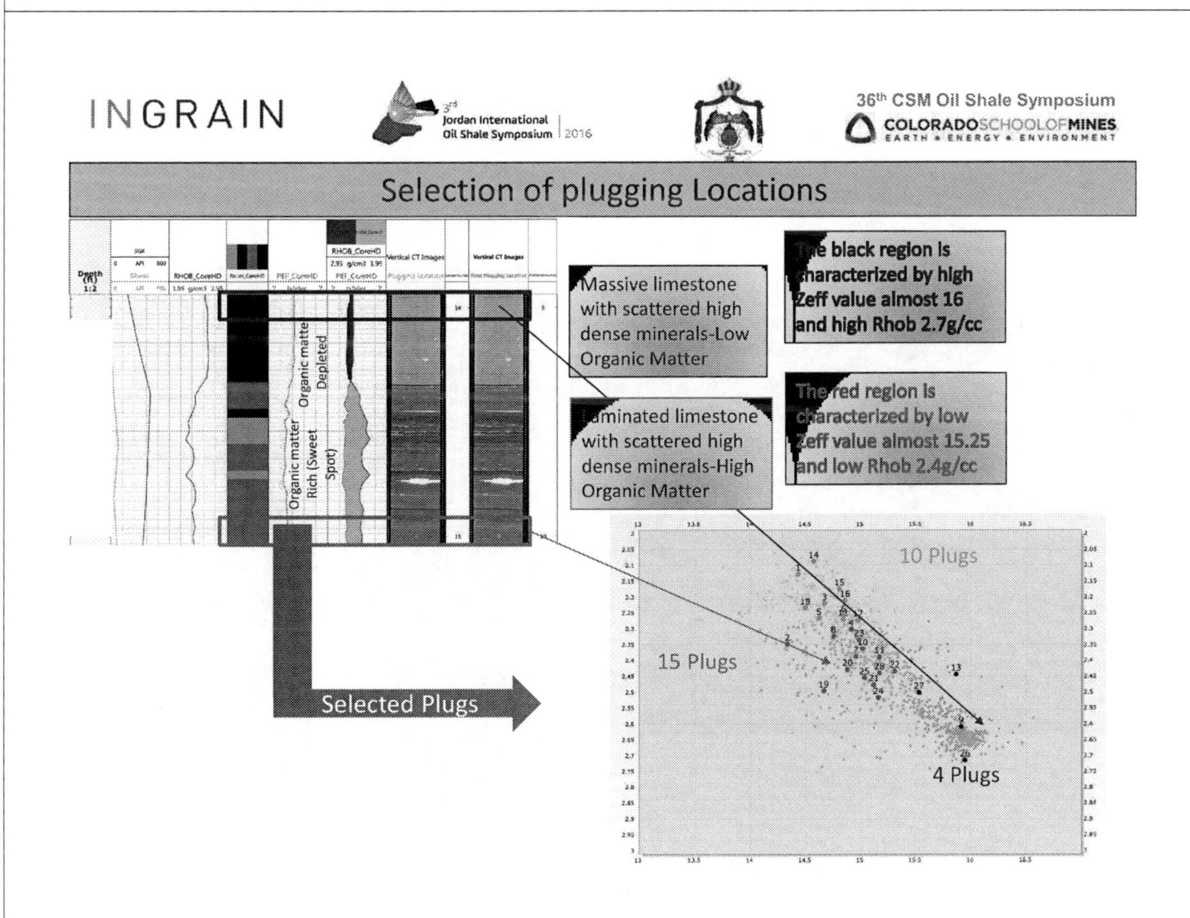
INGRAIN
3rd Jordan International Oil Shale Symposium 2016
36th CSM Oil Shale Symposium
COLORADO SCHOOL OF MINES
EARTH • ENERGY • ENVIRONMENT
Selection of plugging Locations
Massive limestone with scattered high dense minerals-Low Organic Matter
The black region is characterized by high Zeff value almost 16 and high Rhob 2.7g/cc
Laminated limestone with scattered high dense minerals-High Organic Matter
The red region is characterized by low Zeff value almost 15.25 and low Rhob 2.4g/cc
Organic matter Depleted
Organic matter Rich (Sweet Spot)
Selected Plugs
10 Plugs
15 Plugs
4 Plugs

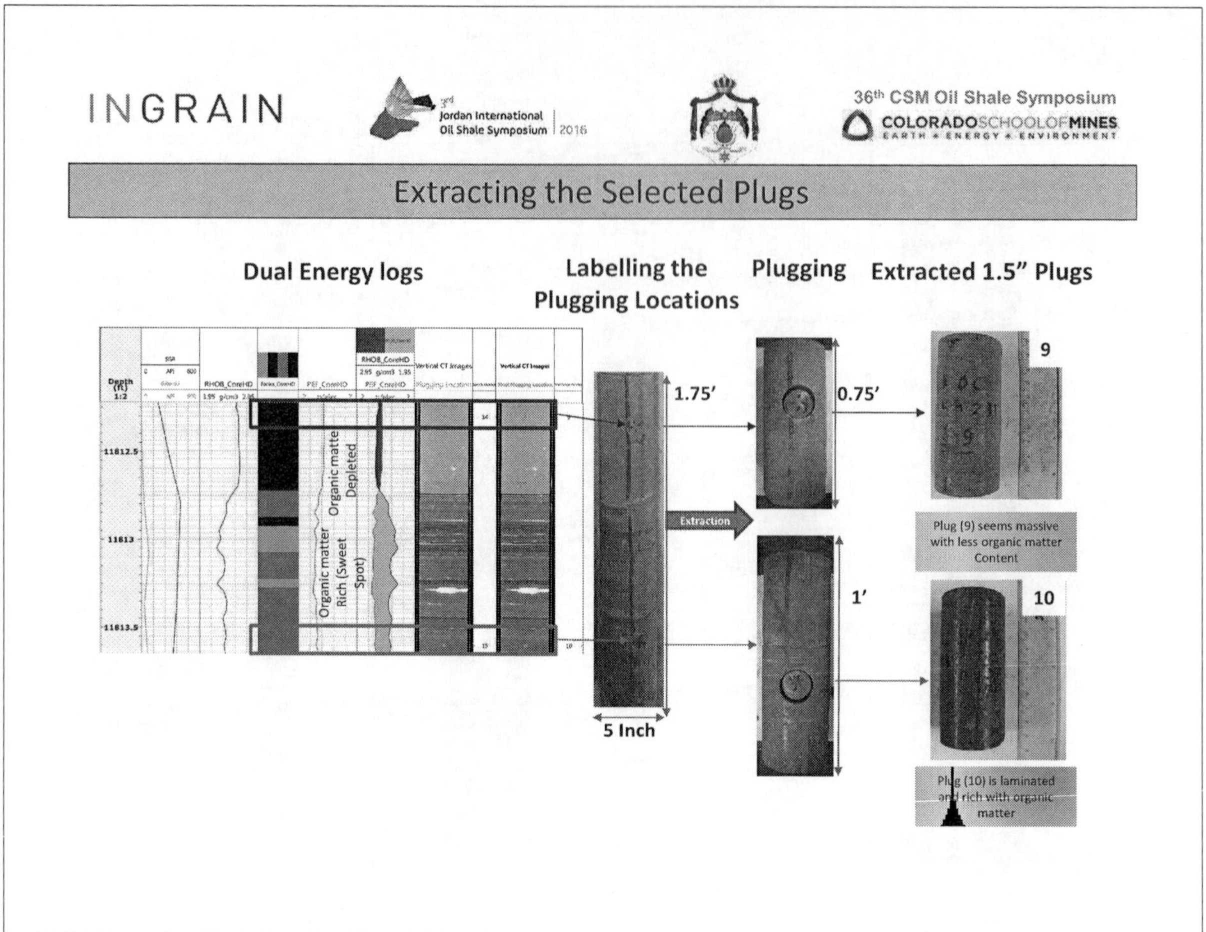

INGRAIN
3rd Jordan International Oil Shale Symposium 2016
36th CSM Oil Shale Symposium
COLORADO SCHOOL OF MINES
EARTH • ENERGY • ENVIRONMENT
Extracting the Selected Plugs
Dual Energy logs
Labelling the Plugging Locations
Plugging
Extracted 1.5" Plugs
Organic matte Depleted
Organic matter Rich (Sweet Spot)
1.75'
0.75'
1'
5 Inch
Extraction
9
10
Plug (9) seems massive with less organic matter Content
Plug (10) is laminated and rich with organic matter

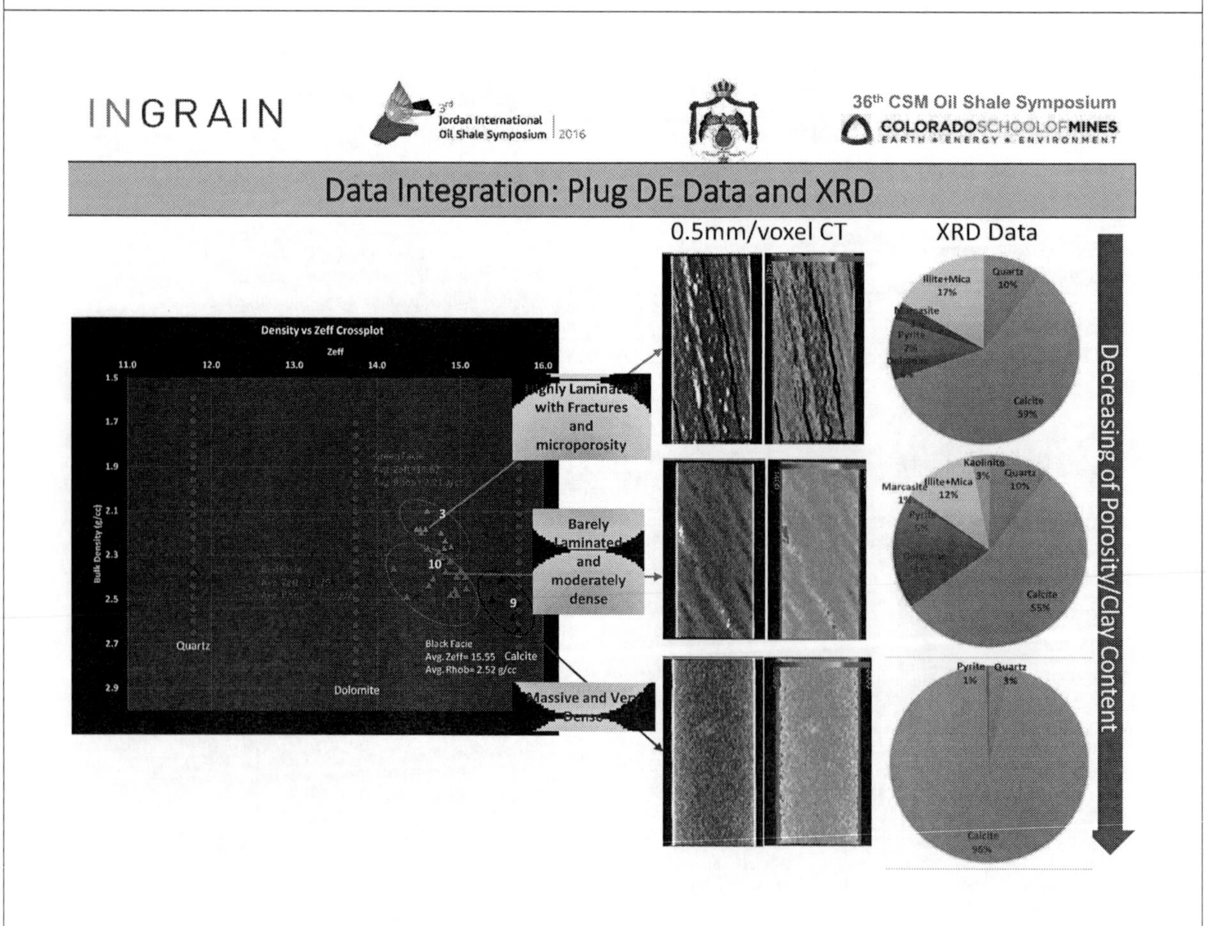

INGRAIN
3rd Jordan International Oil Shale Symposium 2016
36th CSM Oil Shale Symposium
COLORADO SCHOOL OF MINES
EARTH • ENERGY • ENVIRONMENT
Data Integration: Plug DE Data and XRD
0.5mm/voxel CT
XRD Data
Density vs Zeff Crossplot
Zeff
Bulk Density (g/cc)
Quartz
Dolomite
Calcite
Black Facie
Avg. Zeff= 15.55
Avg. Rhob= 2.52 g/cc
Highly Laminated with Fractures and microporosity
Barely Laminated and moderately dense
Massive and Very Dense
3
10
9
Illite+Mica 17%
Quartz 10%
Pyrite 7%
Calcite 59%
Kaolinite 3%
Marcasite 1%
Illite+Mica 12%
Quartz 10%
Pyrite 5%
Calcite 55%
Pyrite 1%
Quartz 3%
Calcite 96%
Decreasing of Porosity/Clay Content

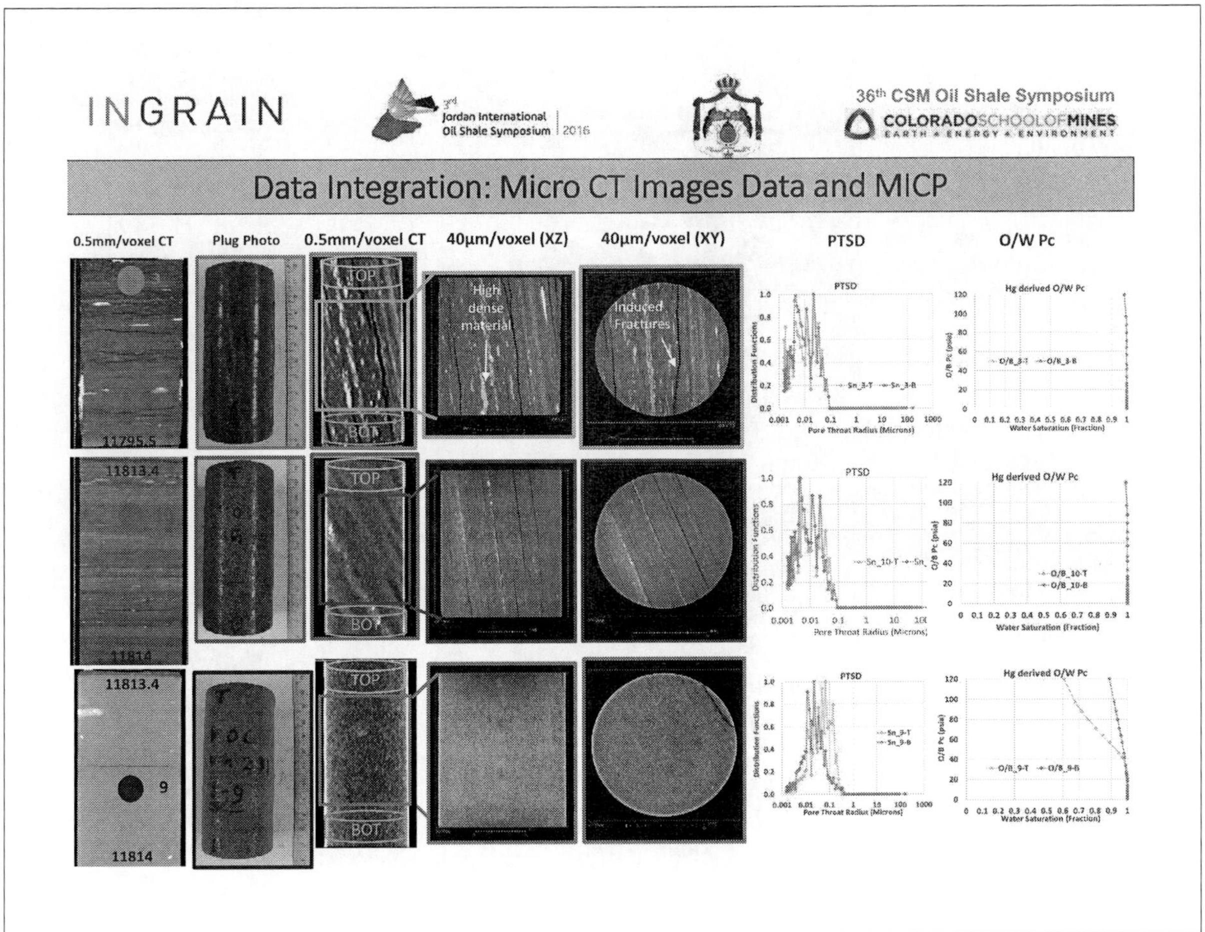
INGRAIN
3rd Jordan International Oil Shale Symposium 2016
36th CSM Oil Shale Symposium
COLORADO SCHOOL OF MINES
EARTH • ENERGY • ENVIRONMENT
Data Integration: Micro CT Images Data and MICP
0.5mm/voxel CT
Plug Photo
0.5mm/voxel CT
40µm/voxel (XZ)
40µm/voxel (XY)
PTSD
O/W Pc
11795.5
TOP
BOT
High dense material
Induced Fractures
Hg derived O/W Pc
Pore Throat Radius (Microns)
Water Saturation (Fraction)
Distribution Functions
O/B Pc (psia)
11813.4
11814
9

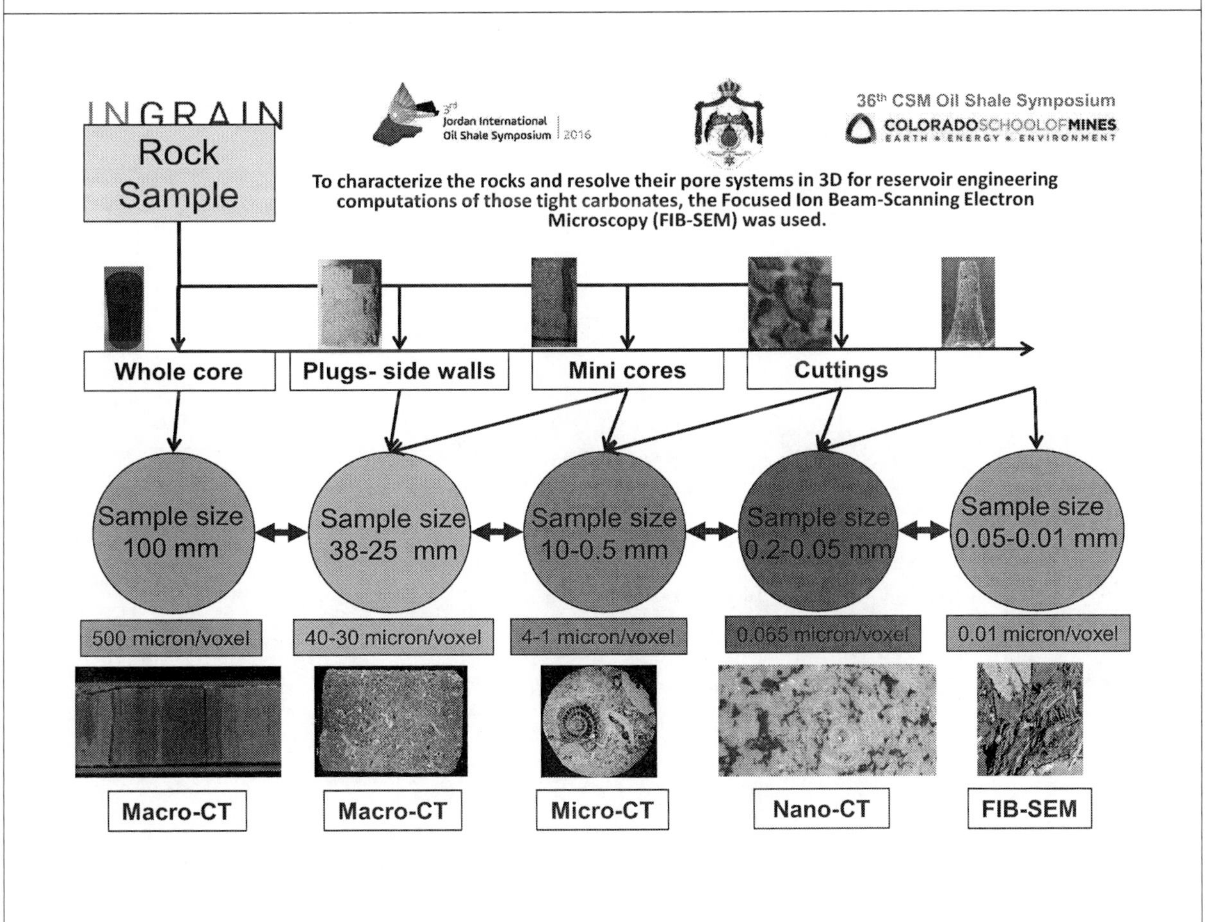
INGRAIN
3rd Jordan International Oil Shale Symposium 2016
36th CSM Oil Shale Symposium
COLORADO SCHOOL OF MINES
EARTH • ENERGY • ENVIRONMENT
Rock Sample
To characterize the rocks and resolve their pore systems in 3D for reservoir engineering computations of those tight carbonates, the Focused Ion Beam-Scanning Electron Microscopy (FIB-SEM) was used.
Whole core
Plugs- side walls
Mini cores
Cuttings
Sample size 100 mm
Sample size 38-25 mm
Sample size 10-0.5 mm
Sample size 0.2-0.05 mm
Sample size 0.05-0.01 mm
500 micron/voxel
40-30 micron/voxel
4-1 micron/voxel
0.065 micron/voxel
0.01 micron/voxel
Macro-CT
Macro-CT
Micro-CT
Nano-CT
FIB-SEM

INGRAIN

3rd Jordan International Oil Shale Symposium 2016

36th CSM Oil Shale Symposium

COLORADO SCHOOL OF MINES

Imaging Technologies

Image Processing

Nano CT scanner – 50 nm

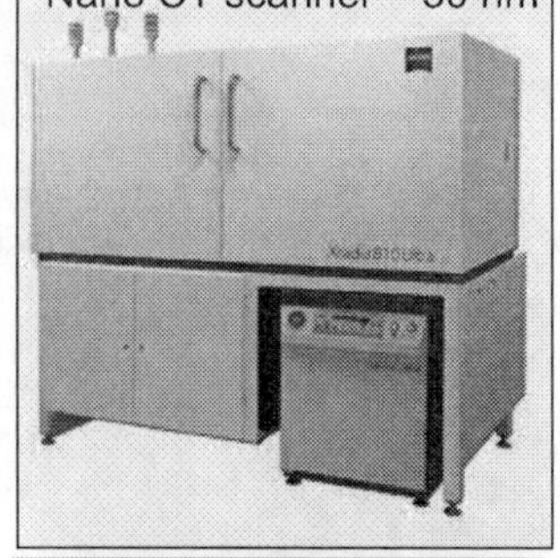

Two versions of micro CT scanners
Resolutions from 40 um to 0.5 um

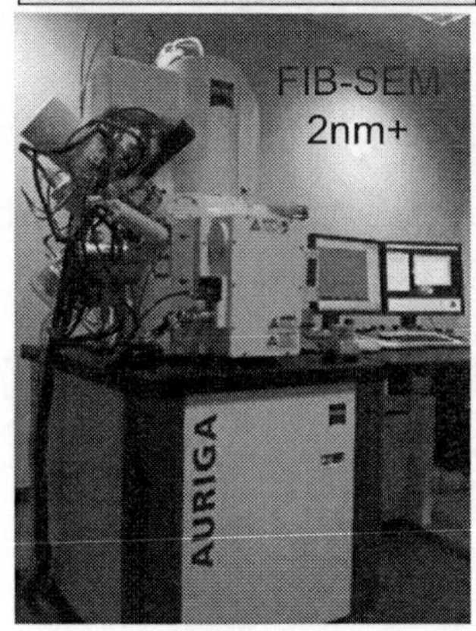

INGRAIN

3rd Jordan International Oil Shale Symposium 2016

36th CSM Oil Shale Symposium

COLORADO SCHOOL OF MINES
EARTH • ENERGY • ENVIRONMENT

High Resolution SEM Imaging: Resolving Pore System

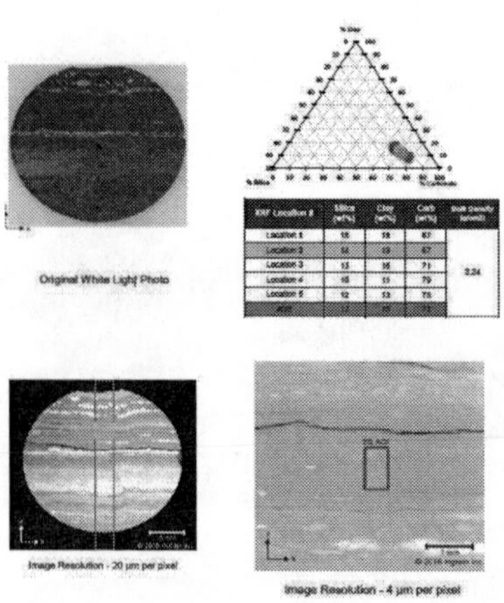

XRF, X-ray Projection Imaging and SEM sample Selection

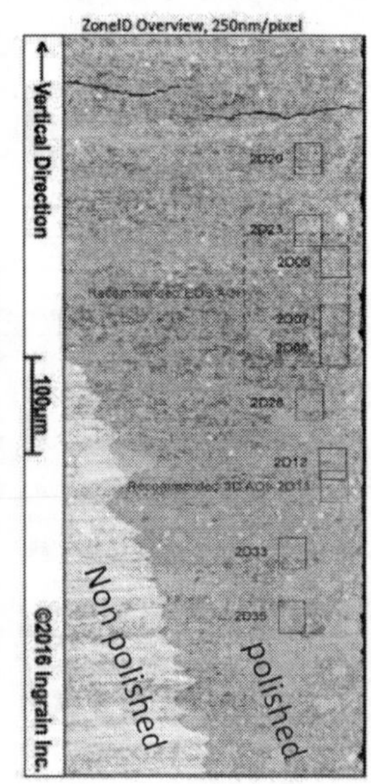

2D SEM Overview

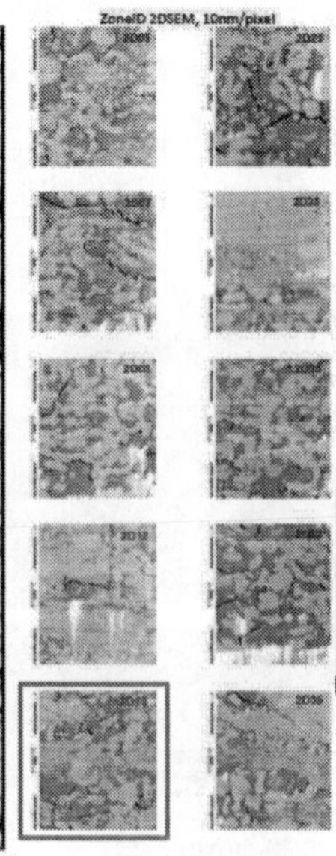

2D SEM Images @ 10nm per pixel

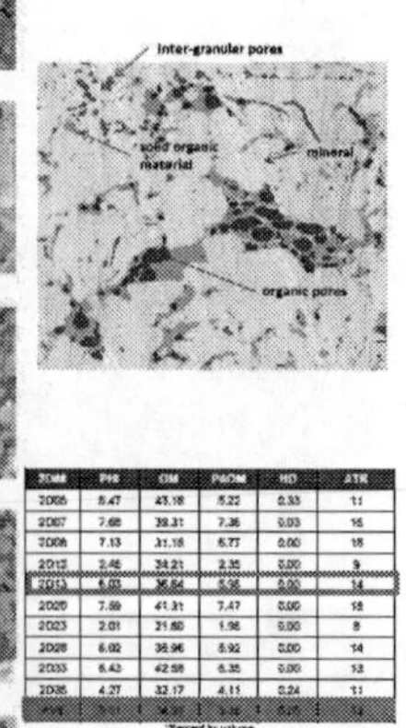

2D SEM Segmentation

INGRAIN

3rd Jordan International Oil Shale Symposium | 2016

36th CSM Oil Shale Symposium

COLORADO SCHOOL OF MINES
EARTH • ENERGY • ENVIRONMENT

High Resolution SEM Imaging: Resolving Pore System

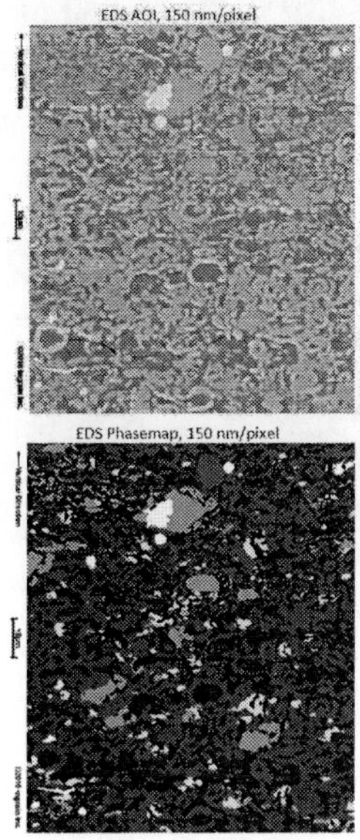

EDS Analysis @ 150 nm per pixel

Mineralogy	Volume %
Quartz	4.3
Plagioclase	0.0
K-Spar	2.5
Clay	4.0
Calcite	40.0
Dolomite	0.0
Pyrite	1.0
Rutile	0.0
Apatite	0.0
Other	48.2
Total	100

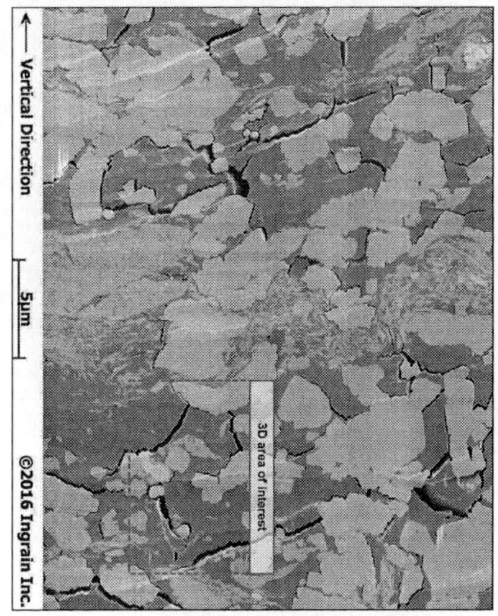

Selected 2D SEM Images @ 10nm per pixel

INGRAIN

3rd Jordan International Oil Shale Symposium | 2016

36th CSM Oil Shale Symposium

COLORADO SCHOOL OF MINES
EARTH • ENERGY • ENVIRONMENT

High Resolution SEM Imaging: Resolving Pore System

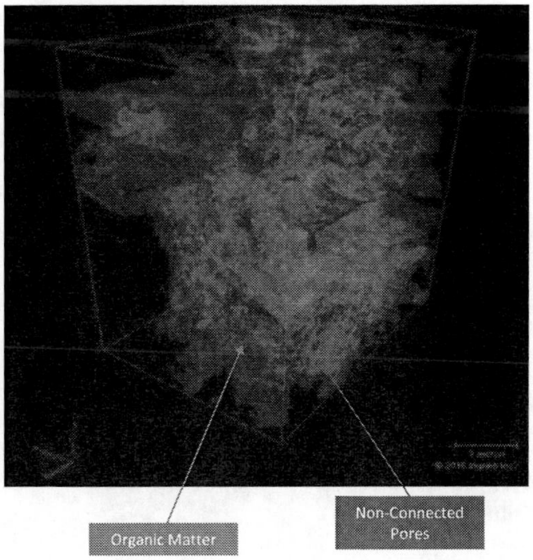

Quantifying Porosity Associated with Organic Material (PAOM) and Apparent Transformation Ratio, ATR

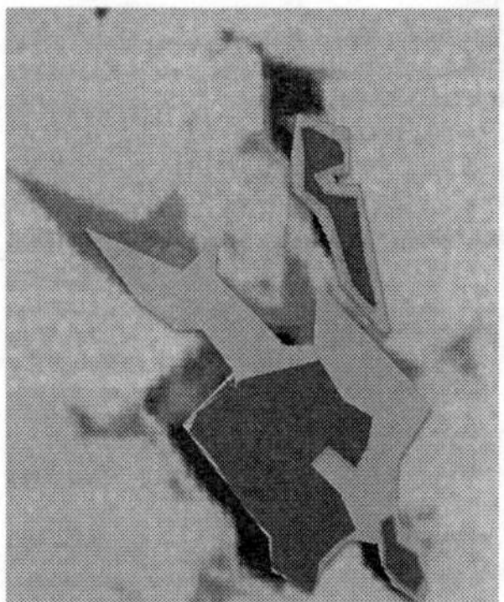

Enlarged image from FIB-SEM

Blue = Porosity associated with organic material (PAOM)
Green = Solid OM

ATR = Blue Area / (Blue + Green Area)
= PA_OM / (PAOM + Solid OM)

FIB-SEM example:
Sample has PAOM = 3.14% and solid OM = 28%
ATR = 10%

INGRAIN
3rd Jordan International Oil Shale Symposium 2016
36th CSM Oil Shale Symposium
COLORADO SCHOOL OF MINES
EARTH • ENERGY • ENVIRONMENT

Computing Absolute Permeability and Brittleness Index using HPC.

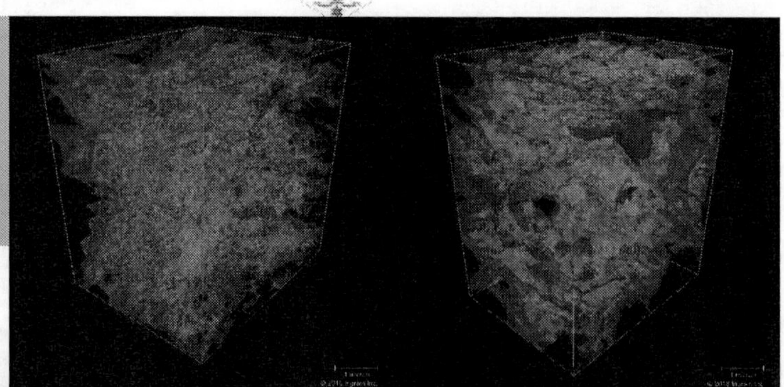

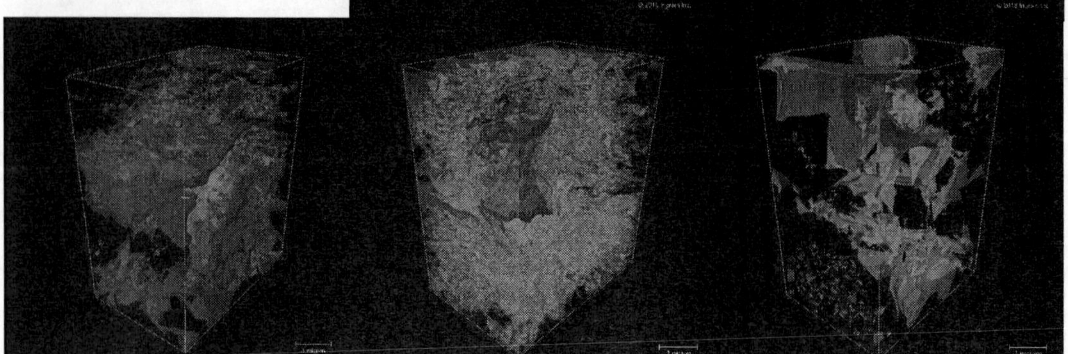

INGRAIN

3rd Jordan International Oil Shale Symposium 2016

36th CSM Oil Shale Symposium

COLORADOSCHOOLOFMINES

EARTH • ENERGY • ENVIRONMENT

Results

Formation A – Siliciclastic

(a)

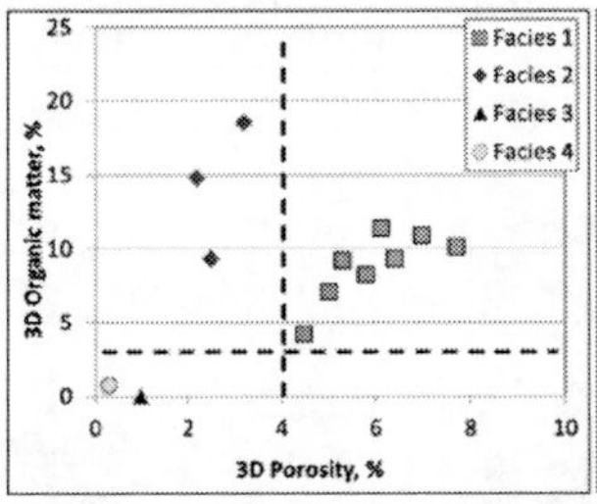

(b)

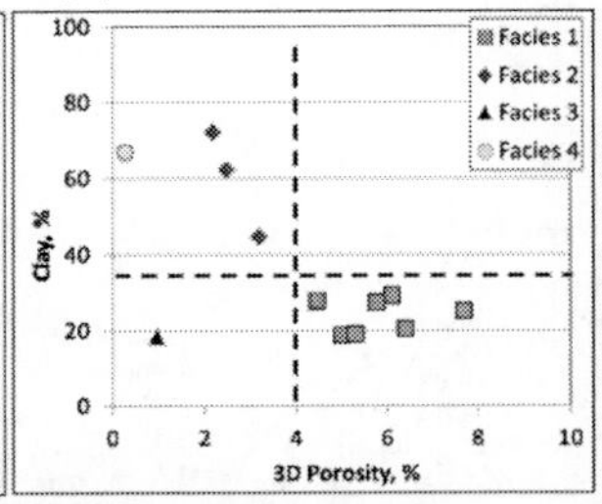

(c)

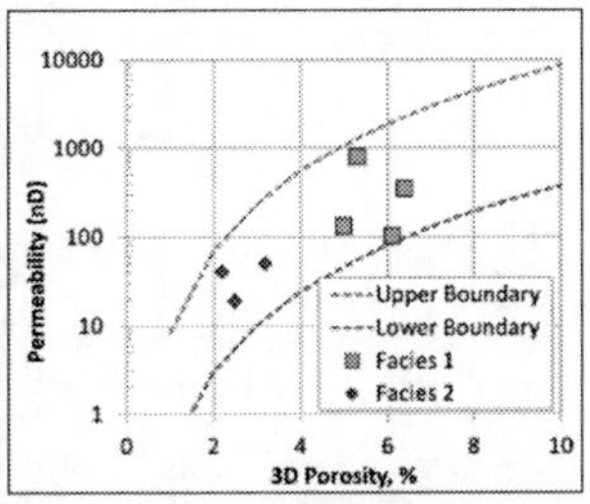

Facies 1

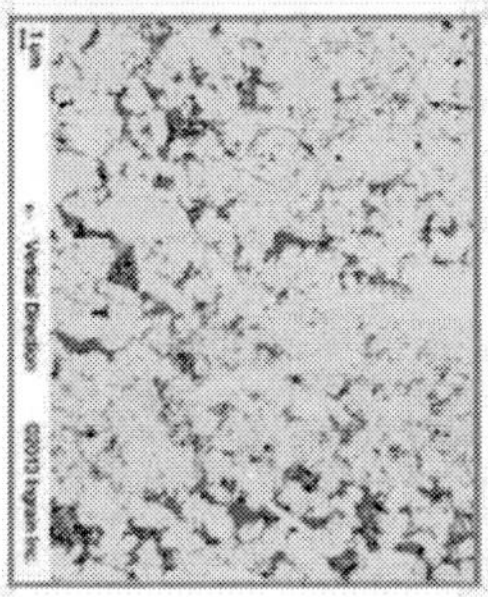

Facies 2 (high clay)

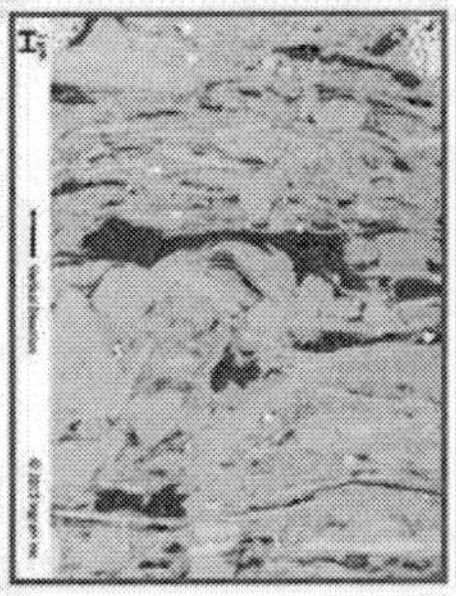

Facies 3

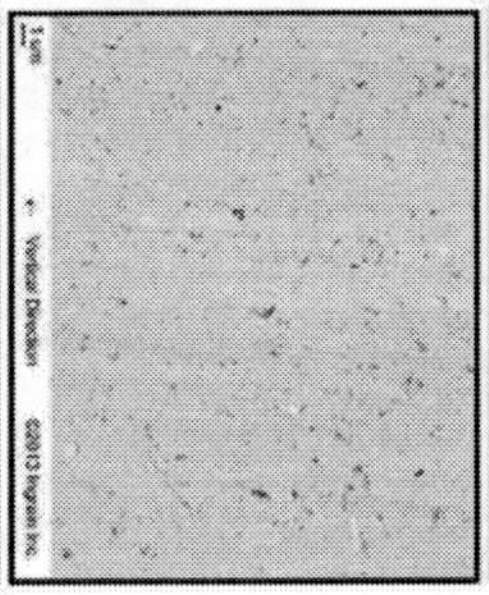

Facies 4 (high clay)

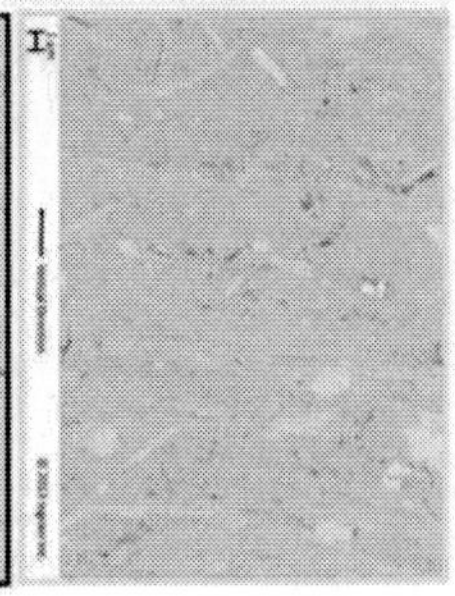

INGRAIN

3rd Jordan International Oil Shale Symposium 2016

36th CSM Oil Shale Symposium

COLORADOSCHOOLOFMINES

EARTH • ENERGY • ENVIRONMENT

Results

Formation B – Carbonaceous (Low clay content)

(a)

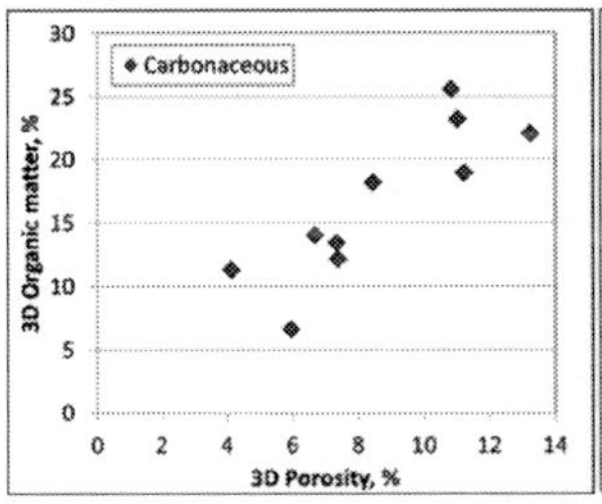

(b)

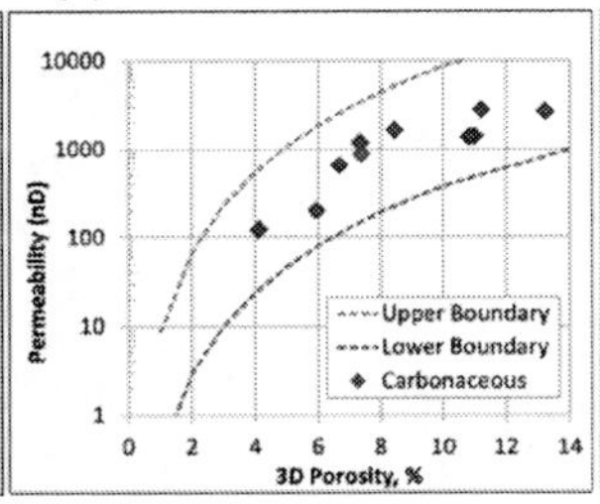

(c)

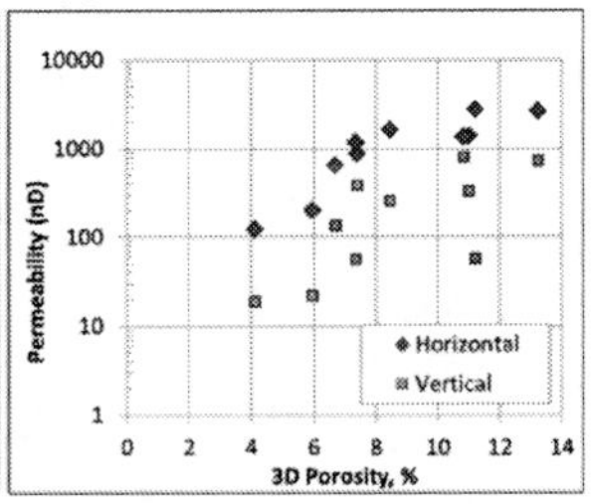

Low porosity/OM

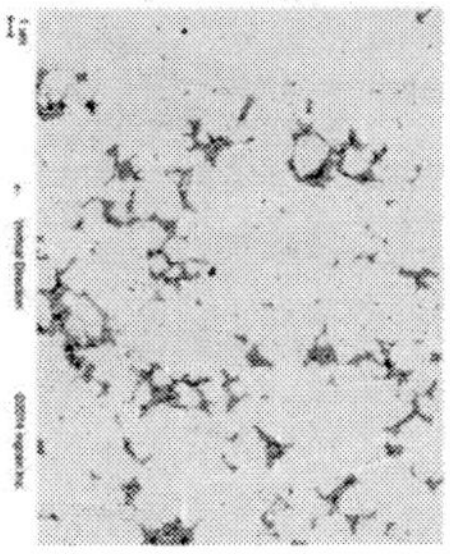

High porosity/OM

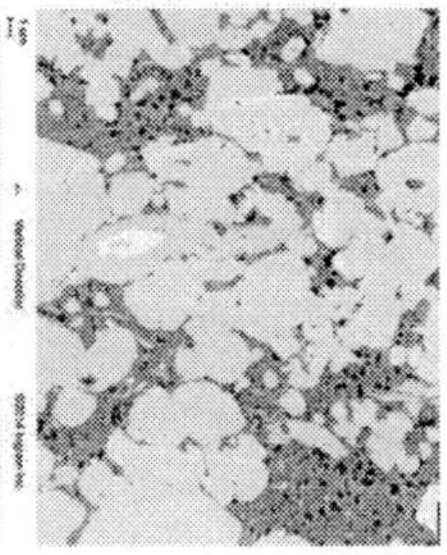

INGRAIN

3rd Jordan International Oil Shale Symposium | 2016

36th CSM Oil Shale Symposium

COLORADOSCHOOLOFMINES
EARTH • ENERGY • ENVIRONMENT

Results

Formation C – Carbonaceous (Low clay content)

(a)

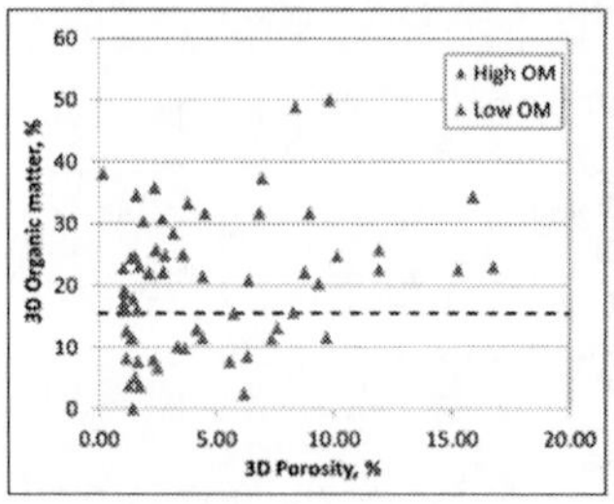

(b)

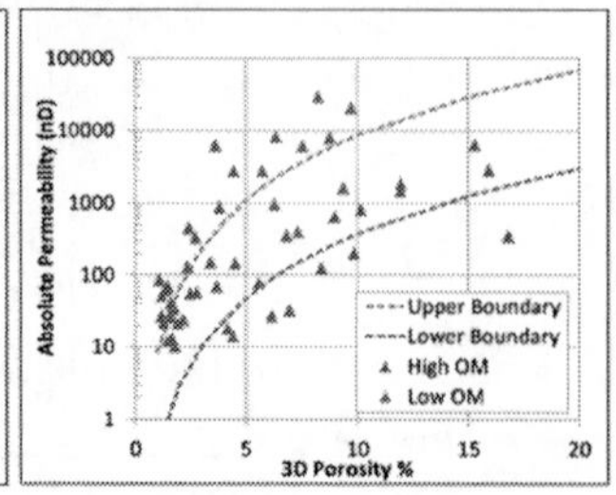

(c)

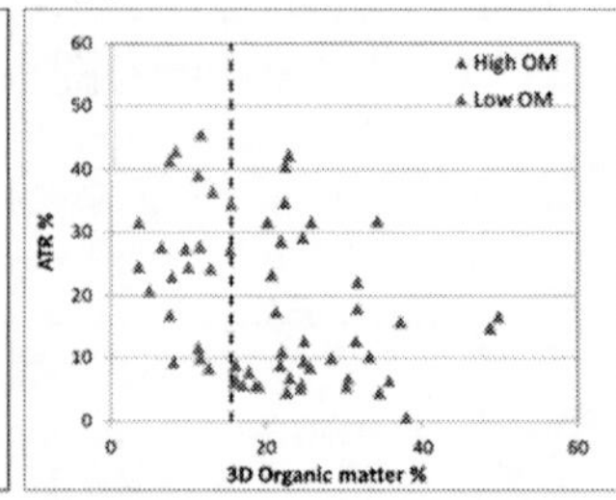

(d)

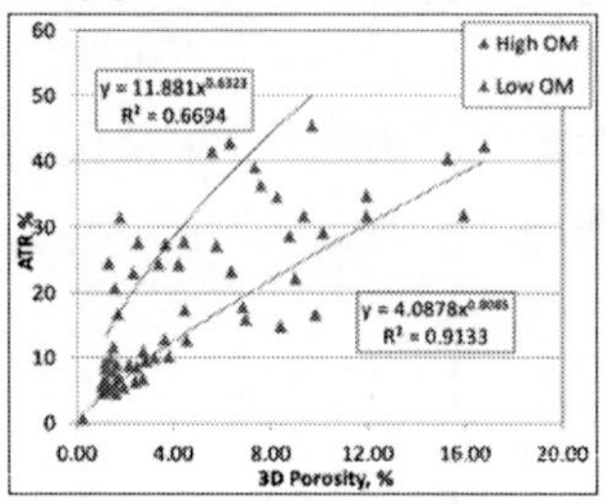

(e)

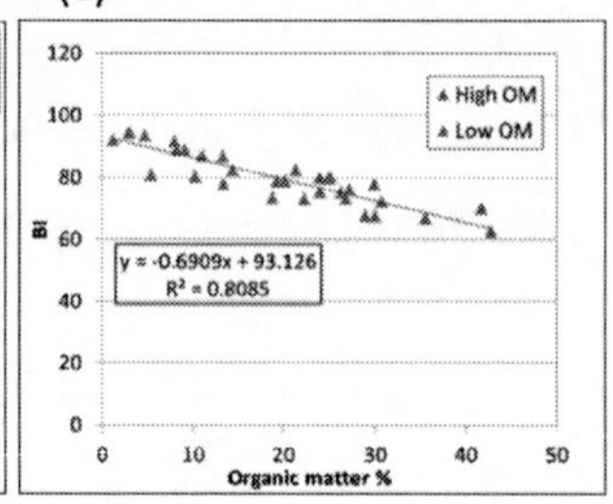

INGRAIN

3rd Jordan International Oil Shale Symposium | 2016

36th CSM Oil Shale Symposium

COLORADOSCHOOLOFMINES
EARTH • ENERGY • ENVIRONMENT

Results

High OM%

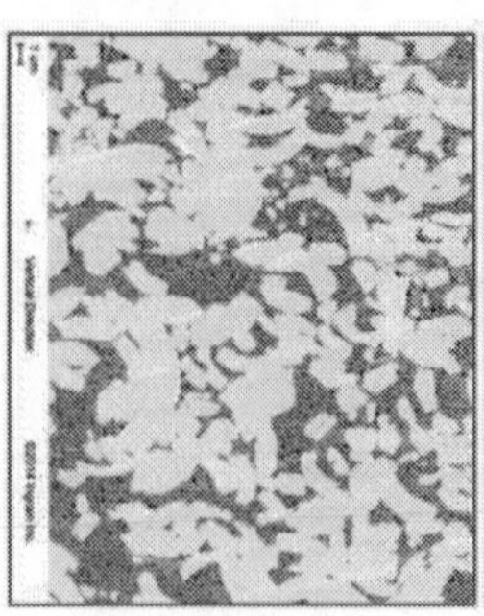

phi = 10.16%

ATR = 29.12%

Low OM%

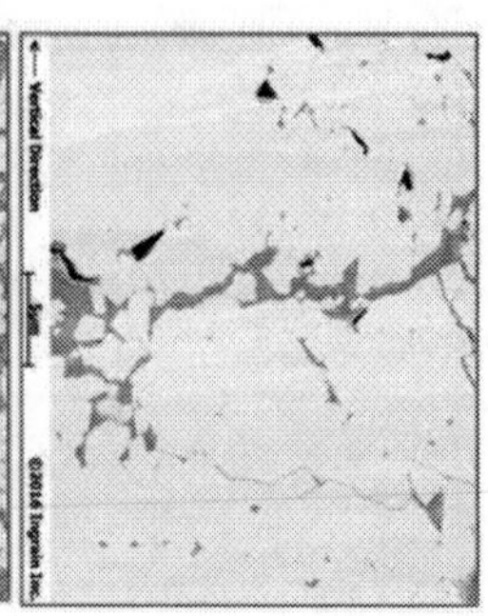

phi = 2.52%

ATR = 27.59%

High OM%

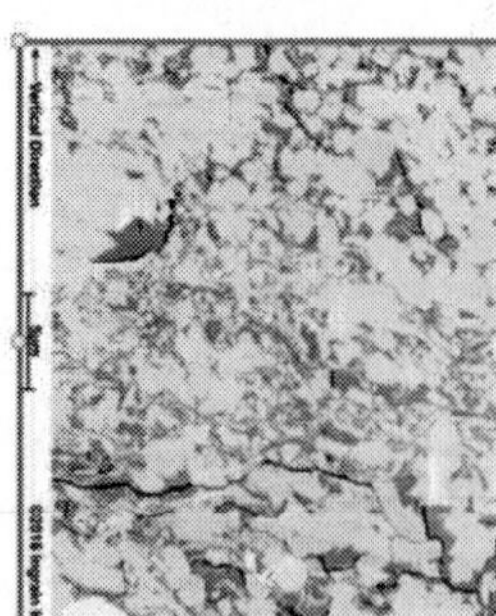

phi = 6.38%

ATR = 23.14%

Low OM%

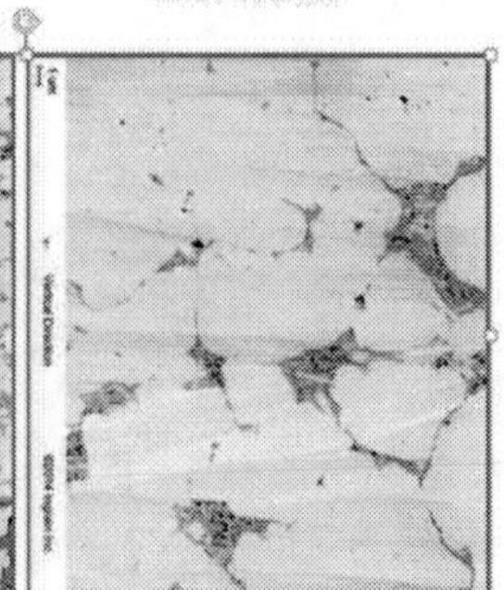

phi = 6.32%

ATR = 43%

INGRAIN

3rd Jordan International Oil Shale Symposium 2016

36th CSM Oil Shale Symposium

COLORADO SCHOOL OF MINES
EARTH • ENERGY • ENVIRONMENT

Key Findings and Summary

- More than one-thousand data sets from 14 wells and three different formations in the Middle East were analyzed in this study.
- Advanced DRP work flow was adopted in this study which take in consideration the high geological, geochemical and petrophysical heterogeneity of shale source rocks.
- Large property variations were seen in the evaluated formations, and each formation showed unique characteristics and different controlling parameters.
- Mineralogy, clay content and organic matter content were three main controlling factors that had direct effects on the porosity.
- Organic matter and pore type were also critical parameters in 3D connectivity and determination of permeability.
- In a single mineralogy system, a strong relation was observed between brittleness index and organic matter.

INGRAIN

3rd Jordan International Oil Shale Symposium 2016

36th CSM Oil Shale Symposium

COLORADO SCHOOL OF MINES
EARTH • ENERGY • ENVIRONMENT

Novel/Additive Information

- The data studied provided statistical information about the dependence of shale properties on many different factors in different formations.

- These data trends can be linked to the varying depositional environments that were found to have a great effect on the pore types, rock fabric and petrophysical properties.

- Such analysis is a critical addition to the understanding of shales in the Middle East, and may be used in direct comparisons with shale plays in North America.

INGRAIN

3rd Jordan International Oil Shale Symposium | 2016

36th CSM Oil Shale Symposium

COLORADO SCHOOL OF MINES
EARTH • ENERGY • ENVIRONMENT

Acknowledgments

The authors would like to express their special appreciation and thanks to Jordan Ministry of Energy and Mineral Resources for giving us the opportunity to present this work in such important event. I would also like to thank Mr. Bahjat Al-Adwan and Ms. Ghussaina Al Hilu from MEMR and the organizers for their support and cooperation.

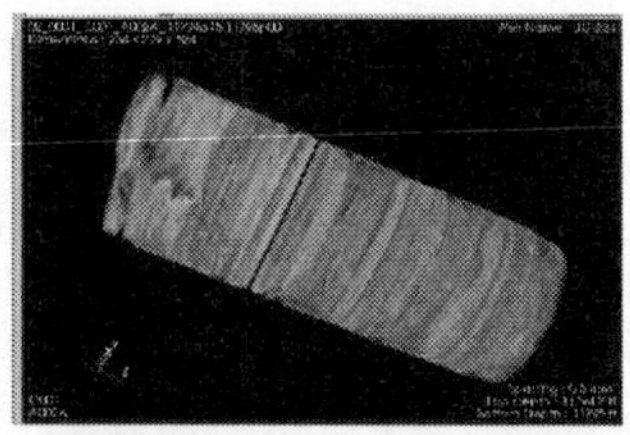

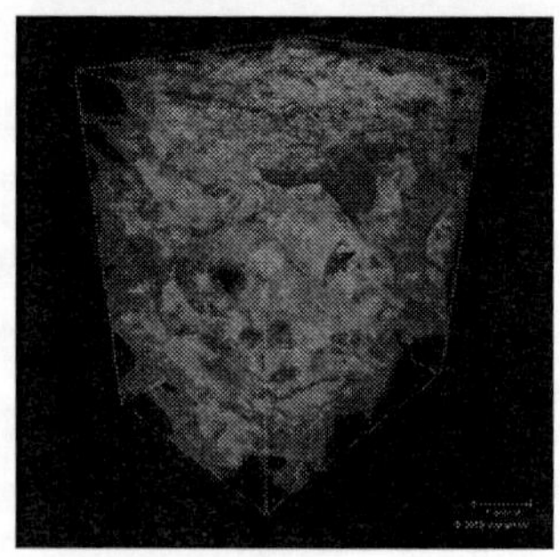

Questions?

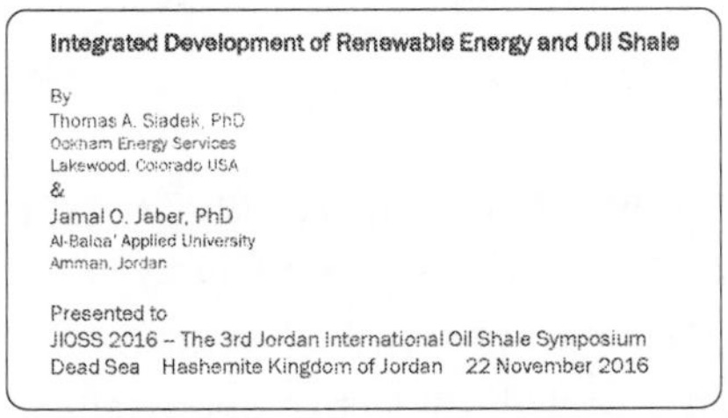

Integrated Development of Renewable Energy & Oil Shale

By Thomas A. Sladek, PhD and Jamal O. Jaber, PhD

Presented to 3rd Jordanian International Oil Shale Symposium
Dead Sea Hashemite Kingdom of Jordan 22 November 2016

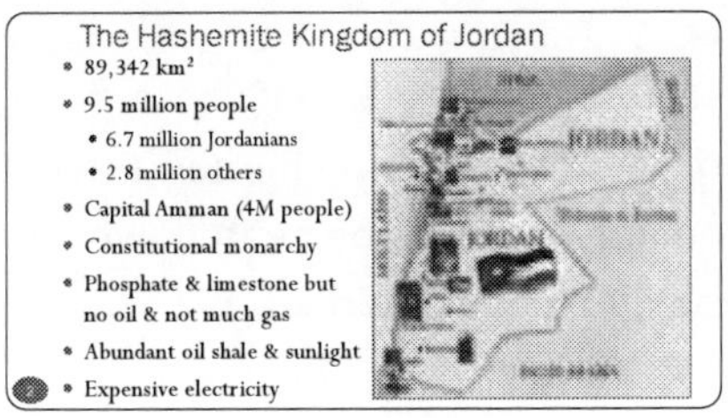

The Hashemite Kingdom of Jordan is located on the western edge of Asia, in the region commonly called the Middle East, south of Syria and west of Iraq and Saudi Arabia. Jordan is 89,342 km^2 in size, about the size of Indiana in the United States or Portugal or Hungary in Europe. Approximately 9.5 million people live in Jordan, including 6.7 million Jordanians and 2.8 million people from other countries, especially Syria, Egypt, and Iraq. Four million people (42% of the population) live in Amman, Jordan's political capital, which is in the northern part of the country, about 100 km south of the border with Syria. Amman's population has more than doubled in the past ten years, largely because of a flood of emigrants and refugees from Syria and Iraq.

Jordan is landlocked except for 26 km of shoreline on the Gulf of Aqaba in the Red Sea. Most of the land is on an arid desert plateau. Precipitation is sparse. Less than 4% of the land is capable of growing crops. The scarce surface water is fully utilized, and the large groundwater resources are being depleted. Jordan has large resources of useful minerals, such as phosphate rock and limestone, but only a small amount of natural gas and essentially no petroleum or coal. The near absence of conventional fossil fuels has caused serious problems for the Kingdom, because its need for energy is large and growing. Jordan imports two-thirds of the natural gas it consumes and essentially all of the crude oil. This reliance on imported energy has warped the Kingdom's economy and endangered its economic, political, and social wellbeing. Electricity has been especially problematic, with high costs and frequent disruptions in the supply of essential fuels.

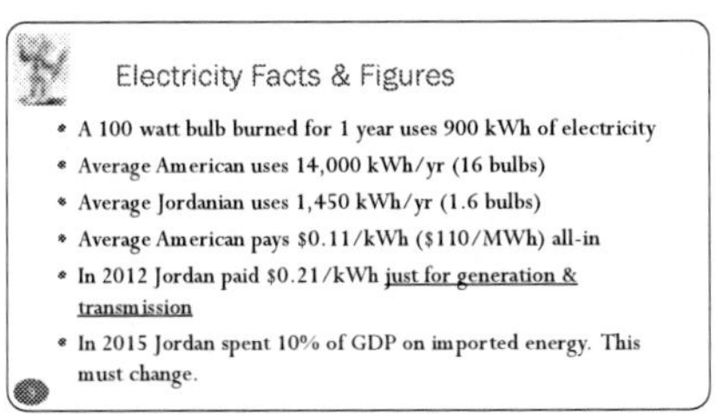

Although nearly all Jordanians have access to electricity, they are very frugal with its use. Per capita consumption of electricity in Jordan is only one-tenth that of the United States. One reason is the high cost. American pay only 11 US cents per kilowatt hour, including the costs of fuel, generation and transmission services, taxes and surcharges, and a bit of profit for each of the players. In 2012, Jordanians spent 21 cents per kilowatt hour just for generation and transmission. A big part of the expense was the cost of the fuel burned in the power plants. In the USA, 96% of electricity is from coal, natural gas, nuclear, and hydroelectric dams. Only 1% is from oil. In Jordan, more than 90% of electricity is generated from imported oil and gas. A power plant burning $50 per barrel oil will spend 8 to 14 US cents per kilowatt hour just for the oil.

Oil prices have decreased in recent years, but they may go up again if OPEC and Russia carry out their plans to restrict supplies. In 2015, despite low oil prices, Jordan spent 10% of its gross domestic product on imported energy. This outflow of capital endangers the security and economic wellbeing of the Kingdom, and it must change.

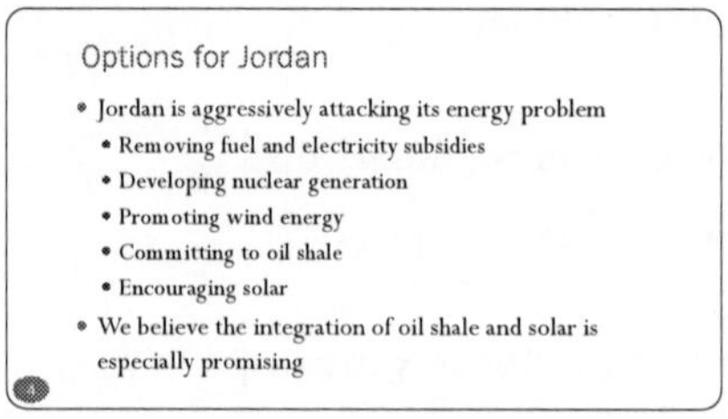

The Government of Jordan is aggressively attacking its energy problem, by removing subsidies and encouraging development of nuclear energy, wind, oil shale, and solar. We believe the integration of the last two resources -- oil shale and sunlight -- is an especially attractive approach.

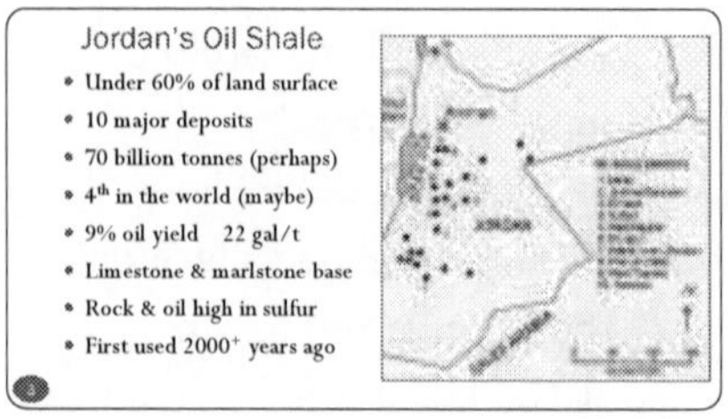

Jordan has huge resources of oil shale -- sedimentary rock that contains kerogen, an immature form of petroleum. Oil shale may be burned as a solid fuel in industrial boilers and power plants, or it can be heated, which causes the kerogen to decompose to form liquid and gaseous hydrocarbon products and a carbonaceous solid residue. The liquid product -- crude shale oil -- can be upgraded by thermal and chemical treatment to form a synthetic crude oil, which can be processed like conventional crude in oil refineries. The gaseous product can be burned to produce process heat or electricity. The solid residue -- called retorted oil shale or spent shale -- also has value as fuel, especially in cement kilns, where its mineral constituents become part of the cement product.

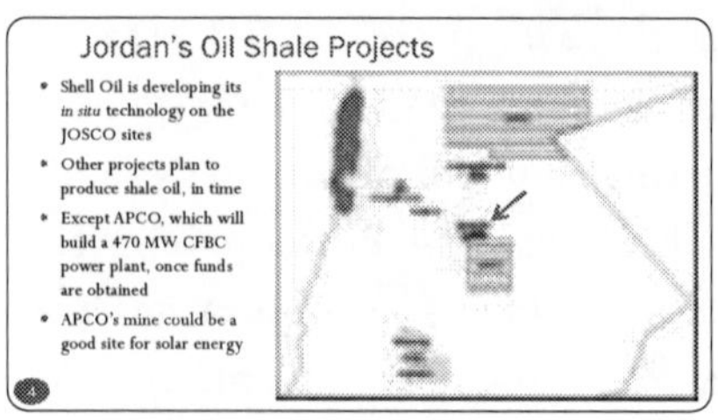

In 2005, the Government of Jordan embarked on a complex, long-term program to develop its oil shale resources, to supply electricity to the national grid and motor fuels for the transportation sector. After ten years of diligent effort, agreements are in place for field testing of Shell Oil's *in situ* retorting technology, for several projects to produce liquid fuels from oil shale by surface mining and aboveground retorting, and for a project to develop a large power plant fueled with solid oil shale. Shell is proceeding with its work, but development of the aboveground retorting projects has been slowed by low oil prices. APCO's power plant project is ready to proceed, once final approvals are provided for the financing. The 470 net MW project will use surface mining to extract oil shale from the Attarat Um Ghudran deposit. Crushed oil will be burned in fluidized bed boilers, and the electricity will be sold to the national grid under a power purchase agreement with the National Electric Power Company. All of the oil shale projects, and especially the APCO project offer a way to accelerate development of another of Jordan's energy resources -- renewable energy from sunlight.

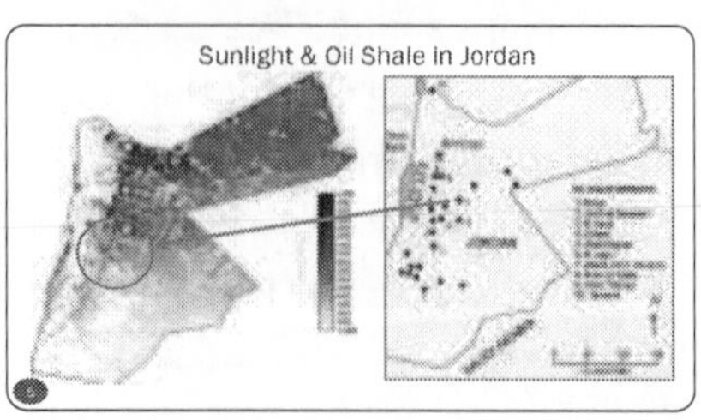

Jordan lies at about 31° north latitude, like Tucson and San Diego in the United States, and like those cities, Jordan enjoys intense, abundant sunlight for most of the calendar year. On the slide, solar irradiation increases as we move down the scale from blue to bright yellow and white. Irradiation in most of central Jordan is very high, at least 2500 kilowatt hours per square meter per year. This compares favorably with the 2700 units of direct normal irradiation in Colorado's San Luis Valley, which is the highest in the USA. The Government of Jordan has already taken steps to encourage solar development throughout Jordan. Although there has been good response, and projects are proceeding, widespread implementation has been impeded by the need for large areas of land, difficult access to the national power grid, distance from the urban areas that constitute the largest power sinks, security, and, of course, cost. Development of photovoltaic facilities in coordination with development of oil shale facilities could overcome all of these impediments.

Jordan's best solar area includes all of the oil shale region and especially the Attarat Um Ghudran oil shale deposit, where Attarat Power Company has proposed to build its electrical generating station. Of APCO's total concession area of 38 km^2, 11 km^2 will be subjected to surface mining, to extract oil shale feed for the 470 net MW power plant and to dispose of the ash product from that plant. The mined areas, after reclamation, could be used to support a large photovoltaic power plant, like the 30-MW Alamosa Solar Project in the San Luis Valley, the largest concentrating photovoltaic power plant in the world.

Photovoltaic electricity has become much more practical in recent years, because the costs of building such renewable energy plants have decreased substantially since the global recession really took hold in 2012. In 2015, projects in Africa had estimated capital costs of as little as $1350 per kilowatt of installed capacity, and the global average for utility-scale projects was $1800 per kW. Large projects attain the lowest investment costs. A 100 MW project in Jordan certainly would qualify as a large PV project. Let's look at the economics of such a project.

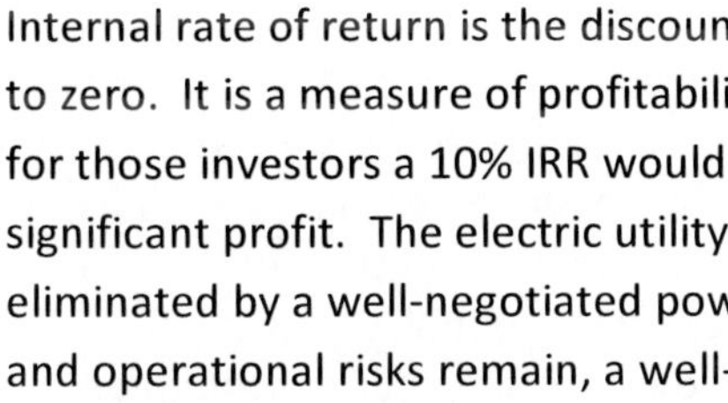

The slide shows the assumptions used in a financial model to forecast the economic performance of a PV power plant with a nameplate capacity of 100 MW. The independent variable is the power tariff - the price charged by the project for its electricity. In the model, the magnitude of the tariff was adjusted to provide a 10% internal rate of return on invested equity from pretax cash flow, over the facility's 20-year operating life.

Internal rate of return is the discount rate which, when applied to annual cash flows, reduces the total cash flow to zero. It is a measure of profitability. Many investors pay at least 10% to obtain the money they invest, and for those investors a 10% IRR would be a breakeven return -- covering the cost of capital but conveying no significant profit. The electric utility business is different, however, because most of the marketing risk can be eliminated by a well-negotiated power purchase agreement with the host grid operator. Although development and operational risks remain, a well-managed utility company can obtain low-cost equity capital by, for example, issuing instruments of corporate debt, such as bonds. Xcel Energy, a large investor-owned utility company that serves my part of Colorado, is guaranteed a 10% return on its invested equity and must make its profits on the margin between that 10% return and its weighted average cost of capital. To make profits, Xcel must raise equity cheaply, by borrowing money at low interest rates, as it has been doing for 107 years. Its present weighted average cost of capital is 2.57%, and its return on invested capital is 5.78%. It would not be offended by a 10% IRR.

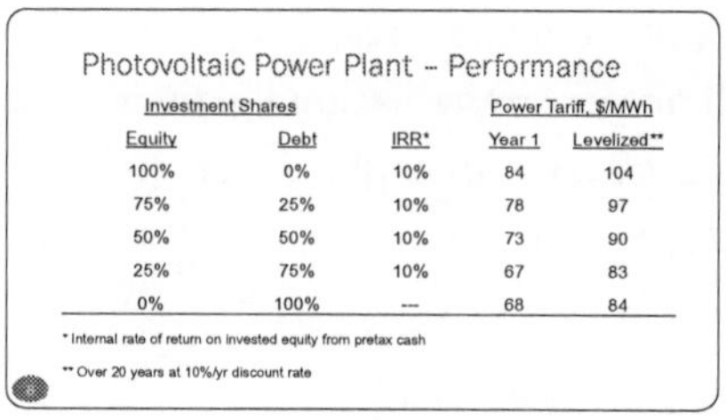

Photovoltaic Power Plant -- Performance

Investment Shares			Power Tariff, \$/MWh	
Equity	Debt	IRR*	Year 1	Levelized**
100%	0%	10%	84	104
75%	25%	10%	78	97
50%	50%	10%	73	90
25%	75%	10%	67	83
0%	100%	---	68	84

* Internal rate of return on invested equity from pretax cash

** Over 20 years at 10%/yr discount rate

Financing with 100% equity requires the highest tariff -- \$84 per megawatt hour in the project's first year. The tariff decreases rapidly as equity is replaced with relatively cheap debt. With 75% debt, the tariff is only \$67 per megawatt hour -- 6.7 US cents per kilowatt hour. This is about twice the typical wholesale price in the United States but less than half of the rates charged in Jordan for mines, bankers, and other large industrial users. The tariff rises slightly with 100% debt financing, to ensure enough cash is generated to service debt in the plant's first operating year. If tariffs in subsequent years were reduced to just cover operating costs and debt service, the levelized tariff would decrease from \$84 to \$81.

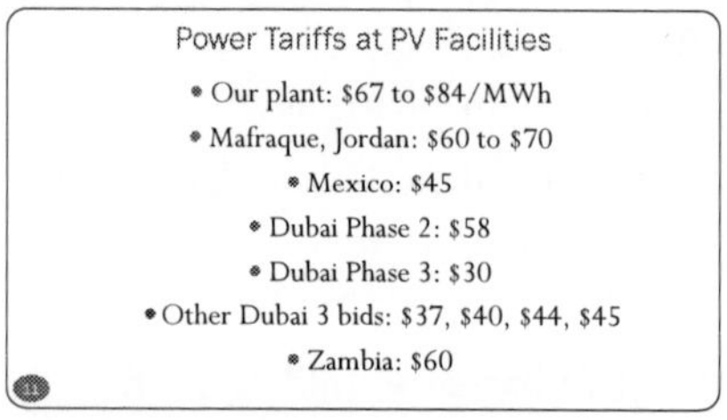

These low prices are similar to those reported for utility-scale PV plants in other locations. The average tariff for a 50 MW PV plant proposed for a site near Mafraque in Jordan was \$60 to \$70 per MWh. A large project in Mexico contracted for PV power at \$45 per MWh. The winning bid for the second phase of a solar energy park in Dubai was \$58 per megawatt hour. Eighteen months later, a contract for the third phase was awarded at \$30 per MWh. Four other bids were received, for \$37, \$40, \$44, and \$45. The Republic of Zambia recently announced a power purchase agreement under which electricity will be purchased from a utility-scale solar PV project at a cost of \$60/MWh. That project was developed with support from the World Bank's Scaling Solar Program, which offers financing, insurance, and advisory services and significantly reduces the exchange rate, offtake, and country risks which have traditionally impeded investment in long-term, capital intensive projects in developing nations.

Such low prices are possible because the cost of photovoltaic systems declined by nearly 80% between 2007 and 2014. Some experts expect the cost to drop another 40% by 2020. Low capital costs, combined with low operating and maintenance costs and no fuel costs, have yielded projects with very low costs for the delivered energy. Electricity made from sunlight, at least in large-scale applications, can now be less expensive than electricity obtained by burning oil and gas.

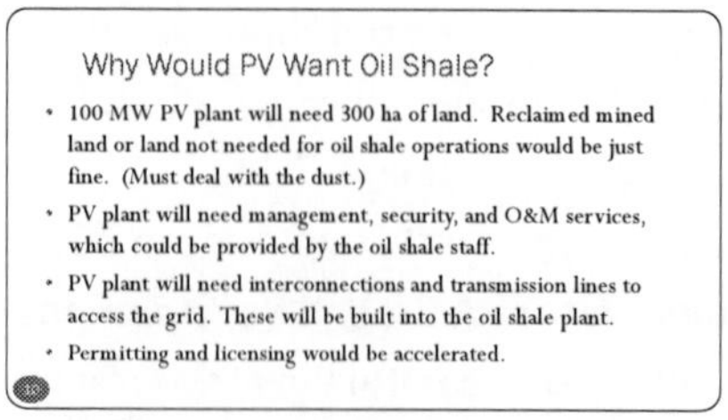

So why would our power plant want an oil shale site? Land is one reason. A large PV power plant requires approximately 3 hectare of land per MW of capacity, for the solar arrays, access roads, substations, service buildings, and other infrastructure. Our 100 MW power plant would need 300 hectare. It is difficult to find such large areas of continuous, accessible, unoccupied land, even in the sparsely populated areas of central Jordan. One outstanding exception is the land committed to oil shale production. Reclaimed mined land has very few short-term uses. Might as well put solar collectors on it. Location near a surface mine does present one operating challenge -- the dust generated by the mining and waste disposal operations. The dust would have to be removed frequently, to maintain collection efficiency. The problem is not unique to oil shale. The PV plant in the San Luis Valley, for example, is surrounded by potato fields, which are bare for most of the year. Wind speeds are high, and dirt moves around all the time.

Ancillary facilities are another reason. The oil shale power plant will be equipped with the lines and interconnection gear needed to deliver electricity to the national grid. The design of that equipment could be altered to handle the output from a PV plant at relatively little cost. The PV plant could also contract with the oil shale plant's managers and operators to handle security and the ongoing operations and maintenance requirements, thus reducing the need for separate staff.

Accelerated development is a third reason. Siting a PV plant requires a range of environmental and social studies, such as establishing air and water quality baselines and completing biological and archeological surveys. Much of this work has already been carried out for the oil shale project. The reports should only require updating, which should substantially expedite the permitting tasks.

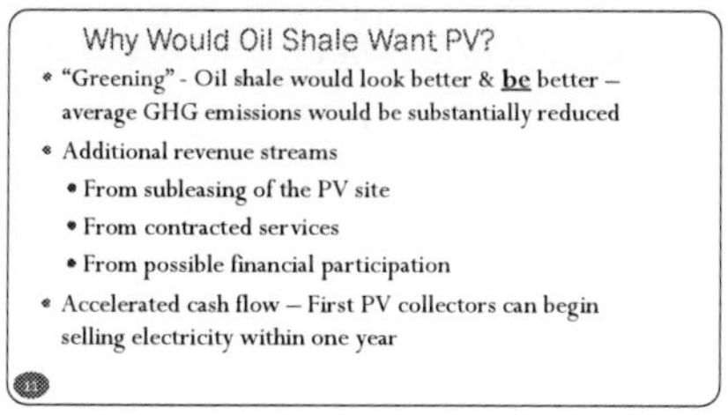

Why would the oil shale plant want a PV plant? The principal social benefit would be a "greening" of the oil shale facility. Adding renewable generation to the electricity obtained from the fossil fuel would reduce the overall emissions per kWh delivered to the grid. This would help make the oil shale project more attractive to potential financiers and more acceptable to the public at large. Also, installation of the solar collectors could begin immediately, on land that will not be required for the oil shale mine or processing facilities. If the interconnection facilities and transmission line are developed in parallel, the project could begin selling electricity within 12 months.

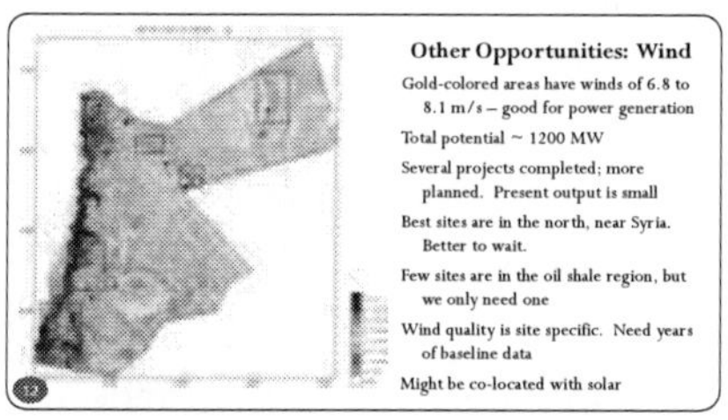

There are other ways to add renewable energy to an oil shale project. Wind turbines might be very practical in certain areas of Jordan. Several wind projects are already on line, and others are in the pipeline. Although wind farms are more expensive to build and operate than solar arrays, their capacity factors can be substantially larger and the electricity cheaper overall. In Estonia, the national energy company Eesti Energia installed wind turbines in an area previously used to dispose of ash from the company's oil shale fired power plants. The wind farm has been an economic success and operates with a capacity factor of about 40%. Similar potential may exist in Jordan's oil shale region. However the practicality of wind energy is site specific, and several years of data gathering would be required to determine feasibility.

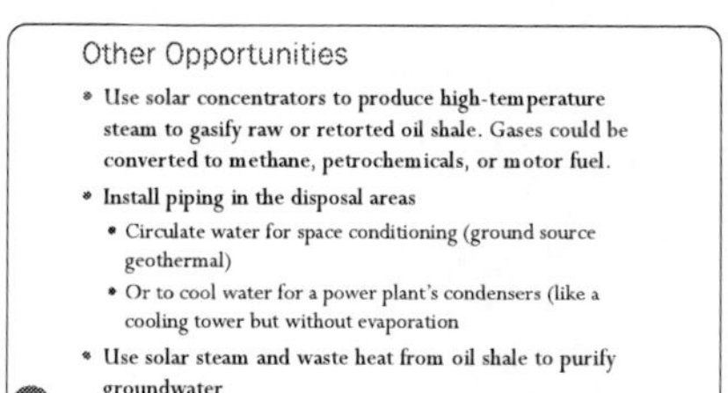

There are other ways to incorporate renewable energy in an oil shale project. Solar concentrators can produce steam at high temperatures and pressures that could be used to gasify raw or retorted oil shale. Water pipes could be installed under the ash and process tailings and used in a ground-source geothermal energy system to heat and cool living spaces. Or the pipes could be used to cool the water used in a power plant's condensers, thus reducing the loads on cooling towers and the associated evaporative losses. Or solar steam and waste heat from the oil shale plant could be used to purify groundwater, to relieve Jordan's large and growing shortage of potable water.

In summary, the benefits to an oil shale project of co-locating with a photovoltaic power plant include:

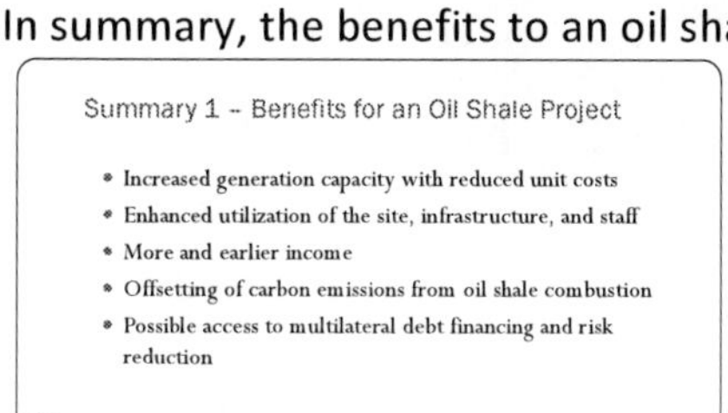

- Increased generation capacity with reduced unit costs
- Enhanced utilization of the site, infrastructure, and staff
- Increased and accelerated revenues

- Offsetting of carbon emissions from oil shale combustion
- Access to multilateral debt financing and risk reduction

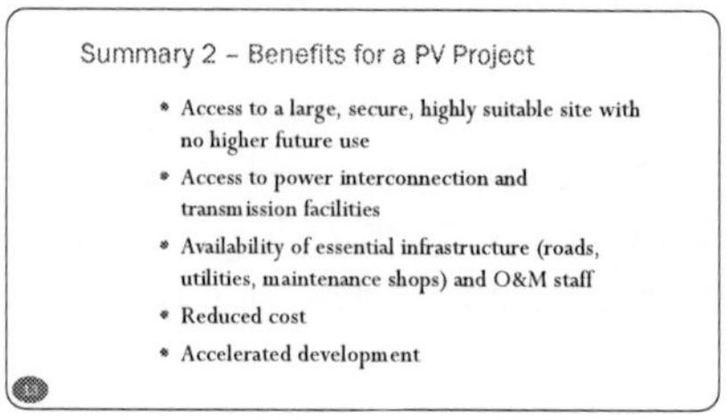

Benefits for the PV project include:

- Access to secure, highly suitable land with no higher future use
- Access to power interconnection and transmission facilities
- Availability of essential infrastructure (roads, utilities, maintenance shops) and O&M staff
- Lower costs   Accelerated development

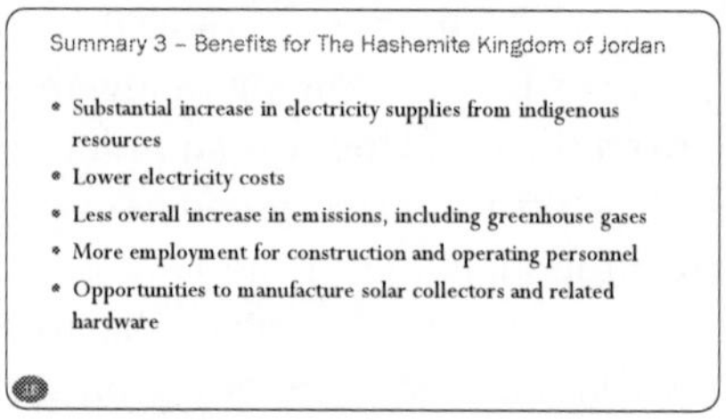

Benefits for the Kingdom of Jordan include:

- Substantial increase in electricity supplies from indigenous resources
- Lower electricity costs
- Lower overall increase in emissions, including greenhouse gases
- More construction and operating employment
- Opportunities to build solar collectors & related hardware

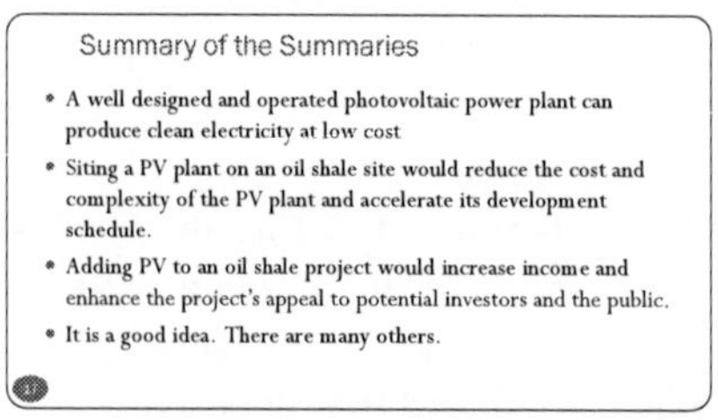

- A well designed and operated photovoltaic power plant can produce clean electricity at low cost
- Siting a PV plant on a reclaimed oil shale mine would reduce the cost and complexity of the PV plant and accelerate its development schedule.
- Adding PV to an oil shale project would enhance that project's appeal to potential investors and the public.

In conclusion, substantial decreases in the installation and operating costs of solar collectors have made photovoltaic power plants competitive with plants producing power from fossil fuels. Establishing a PV plant on a reclaimed oil shale mining site could enhance the efficiency, appeal, and economic performance of both generating stations. It is an opportunity worthy of further consideration.

Ministry of Energy & Mineral Resources of the Government of Jordan (MEMR)
PO Box 140027
8th Circle, Abdel-Rahman
Al-Jamal Street / Passage, Building No. 18
Amman, Jordan 11814

ISBN 978-1-5108-5395-9

CURRAN ASSOCIATES INC.
proceedings.com